The Astronomical Clock of Strasbourg Cathedral

Scientific Instruments and Collections

STUDIES PUBLISHED UNDER THE AUSPICES OF THE SCIENTIFIC INSTRUMENT COMMISSION

VOLUME 8

The titles published in this series are listed at *brill.com/sico*

The Astronomical Clock of Strasbourg Cathedral

Function and Significance

By

Günther Oestmann

English translation by

Bruce W. Irwin

BRILL

LEIDEN | BOSTON

Revised and updated translation from the German edition *Die Straßburger Münsteruhr: Funktion und Bedeutung eines Kosmos-Modells des 16. Jahrhunderts.* Published by GNT-Verlag in 1993. All rights reserved.

Cover illustration: Detail of the "The Strasbourg Cathedral Clock," woodcut by Tobias Stimmer, 1574 (smaller version). Brunswick, Herzog Anton Ulrich Museum, TStimmer AB 3.9.

The Library of Congress Cataloging-in-Publication Data is available online at http://catalog.loc.gov
LC record available at http://lccn.loc.gov/2020007395.

Typeface for the Latin, Greek, and Cyrillic scripts: "Brill". See and download: brill.com/brill-typeface.

ISSN 1877-7066
ISBN 978-90-04-42346-6 (hardback)
ISBN 978-90-04-42347-3 (e-book)

This book is printed on acid-free paper and produced in a sustainable manner.

Contents

Preface

This book is a revision of my dissertation, which was first published a quarter of a century ago, with Prof. Bruno Reudenbach (Hamburg) and the late Prof. Matthias Schramm (Tübingen) as my advisers.[1] That work followed a master's thesis which was submitted to the Department of Cultural History and Culture Studies of the University of Hamburg in 1990. The thesis was awarded the Philipp-Matthäus-Hahn Prize by the City of Kornwestheim three years later.[2]

At that time my goal was to carry out a source-critical investigation that was as comprehensive as possible and to produce an overall interpretation of the clock based on it. I can—*absit iactantia*—consider that that project was successful. Of course, the research has not stood still since then, and the literature which has appeared in the meantime has been incorporated into this work on a case-by-case basis; the original structure of the text has, however, been retained. Some of the pictures from the "chemical age" of photography that were used in the original dissertation have been replaced by digital reproductions and illustrations have been added to the total set of figures.

It would have been a particular pleasure for me to have been able to present this book to the former director of the Archives municipales de Strasbourg, Dr. François Joseph Fuchs, who unfortunately died in 2012 at the age of 91.[3] At the time of my initial research he was constantly willing to provide assistance, and I am indebted to him for numerous tips on local history, valuable information on archive sources, and for opening the doors of various institutions in Strasbourg for me. I look back fondly on my visits to the Upper Rhine region as a young doctoral candidate, which was especially pleasant, above all because of the friendly hospitality in the Fuchs home.

The English translation provided by Bruce W. Irwin (Hildesheim) with great care and expertise was only possible thanks to generous support from the Antiquarian Horological Society, Dr. Günter K. Strauß (Madrid), and an anonymous benefactor. Giorgio Strano (Curator at the Museo Galileo, Florence) and Alison Morrison-Low (retired Principal Curator at National Museums Scotland, Edinburgh) diligently revised the translation, and notwithstanding his terminal illness Andrew James (in 2018 Master of the Worshipful Company of Clockmakers, London) read the manuscript attentively and made very

1 Oestmann 1993/2000.

2 Oestmann 1990.

3 See the two obituaries by Jordan 2012 and Thomann 2013, as well as Thomann 1996.

valuable comments and suggestions. He sadly did not live to see the finished book and passed away on 24 October 2019.[4]

I have the pleasant duty of heartily thanking all of those involved, not least of all Dr. James Nye, Chairman of the Antiquarian Horological Society and Court Assistant of the Worshipful Company of Clockmakers, for their active contributions in making this publication possible.

Once again the quote from a poem by Goethe applies:

Das Beste schaffet unverdrossen;
Wohlwollen unsrer Zeitgenossen, das bleibt zuletzt erprobtes Glück.

(Untiringly create the best;
The goodwill of our contemporaries will meet the test, with good fortune to the end.)

Den Freunden (To Friends), 1826

Günther Oestmann
Bremen, March 2020

4 See the obituary by Nye 2020.

Illustrations

Abbreviations

AAHA	*Archives alsaciennes d'histoire de l'art*
ADB	*Allgemeine Deutsche Biographie*, München-Leipzig, 1875–1912
AMS	Archives Municipales de Strasbourg
BMHM	*Bulletin du Musée historique de Mulhouse*
BNUS	Bibliothèque Nationale et Universitaire de Strasbourg
CR	*Corpus Reformatorum*, vv. 1–28; Carl Gottlieb Bretschneider, Heinrich Ernst Bindseil (eds.), *Philippi Melanthonis opera quae supersunt omnia*, Halle-Braunschweig, 1834–1860
DSB	Charles Coulston Gillispie (ed.), *Dictionary of Scientific Biography*, New York, 1970–1980
EA	*Encyclopédie de l'Alsace*, Strasbourg, 1982–1986
Hain	Ludwig Hain, *Repertorium bibliographicum, in quo libri omnes ab arte typographica inventa usque ad annum MD. typis expressi* […], Stuttgart, 1837; reprint Milano, 1948
KGW	Johannes Kepler, *Gesammelte Werke*, München, 1938–
KJV	*King James Version of the Bible*, Oxford, 1769
LH	*The Lutheran Hymnal*, St. Louis (MO), 1941
LBW	*Lutheran Book of Worship*, Minneapolis (MN), 1978
LCI	Engelbert Kirschbaum (ed.), *Lexikon der christlichen Ikonographie*, Freiburg im Breisgau, 1968–1972
LM	*Lexikon des Mittelalters*, München-Zürich, 1980–1998
NDB	*Neue Deutsche Biographie*, Berlin, 1953–
NDBA	*Nouveau dictionnaire de biographie alsacienne*, Strasbourg, 1982–2007
OND	Musée de l'Œuvre Notre-Dame, Strasbourg
ÖNB	Österreichische Nationalbibliothek, Vienna
RAC	*Reallexikon für Antike und Christentum: Sachwörterbuch zur Auseinandersetzung des Christentums mit der antiken Welt*, Stuttgart, 1950–
RE	*Paulys Realencyclopädie der classischen Altertumswissenschaft*, Stuttgart-München, 1893–1980
TBOO	John Louis Emil Dreyer (ed.), *Tychonis Brahe Dani Opera Omnia*, Copenhagen, 1913–1929; reprint Amsterdam, 1972
Thieme/Becker	Ulrich Thieme, Felix Becker (eds.), *Allgemeines Lexikon der bildenden Künstler von der Antike bis zur Gegenwart*, Leipzig, 1907–1950
Zedler	Johann Heinrich Zedler (ed.), *Grosses vollständiges Universallexicon Aller Wissenschaften und Künste, Welche bißhero durch menschlichen Verstand und Witz erfunden und verbessert worden*, […], Halle-Leipzig, 1732–1750

Introduction

> The bishop's church which is called the 'Münster' was wonderfully built of hewn stone and rises forming an imposing building. It is adorned with two towers, the one which has been completed, an impressive work of art, buries its head in the clouds.[1]

Thus Strasbourg Cathedral was described by the later Pope Pius II, Enea Silvio Bartolomeo Piccolomini, in his *Germania* of 1457–1458. For the humanist Jakob Wimpfeling (1450–1528) the Cathedral was comparable to the great structures of antiquity and he considered the Cathedral tower the eighth wonder of the world.[2]

Barely seventy years later, a new work of wonder appeared, this time inside the Cathedral: the astronomical clock built in 1571–1574 to replace its predecessor from the middle of the fourteenth century. It is glorified by Nicodemus Frischlin (1547–1590) in his poem of praise for the city of Strasbourg and its tower,[3] and the same Frischlin dedicated another lengthy poem of about 1,600 Latin hexameters to its description.[4]

As a result of numerous literary tributes following Frischlin's lead and not least because of the descriptions of its intellectual creator, Conrad Dasypodius, the clock in Strasbourg Cathedral is probably the best known example of the type of clock that is frequently referred to as 'monumental' or 'display' clocks

1 "*Ecclesia pontificalis, cui monasterio nomen est, secto lapide magnifice constructa, in amplissimam fabricam assurexit duabus ornata turribus, quarum altera, que perfecta est, mirabile opus, caput inter nubile condit*"; Piccolomini 1962, 51; on the literary reception of Strasbourg Cathedral see Berns 1989.

2 Wimpfeling 1505, chap. LXVII, fol. 39^{v}.

3 Schadaeus 1617, 110–112; additionally about the clock (using the description of Dasypodius) 38–39 and 39–43 where the descriptive poem by Johannes Fischart can be found in the smaller edition of the woodcut by Tobias Stimmer, as well as German translations of the Bible verses on the clock on the bottom of page 43.

4 Frischlin 1575; on which see Strauß 1856, 48–50, Schade 1984, and in detail Kühlmann 2016. Incidentally, Frischlin was an opponent of Copernicanism (Strauß 1856, 331–333) and also campaigned against astrology in his *De astronomicae artis cum doctrina coelesti et naturali philosophia congruentia* [...], Frankfurt M., 1586, which was harshly criticized by Michael Maestlin (*ibid.*, 328–329).

 | DOI:10.1163/9789004423473_002

(*Schauuhren*) by researchers.[5] Whereas the first term emphasizes the size of the object, stressing its public, representative installation and architectural integration in public buildings such as churches or town halls, the second term notably refers to its assigned function of displaying astronomical phenomena.

This book investigates what this clock represented and what was displayed by it, as well as whether its function was limited to its displaying character, that is, to the spectacular movements of its automata and its depiction of astronomical events.

To begin with, the first clock from the fourteenth century and the project for building a new clock in the first half of the sixteenth century, as the direct predecessor of the second Cathedral clock built in 1571–1574, are dealt with in *Part 1*. In the descriptions of the artistic and technical features of the clock, which follow in *Part 2*, extensive secondary literature can be consulted, treating historical and art-historical questions as well as problems dealing with the history of astronomy and technology. The research on this object was facilitated by the availability of excellent sources, as—despite the considerable, irreparable losses resulting from the fire in the Strasbourg Municipal Library in 1870 during the Franco-Prussian War—numerous important documents still exist, besides which there is a wealth of more or less well informed guide books. Some of these sources were documented by the pioneering work of the ecclesiastical and art historian Franz Xaver Kraus (1840–1901),[6] as well as of the tower-clock manufacturers Alfred and Théodore Ungerer (1861–1933 and 1894–1935, respectively). The Ungerers published their results in a monograph in 1922,[7] and several papers after that. The seminal work of the Ungerers, which is still of fundamental importance today, even if it does deal with the 1571–1574 clock and its predecessor, places its emphasis on the third clock, which was built by Jean-Baptiste Schwilgué in 1838–1842 and installed in the old housing. The technical components of the sixteenth century clock all fell victim to the new clock, and were removed; fragments of them are displayed today in the Musée des Arts Décoratifs in Strasbourg.

The paintings on the clock housing are among the main works of the painter from Schaffhausen, Tobias Stimmer (1539–1584), and have received appropriate attention in art history research.[8]

A work on the three Strasbourg Cathedral clocks that appeared in 1992 and was primarily based on the investigations and knowledge of the engineer

5 Zinner 1939, 41, 57; Zinner 1979, 23ff; King/Millburn 1978, 42ff.

6 Kraus 1876.

7 Ungerer 1922.

8 Stolberg 1898; Bendel 1940, 58ff; Thöne 1936, 33.

Henri Bach (1909–1991) includes a number of good illustrations of the fragments of the second clock and very clear technical drawings.[9] Bach worked for the Ungerers for a long time, but his expositions rarely go beyond results that had already been published elsewhere, and his work does not answer the question as to Dasypodius's fundamental plan.

A singular set of sources that are kept in the municipal archives of Strasbourg should also be mentioned here: a collection of primary and secondary manuscript sources on the Cathedral clock which was accumulated methodologically and with great care between 1929 and 1936 by the musician Charles-Eugène Wagner.[10] Presumably Wagner was initially only interested in the melodies of the carillon of the Cathedral clock and successively collected more and more material on the clock. This collection forms an indispensible foundation for all further research on the astronomical clock in Strasbourg Cathedral.

Despite all the detailed research, the question as to the connection between all of the components of the Strasbourg clock had not been satisfactorily resolved, as in most cases the arguments were based primarily on the painted and sculptural features, while the astronomical indications were only included as an aside at most.[11] The unevaluated inclusion of the descriptions of Alfred and Théodore Ungerer had led to skewed interpretations of the Cathedral clock—in particular in terms of Conrad Dasypodius's views on the teachings of Copernicus.[12] Texts written by Dasypodius himself exist, however, which

---

9 Bach/Rieb 1992.

10 Strasbourg, AMS: Wagner 90 Z 22–24. Wagner (1874–1947) was a musicologist and brilliant pianist, performed in Paris salons, and enjoyed an excellent reputation as an accompanist for numerous virtuosos, including the violinist Fritz Kreisler. He was particularly interested in the musical history of Strasbourg, placing his emphasis on church music and songs from the period between the Reformation and the French conquest of the city in 1681. Wagner amassed a sizeable collection on the church musician Christoph Thomas Walliser (1568–1648), which is also in the municipal archives. Together with Théodore Gérold, he published *Les plus anciennes mélodies de l'Église protestante de Strasbourg et leurs auteurs* in Paris in 1928, compiled a bibliography of Strasbourg hymnals from 1524 to 1681, and planned a series of articles under the title *Denkmäler der Tonkunst im Elsaß* (Monuments of Musical Art in Alsace), which was never realized. This information was taken from a letter written by the retired pastor Ernest Muller in Marlenheim to Dr. François Joseph Fuchs on 20 August 1982; see the more recent article by Jean-Luc Gester in *NDBA*, v. 38, 4057.

11 Even the catalog for the Basel exhibition on Stimmer's œuvre only dealt with the paintings on the clock housing, despite the fact that studies were mentioned on the technical components of the clock; Tanner 1980, 97–106.

12 Sumner 1989, 71–77; Harbison 1976, 194–197. Even Francis C. Haber's interpretation of the Cathedral clock, which referred to Dasypodius's *Heron Mechanicus*, was affected by some fallacies adopted from the secondary literature; Haber 1975.

make a coherent reconstruction of the concept behind the whole clock possible. One of these sources, which had so far not been taken into account and which even exists in published form, was discovered by the author while carrying out research in the library of the University of Göttingen. This was no less than an accountability report of an astronomical clock, submitted by the designer himself, in which the basic idea behind the clock based on astrological demands is presented. This encompasses both the artistic embellishments of the Strasbourg Cathedral clock as well as the astronomical displays and their technical realization.[13] The discussion of this singular source in connection with the treatise *Heron Mechanicus* by Dasypodius—which as the framework for this basic idea provides important insights into the self-perception of the Strasbourg mathematician and his reception of the ancient scholars on mechanics—makes up the nucleus of *Part 4*.

Finally, the *Conclusion* deals with the question of the functions of an astronomical clock inside a church and its reception.

This report on the state of the research about the Cathedral clock in Strasbourg was completed in 1991. The literature on the Strasbourg Cathedral clock has been augmented meanwhile by two dissertations.

In a transdisciplinary study Andreas Gormans examined the mnemonic significance of medieval and modern illustrations which are composed of a circle and a square and discussed their origin, meaning and history. In this respect he has also treated the Strasbourg clock and connected it to the diagrammatic *ars memorativa* in the Middle Ages.[14]

The second study by Michael Dengler has been submitted in 2011 to the Department of History and Sociology of the University of Konstanz. In this work, bearing the title *Zeitmaschinen, Sakralautomaten, Frömmigkeitsapparate: Die Produktion sakraler Zeiten im Kirchenraum der Vormoderne* (Time machines, Sacral Automata, Pious Apparatuses: The Production of Sacral Time within Churches in Pre-Modern Times),[15] the Strasbourg Cathedral clock is understood to be a complex orchestration of manifold things. However, Dengler's investigation suffers in many of its sections from considerable terminological overloading, and whether the suggestive, at times somewhat seemingly fashionable machine metaphors are appropriate for the object investigated remains to be seen. The author obviously zealously consulted the works of

13 Incomprehensibly, Volker R. Remmert, still asserted in 2009: "Nothing seems to be known about the origins of the original iconographic programme"; Remmert 2009, 43, n. 3.

14 Gormans 1999, 317–329.

15 http://nbn-resolving.de/urn:nbn:de:bsz:352-263349 (accessed on 24 July 2019).

Wagner mentioned above and established connections to the 'machine books' from the Renaissance, but unfortunately does not go into detail on the astronomical functions of the Strasbourg clock. Above all, he completely ignores the astrological levels of meaning on which Dasypodius based his concept and their interconnection with the architectural theory of Vitruvius.

The Strasbourg clock is also dealt with in Pascal Weitmann's postdoctoral habilitation thesis, the main topic of which was the ancient Greek and Roman automata and their reception in early modern times, but his expositions (at least those on the Cathedral clock) are based on insufficient sources.[16]

John David North (1934–2008) had provided me most generously with unpublished translations of the text *Heron Mechanicus* by Dasypodius and of *Carmen de Astronomico Horologio Argentoratensi* by Nicodemus Frischlin. These had been produced at the beginning of the 1970s by the historian and philologer Bernard Aratowsky (1922–1986), who had formerly been a professor at the State University of New York at New Paltz. I prepared an edition of Aratowsky's partial translation of *Heron Mechanicus* together with a facsimile of the original, which was published in 2008.[17]

A diploma-degree thesis submitted in 2008 at the University of Teacher Education in St. Gallen (PHSG) examines the life and work of Conrad Dasypodius. Its author, Roger Jud, does not limit himself to historical discussions (a lengthy section is devoted to the Cathedral clock),[18] but formulates ideas based on Dasypodius's edition of Euclid's *Elements* on how this work could be used in upper-level high school mathematics instruction.

In connection with a publication on the contemporary astronomical clock in the foyer of the headquarters of Festo AG & Co. KG in Esslingen-Berkheim, Germany, designed by Hans Scheurenbrand (1938–2015), the science journalist Hermann-Michael Hahn discusses the gearing for the astrolabe of the Strasbourg Cathedral clock of 1571–1574 in some detail.[19]

Shortly before he died, Karl Heinz Burmeister (1936–2014), the former director of the Vorarlberg State Archives and the author of a three-volume work on the astronomer Georg Joachim Rheticus (1514–1574), was able to complete a substantial bio-bibliographical compilation of Rheticus's contemporaries, which also includes some interesting and further developed information on persons involved with the construction of the Strasbourg Cathedral clock.[20]

---

16 Weitmann 2011, 67–70.

17 Dasypodius 2008.

18 Jud 2008, 31–46.

19 Scheurenbrand/Schwarz/Hahn 2005, 112–114 and 200.

20 Burmeister 2015.

For the edition of the works of Nicolaus Copernicus, the art historian Gudula Metze recently compiled a catalog of the illustrations of the astronomer,[21] and submitted a dissertation based on this to Ludwig-Maximilians University in Munich.[22] In these two publications, which represent a valuable contribution to research, the author discusses the portrait of Copernicus painted by Tobias Stimmer on the Strasbourg Cathedral clock in detail.[23]

And finally, Elly R. Truitt, associate professor of history at Bryn Mawr College, should be mentioned together with her study on medieval automata, which presents a brief interpretation of the Strasbourg Cathedral clock that is primarily based on Dasypodius's *Heron Mechanicus*. Unfortunately, no mention is made here either of the fundamental astrological concepts, which are central for understanding both the artistic furnishings as well as the technical features of the clock.[24]

---

21 Kühne/Kirschner 2004, 330–415.

22 Metze 2004.

23 Kühne/Kirschner 2004, 331–337; Metze 2004, 20–32.

24 Truitt 2015, 149–153; see the review on this by Deutsch 2017, and also Truitt 2013.

PART 1

The First Clock in Strasbourg Cathedral and the Project for a New Clock in the Sixteenth Century

CHAPTER 1

The First Clock in Strasbourg Cathedral

As can been seen in the chronicle of the city recorded by Fritsche (Friedrich) Closener in 1362, the first clock in the Strasbourg Cathedral was built between 1352 and 1354: "One counted the year 1352 when the work was started on the clock in the Cathedral and it was well completed two years after that".[1]

Nothing is known about the builder of the clock—as no invoices for the 'Frauenwerk' (*opus sancte Marie*, Works of the Cathedral of Our Lady of Strasbourg) before 1414 have been saved except for a few individual notes from the end of the fourteenth century—and the clockmaker Jean Boernaave, who appears in some recent literature and who, under the name Ben al-Benzar, supposedly learned his craft from the Arabs, is a completely fictional figure.[2] As with every other somewhat famous monumental clock, the tale was circulated according to which the city council had the master builder blinded to ensure that he was not able to build a more impressive work for another city. Such tales are not based on facts and should be disregarded as legends about artists.[3]

The city government surely issued the commission, in all likelihood making use of the 'Frauenwerk', which it controlled. The bishop and the Cathedral chapter only played the role of builder on paper at that time, the real responsibility for building and furnishing the Cathedral lying with the city. The historical development of this situation will be described in general in the following.

1 "*Do man zalt 1352 jor, do wart daz urlei zum münster angevangen zu machende, und wart dernoch wol uber 2 jor vollebroht*"; Hegel 1870/71, 133. Closener wrote the first German chronicle of Strasbourg, which was conceived of as a history of emperors and popes, and was based on older sources (*Chronicon pontificum et imperatorum* by Martinus Polonus, Eike von Repgow, Ellenhard's collection of Strasbourg Histories and Memorabilia (*Historien und Denkwürdigkeiten*) from the end of the thirteenth century. Jacob Twinger von Koenigshoven made use of Closener's works—which he does not admit—and influenced the entire German historiography of the fifteenth century with the chronicles that he began in 1382. He included the fragmentary history of Strasbourg by Closener in a world chronicle. Koenigshoven's chronicle was critically edited by Johann Schilter and published in 1698; Schilter 1698, 574–591. Koenigshoven's reference to the construction of the Cathedral clock in Hegel 1870/71, v. IX, 725.

2 LM, v. 1, col. 1134, s.v. "Astrarium" by Heribert M. Nobis. Here the description that was probably freely imagined by Theodor Schweighaeuser with fairy-tale-like embellishments was presented as historical fact; Schweighaeuser 1876, 10–24.

3 Kris/Kurz 1980, 118.

 | DOI:10.1163/9789004423473_003

The transfer of the 'Frauenwerk' administration initially to the Cathedral chapter in 1262 and then to the city council in 1290 are dates recorded by the chronicler Jacob Twinger von Koenigshoven. He connected the victory of the city over Bishop Walther von Geroldseck in the battle of Oberhausbergen to the transfer of the administration of the 'Frauenwerk' to the city council.[4] However, no sources can be found to support this assumption; according to the investigations of Schulte[5] and Wiek,[6] the transfer of responsibility to the city took place in between 1282–1284 and 1286, without any conflict with the bishop and the Cathedral chapter. The beginning of construction of the west facade of the Cathedral must be assumed to be the reason for the shift of responsibility to the council.[7] Individual citizens acted as Cathedral caretakers earlier, which can be documented as early as 1261.[8] In 1290 the mayor and the council installed two lay caretakers (*gubernatores*), and a supervisor (*procurator*) can be documented starting in 1289.[9]

The transfer of the administration of the Cathedral 'Frauenwerk' was originally only a temporary measure, but later, based on established rights, became permanent. At the beginning of the fourteenth century, the council decided that the 'Frauenwerk' should remain under the administration of the City of Strasbourg, which was also responsible for installing the caretakers (administrators).[10] The city laws of 1322 finally indicate that audits were to be submitted to the mayor and the council, placing the Cathedral 'Frauenwerk' completely under the authority of the municipal administration:

> The seventh [article] is that the mayor and council and no one else shall have power over the Frauenwerk, and that they shall install caretakers for it, who are trustworthy, and who are useful and good for the Frauenwerk, and shall also be called upon by the council twice a year, at Christmas and at the summer solstice, to present the invoices of the caretakers of Our Lady.[11]

4 Hegel 1870/71, v. IX, 663, 726.
5 Schulte 1927, 35–36.
6 Wiek 1958, 21–95; also in Wiek 1959, 43–93 and Schöller 1989, 210–212.
7 Wiek 1959, 90–92; Klein 2008.
8 Wiek 1959, 56ff.
9 *Ibid.*, 71–72 and 82–83.
10 "*Das unser frawen werck in der stat Strassburg gewalt sol bleiben, die ouch pflegere daruber zu sezen* (The Frauenwerk shall remain subject to the power of the City of Strasbourg, which shall also name caretakers for it)"; from a collection of council statutes drawn up between 1300 and 1322, reprinted in Schulte/Wolfram 1888, v. IV.2, 23, n. cc.
11 "*Der sybende* [Artikel] *ist, das Unsser frowen wercke meister und rat sollent in ir gewalt haben und nieman anders und sollent pfleger daruber setzen, die sie truwent, die nutze und*

As a result of these general conditions, the construction of the first astronomical clock in the Cathedral was clearly an undertaking of the townspeople, as was also the case for the new clock project in the sixteenth century. The southern transept was chosen as the location, and the clock stood on the wall to St. Catharine's Chapel, opposite the location for the new clock of 1571–1574.

The figures on the pillars of the southern portal—a king sitting in judgment (probably Solomon) with Christ as Pantocrator on the central pillar, flanked by sculptures of Ecclesia and Synagoga—are interpreted as a judgment scene.[12] In front of this portal was the *Fronhof*, where court actions such as vows of union and disputes between citizens were settled; persons who wished to be freed from a ban were also received by the mayor here.[13] During their homage journeys, the German kings received tribute from the bishops as the rulers of the region in front of the southern portal.[14] Court actions also took place inside the southern transept with its Pillar of Angels, besides other locations in the Cathedral.[15]

Thus the astronomical clock was built in a part of the Cathedral that was particularly closely connected to city life, and this circumstance also justifies the conclusion that the project was an expression of the city's interest in building the clock in the bishop's church.

The clock's dimensions and its approximate outer structure can be reconstructed from the extant holes with lead wall plugs for attaching the housing.

There was a calendar disk with a diameter of 2.8 m in the bottom tier (width 4.1 m, height 4 m); the two tiers above this had a width of about 2.9 m and the clock's total height was 11.7 m. Two galleries, whose supports which projected from the wall are still recognizable, provided access to the upper two tiers.[16]

The circle carved in the wall (approximately at the level of the calendar disk, indicated with a solid line in *Fig. 1.1*) supposedly had the diameter of a bell that was cast in 1519 and hung in the newly constructed bell tower in 1521, but was cracked by severe frost and had to be broken up.[17] Here the question arises as to when the circle was chiseled into the wall, because the first clock was still

---

gut syent dem wercke, und sollent ouch botten von dem rate zwurent ime jore dartzu schicken zu wynahten und zu sungihten, die die rechenunge verhörent von den pflegern umb Unsser frowen gut"; Schulte/Wolfram 1888, 144–145.

12 Erler 1954, 2ff.

13 *Ibid.*, 24–25 and 27.

14 *Ibid.*, 26–27.

15 *Ibid.*, 34ff, 40 and 43–46; Bengel 2011, 182–184.

16 Ungerer 1922, 7–8.

17 According to Grandidier 1782, 261: "*Près de ce pilier* [the Pillar of Angels], *on voit à la muraille, vis-à-vis de l'horloge, un tour de cercle, qui marque la circonférence de cette grosse cloche, dont nous avons parlé dans le prémier livre de ces Essais*".

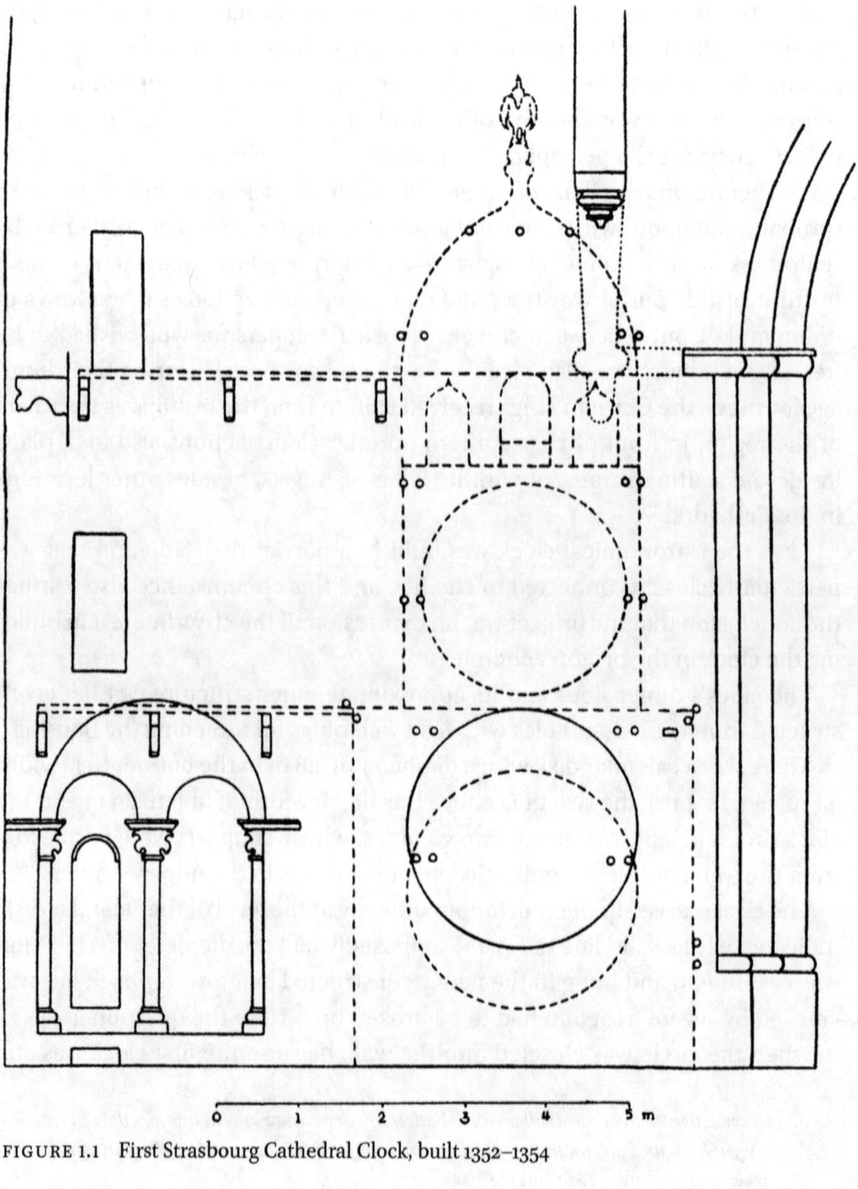

FIGURE 1.1 First Strasbourg Cathedral Clock, built 1352–1354

standing in this location in 1521 and was not dismantled until the second half of the sixteenth century shortly before the new clock of 1571–1574 was completed.[18]

The intellectual creator of the new clock, Conrad Dasypodius, had studied the old clock extensively and recorded further details in his description.[19] According to him, the housing was made of wood and had a three-part arrangement of the displays. The wooden calendar disk at the bottom completed one revolution in a year and listed the movable feasts among other things. Above it was an astrolabe with hands for the Sun and Moon, which was driven by clockwork behind it. Above the astrolabe there was a semicircular balcony on which the Three Kings or Magi moved past a statue of Mary at certain times (perhaps at fixed hours) and bowed. The drive mechanism for this also activated the hammers for the carillon (*Cimbalen*), following which a mechanical cock, probably at the top of the housing, moved and crowed.

On the side (most probably next to the calendar disk or on a sidewall of the clock) there was a table with the astrological properties of the planets in verse.[20] The qualities intended here were those attributed to the planets, corresponding to the four elements: Mars and the Sun were considered warm and dry, Jupiter and Venus warm and moist, Saturn cold and dry; Mercury oscillated between dry and moist, and the Moon, although heated by the Sun, was in general moist.[21] In a rather free interpretation, Alfred and Théodore Ungerer assumed that the image of a little phlebotomy man was also included and that the tables mentioned by Dasypodius thus indicated the influence of the planets on bloodletting, or rather the favorable and unfavorable times for this procedure.[22] Ever since, this unsubstantiated theory has been treated as

---

18 The fragment of a chronicle (called the 'Kleine Münsterchronik') states: "*Anno domini 1352 ward das Urley im Münster mit den drey Königen angefangen worden zumachen und inn zweyen Jaren hernach vollendet. Ist anno 1572 abgebrochen worden und ein anders gemacht worden wie in anno 1573 zu sehen* (In anno Domini 1352, construction was begun on the clock in the Cathedral at Epiphany and completed within two years. It was dismantled in 1572 and another was built as could be seen in 1573)"; Dacheux 1887, 10.

19 Dasypodius 1580a, 2–3. This description of the second Cathedral clock is from the second, expanded edition of the first edition of 1578.

20 "*Auch ist darbey ein taffel auffgehenckt in welcher der siben Planeten eygenschafft rheimens weiß geschrieben gewesen* (A table was also mounted nearby in which the properties of the seven planets was written in rhyme)"; *Ibid.*, 3.

21 Ptolemy 1980, I.4, 35–39; Uhlemann 1857, 69–73. The only remaining fragment of the astronomical clock of 1561–1566 from St. Mary's Church in Lübeck (ring of the zodiac with a gear and planet hands, St. Annen-Museum, inv. no. 1892/145) has corresponding inscriptions. The inscriptions on the planet hands (which have flaked off to a large extent) indicate the respective orbital periods and properties as well as the names of the planets.

22 Ungerer 1922, 8; see also Ungerer 1931, 163.

fact in the literature,[23] and, moreover, a reputable reference work includes the surprising report that, according to the description of Dasypodius from 1578, the clock also had a hand for indicating the points on the human body for phlebotomy relative to the course of the stars (*Zeiger für die vom Sternenlauf abhängige Aderlaßstelle des menschlichen Körpers*).[24] This must be rejected as totally unfounded speculation as the text by Dasypodius does not include the slightest indication of such a feature. Of course that does not eliminate the possibility that the verses included a depiction of the ancient doctrine of *Melothesia*, i.e. the influence of the signs of the zodiac on certain parts of the body.

The clock's astrolabe probably had a northern projection ('southern astrolabe'),[25] which can be assumed indirectly from a source that deals with the astronomical clock in the Church of Our Lady in Villingen.

In 1398 the city council of Rottweil sent the clockmaker Claus Gutsch to Strasbourg with the commission to inspect the *jar rade* (year wheel; most probably the calendar disk), so that a similar work could be completed in Rottweil.[26] However, this project never seems to have been carried out; at least the existence of a monumental clock in Rottweil cannot be documented. The only probable location for such a clock was the Church of the Holy Cross, which was completely rebuilt, including the steeple, in the fifteenth and sixteenth centuries.[27]

Shortly after that, Gutsch built a clock for the Church of Our Lady in Villingen, which was completed at the beginning of 1401 and was similar to the first clock in the Strasbourg Cathedral. A document from 6 May 1401 states:

> We conjointly, the Schultheiss, the Mayor, and the council of the City of Villingen, make multifariously known with this letter and publicly announce that the honorable man Master Claus Gutsch of Rottweil, carrier of this letter, has constructed and made a sphere [astrolabe dial] for us in the Church of Our Lady in Villingen, which indicates and denotes, as is written in this letter: First of all a wheel that moves in the course of a

23 Maurice 1976, v. 1, 28; Bach 1978, 29.

24 LM, v. 1, col. 1134, s.v. "Astrarium" by Heribert M. Nobis. This article also has numerous other mistakes and ambiguities.

25 The north or the south pole can be chosen for the stereographic projection of the celestial vault of an *astrolabium planisphaerium*—if we ignore special designs. The first case is referred to as a 'southern astrolabe', the second is called a 'northern astrolabe'.

26 Letter of recommendation by Rottweil's city council, addressed to the city council of Strasbourg, AMS, Sér. II, 46, no. 32; reprinted in Kraus 1876, 385.

27 Information kindly provided by Dr. Winfried Hecht (Municipal Archive, Rottweil).

year and has a calendar on it and the twelve months painted with the labors commonly done; and one sees the length between Christmas and Carnival every year, and a little man sits in front and points to each day throughout the year, and one finds all the feast days in the year. Above the wheel the sphere is arranged, on it are marked air, water, and earth, and the Sun and Moon and the twelve signs, and the Sun rises and sets and is high and low in accordance with its course in the year; the moon rises and sets as well and is high and low in each phase as the Sun does in a year, and its shine waxes and wanes and it becomes new in the hour as one reads that it appears new in the heavens; and both the Sun and Moon stand every day in their courses and signs, as one can read that they are standing in the heavens; and one sees in the Sun [the Sun pointer] all the hours of the day and the night, and above that are the Three Holy Kings and three servants that come out of one door and go in another; pay homage to Our Lady and turn and bow to her, and the little child turns to them, and an angel flies over them with a star, and two angels on the case blow horns, as the Magi move; and there is a positif organ that plays while they move. This arrangement and this work have been well embellished with woodwork by the above mentioned Master Claus. And in order to recognize this truly and forever, we have publicly attached the seal of our City to this letter, written in Villingen on the first Friday after May Day, to which one counts fourteen hundred years and after that the first year after the birth of Christ.[28]

---

28 *"Wir der Schulthaiss, der Burgermaister und der rautt gemainlich der Statt ze Vilingen kunden meniglichem mit disem brieff und vergehen offenlich, daz der erber man maister Claus Gutsch von Rotwil, zöger dis briefs uns in unserm münster ze Vilingen ain Sper geordnot und gemachot hat, die also wiset und betutet, alz hie nauch an disem brieff geschriben stat: Dez ersten ain rade, daz des jars ainost umb gat, und ist daz kalendarium daran und die zwölf manot gemalot mit den geberden, alz denn ze mal gewonhait ist ze tund; und sicht man die lengi zwuschent wihennechten und vasnaht allu jar, und sitzt ain mennly davor, daz zaiget uff alle tag durch das jar, und findet man alle virtage in dem jar. Ob dem rad ist die sper geordnot, daran ist bezaichent luft, wasser und ertrich und Sunne und man und die zwölf zaichen, und gat die sunne uff und ze gnaden und hohet und nidret nauch dem, alz es in dem iare ist; der man gat ouch uff und ze gnaden und hohet und nidert in ie dem schin alz vil, alz die sunne ains jars tut, und lait zu und nimpt ab an dem schin und wirt nuwe in der stund, alz man liset, daz er an dem himel nuwe sig; und stand beide, Sunne und man, taeglichen in irem staffel und zaichen, alz man liset, daz si an dem himel standin; und sicht man in der sunnen alle stunden dez tags und der naht, und sint darob die hailgen drig kung und drig kneht, die gand umb zu ainer tur uss und zu der andren tur wider in; wisent unser frowen und kerent sich umb gen ir und nigent ir, und kert sich daz kindli umb gen inen, und schwept ain engel ob inen mit ainem sternen, und stand zwein engel uff dem gehus und blasent, so die kung wellent gan; und ist ain positiff daran, daz singt, die wil si gant. Dis ordnung und*

In an astrolabe with a northern projection a solar hand describes a larger day's arc over the ecliptic ring in summer than in winter, as corresponds to the actual situation. This situation was probably intended in the formulation: "*und gat die sunne uff und ze gnaden und hohet und nidret nauch dem, alz es in dem jare ist* (and the Sun rises and sets and is high and low in accordance with its appearance in the year)". As a result of the similarity supported by this document between the clock in Villingen and its model in Strasbourg, which Gutsch had seen with his own eyes, we can assume with considerable certainty that both clocks had an astrolabe that had a stereographic projection of the celestial vault with north-pole orientation (southern astrolabe). Thus these two were among the group of early astronomical clocks whose astrolabe dials differed from those of portable astrolabes, which typically had a southern projection (northern astrolabes).[29]

Further details about the first Strasbourg clock cannot be deduced even indirectly from the Villingen clock because the Three Magi works had disappeared by 1714, and the calendar disk was disassembled in 1839.[30] Around the middle of the nineteenth century, the clock no longer existed and no illustrations survive.[31]

Concerning maintenance work on the Strasbourg Cathedral clock, repairs were possibly carried out on the housing in 1382; invoices from the 'Frauenwerk' indicate expenditures for lead "*um das urley hus zu deckende* (in order to cover the clock housing)" as well as for paying a metal founder (*Kannengießer*). At that time a certain Master Bartholomaeus was also employed in connection with the clock.[32] However, it is possible that this entry did not have anything to do with the astronomical clock, but referred to the roof of the clock room on the platform of the tower. In 1414–1418, frequent mention was made of a "*Herr Cunrat uff dem Huse*" (Mr Cunrat on the house; i.e. the mason's lodge), who adjusted the clock. He was probably an employee of the lodge and a smith. Between 1419 and 1450 a Swiss clockmaker from Bern (*horleymacher von berne*)

dis werk hat der obgenant maister Claus mit holtzwerk wol geziert. Und dis alles zu ainem waren und stätem urkund so haben wir unser Statt Insigel offenlich gehenkt an disen brieff, der ze Vilingen geben ist an dem nechsten fritag nauch dem maye tag, do man zalt nauch Cristy geburt vierzehenhundert Jare und darnauch in dem ersten Jare"; Municipal Archive, Villingen-Schwenningen, MS GG 4a. The document is reprinted in Roder 1880, H. 3, 291–292; Kraus 1890, 124; Steinhauser 1939, 124; Fuchs 1970/71, 59–62, with a photograph of the original.

29 For information on the typology of astronomical clocks see Horský 1967, 29–30 and Oestmann 2017.

30 Kraus 1890, 124.

31 Information kindly provided by Dr. Joseph Fuchs, city archives of Villingen-Schwenningen.

32 AMS, Sér. II, 45a/4, fol. 8^{v}.

serviced the Cathedral clock, who remained in Strasbourg and repaired the mechanical clock in 1450–1451.[33]

1.1 The Automata of the First Cathedral Clock

As already mentioned, both the first Strasbourg Cathedral clock and the clock in Villingen had figures of the Three Magi who moved past a statue of the Virgin Mary. The following clocks were also equipped with such automata:

1. Venice (clock tower at Piazza San Marco), built between 1496 and 1499 by Giampaolo and Giancarlo Rainieri.[34]
2. Lund (Cathedral), first mentioned in 1442; date of origin unknown, but probably of the late fourteenth or early fifteenth century.[35]
3. Frankenberg in Hessen (Parish Church), built in 1359 and destroyed in a great fire of the city on 9 May 1476.[36]
4. Münster in Westphalia (Cathedral), built in 1540–1543;[37] its builders were the printer Dietrich Tzwyvel, the Minorite and Cathedral dean Joannes Aquensis, and the clockmaker Nikolaus Windemaker.[38]
5. Osnabrück (Cathedral), built in 1587 by Vicar Jost Bodeker von Wartbergh (no longer extant).[39]
6. Reggio Emilia (town hall), built in 1481–1483 by Giovanni Paolo Rainieri.[40]

---

33 For transcriptions of the relevant invoice documents see Oestmann 1993/2000, 245–247.

34 Ungerer 1931, 379; Rocca 1719, v. 1, 188.

35 Wåhlin 1923, 179.

36 "*Do man schreib nach Christus geburt 1359 jare, du wart der urtzouger in der parkirchin mit den dryen kunnigen gemacht; der kostede wol tusent pund heller* (As one wrote the year 1359 after the birth of Christ, the clock was made in the parish church with the Three Kings; it certainly cost a thousand pounds of hellers)", with the marginal note: "*ist noch bey handen, aber zerbrochen* (it still exists, but is broken)"; from the chronicle of Wigand v. Gerstenberg (1457–1522), completed in 1506; Diemar 1909, 433; see also Zinner 1939, 45.

37 During preliminary investigations for the restoration of the astrolabe dial in summer 2017, the year 1543 has been discovered.

38 Wieschebrink 1983, 8–11 and 22–23.

39 Veltman 1890, 238–239 and 269.

40 "*Et ulteriorius imagines trium Regum, sive Magorum: qui semper, paulo ante sonum horarum, exeuntes ex ordine e turri praedicta, procedant, et sese inclinantes, et e capite suo coronam declinantes, conspectui praelibate virgines sese visibiliter praesentent, et illam revereantur, quasi adorantes filium, et eidem munera offerentes: ita, qud haec omnia publice spectari possint: et qui Magi deinde in eadem turri ab alter aparte sese recondant, et etiam imaginem unius Angeli, qui semper praecedant dictos Magos, cum tuba in manibus, quam sonet ita, quod per totam plateam, et in confinibus eius audiri possit: et qui exeat, et*

7. Danzig (St. Mary's Church), built in 1464–1470 by the clockmaker Hans Düringer.[41]
8. Augsburg (Cathedral), built in 1594 by the master clockmaker Marquart, the clock- and compass-maker Tobias Klieber, as well as by the physician Georg Henisch (Hoenisch), dismantled in 1609.[42]
9. Macerata (city tower), built in 1569 by Lorenzo Maria and Ippolito Rainieri.[43]
10. Bologna (town hall), built in 1444–1451 by Giovanni del fu Evangelista from Piacenza and Bartolomeo di Gnudolo del fu Ser Rustigani.[44]
11. Olmütz (town hall). The date of origin (1419–1422) and the name of its builder (Anton Pohl) frequently mentioned in the literature cannot be verified in sources. Important reconstruction and extensions were carried out in 1572–1575 by the mathematician and physician Paul Fabricius, as well as by the clockmaker Hans Pohl.[45] As of 1661 the existence of a statue of Mary with mechanical figures of the Three Magi that moved can be documented;[46] whether they already existed in the sixteenth century or earlier is possible, but cannot be verified.
12. Cologne (choir ambulatory in the Cathedral). The date of origin and the builder of this clock are unknown; it was removed in 1760.[47]
13. Paris (St. Jacques' Pilgrim Hospital; no longer extant).[48]

se recondat, prout exibunt, et se recondent dicti Magi", from *De officio Regulatoris horologii* (1500) in Siliprandi 1915, 48; see also Ungerer 1931, 369–370.

41 Schmidt 1927, 390; Drost 1963, 141–142.

42 Munich, Bayerische Staatsbibliothek, Cgm 4189.

43 Ungerer 1931, 359–360.

44 Podestà 1869, 147–148, 159.

45 Kux/Kreß 1904, 9ff.

46 Michalik 1943, 11.

47 First briefly described by Aegidius Gelenius in 1645 (241) and shortly after that by the Jesuit priest and historian of the Three Magi, Hermann Crombach (1598–1680): "*Prae foribus huius vestiarii ad dexteram horologium cernitur arte mirabili, cuius pondera semel in anno elevata, non solum singulos anni dies, sed solis etiam, lunae ceterorumque errantium siderum, cursum et oppositionem artificiose repraesentant: Supra horologium rota versatilis in editiore loco consistit, qua tres Reges Deiparae filioque munera deferentes et adorantes, cum aeris campani festiva harmonia, et tubarum clangore circa Epiphaniam et translationis festum progressu gravi circumaguntur*" (Crombach 1654, v. III, 801). According to Schallenbach 1771, 52, the clock was removed several years ago because of a lack of repairs.

48 "*Item pour changer les aloges a autres 4 l. 14 s. pour iii plates de coivre de quoi les iii roys sont faiz et pour facon 12 s.; pour une ouvrier qui fist les vi. petiz ymages des aloges 22s.*"; Archives Assistance publique Hôpitaux de Paris, fonds St.-Jacques-aux Pelerins, invoices, no. 231, côte 2251 (rouleau), fols. 24–25, from 1326. This seems to be the earliest documented mention of a clock with the Three Magi; the author thanks Gerhard Dohrn-van Rossum for pointing this out.

The theme of the procession of the Three Magi is therefore a common element in the tradition of public clocks, and the question is, why it was chosen. Theodor Wåhlin assumed that the Three Kings personified the three continents Europe, Africa, and Asia, which pay homage to Christ.[49] On the other hand, E. v. Bassermann-Jordan and H. Hofmann connected the automata in the monumental clocks with the religious mystery plays.[50] This assumption can be supported by the example of the Cologne clock, as Cologne Cathedral has housed bone relics of the Three Kings since 1164, when they were brought to the city by the Archbishop and Chancellor Rainald von Dassel as a present from Emperor Friedrich Barbarossa after his conquest of Milan.[51] The theme of the Three Magi not only dominated the sculptural, painted, and other embellishments of Cologne Cathedral, it also played a certain role in the liturgy. Three Magi plays were acted out at the reliquary shrine.[52] Possibly the astronomical clock in Cologne Cathedral, colloquially referred to as the *Umläufer*, was a source of inspiration for the builder of the astronomical clock in Strasbourg; after all, Cologne had developed into a center for the veneration of the Magi since the Milan relics were moved there. Considering the completely undocumented date of origin of the *Umläufer*, this assumption is nothing more than speculation.

A Three Magi play is also documented for Strasbourg in an antiphonary from 1200,[53] in which Mary does not appear.[54] Whether the clock played a role in this theatrical performance is not known. The impulse for this theme for the automata, however, has to be sought in the realm of the liturgical plays.

In the fourteenth century, the Three Magi plays, which originally were firmly associated with the liturgy, increasingly became secular; the plays were no longer exclusively performed in sacred locations, they were increasingly performed in public places, such as the squares in front of town halls and churches.[55] The transfer of this theme to the 'automat theatre' of the monumental clocks can surely be understood to have been part of this secularization.

---

49 Wåhlin 1923, 189. This interpretation was probably based on a passage from a commentary on St. Matthew attributed to Beda Venerabilis: "*Mystice autem tres Magi tres partes mundi significant, Asiam, Africam, Europam, sive humanum genus, quod a tribus filiis Noe seminarium sumpsit*"; Migne, *Patrologiae cursus completus: Series latina*, v. 92, col. 13.

50 Bassermann-Jordan 1921, 19–20.; Hofmann 1975, 159–160.

51 Hofmann 1975, 96ff.

52 *Ibid.*, 154.

53 Text in Wilmart 1928, 8–10, as well as in Lange 1888, 412–415. For the development of the play text, see Anz 1905, passim, and in particular 124–129.

54 Sig 1906, 16–18; Walter 1929, 48.

55 Hofmann 1975, 155–157.

FIGURE 1.2 Automaton cock

The 1.17 m tall mechanical cock is the only thing that remains from the first Cathedral clock (*Fig. 1.2*).[56] Its body and head with the top of its beak were made of wood; both the bottom of the beak with its tongue and the three-part comb, which was originally painted red, as well as the wattles and the feathers nailed to the body were made of metal. The comb consists of wrought copper, the feathers of wrought iron with hammered quills. Gib-head keys were the only fasteners used in the mechanism. The movement (*Fig. 1.3*) was carried out as follows:[57]

56 Musée des Arts Décoratifs, Strasbourg; Beyer 1955, 71; Späth 1909; Deluz 2019.

57 Based on a description by Ungerer 1927b, 9–11.

FIGURE 1.3 Schematic illustration of the mechanical cock. I. Side view of a wing from the inside; II. Cross section of the body

The two rods *1* and *2* reciprocated up and down in the feet; the first of which moved the head and tail. When rod *1* moved down, lever *a*, which was attached to it, rotated around the pivot *A*, so that the head and the three-part articulated neck were raised. A second bent lever *b* (pivot *B*) was connected to lever *a* and moved the rod *e* via the connecting rod *c* and the rocker arm *d* (pivot *D*), so that the bottom of the beak (hinge *F*) moved down and the split tongue *g* moved out. When the head was raised, the tail, which could pivot around *K*, was moved down by the rods *h* and *k* and the connecting rod *i*. Rod *2* made the wings flap (moved them up and down) and was activated by lever *l* with the connecting rod *l′*, which moved the two levers *n* via the connecting rods *m*, as the wing axes *NN* were firmly attached to the levers *n* with rivets. An additional mechanism spread the feathers of the wings while they were flapping: when the wings pivoted around *NN*, the rod *o* connected at *n′* moved the rocker arm *p* (pivot *P*), so that the slat *q* carried out an almost parallel movement. The ends of the wing feathers were attached to the slat *q*. The front feather r_1 pivoted around *R* in the bridge *S* and extended further than the other feathers, as the lever arm of the slat *q* was shortest there. The extension of the spread decreased for the feathers r_2, r_3, r_4, r_5 as the length of the lever arm increased;

they were not mounted in pivots in the bridge *S*, but moved freely within the space between the cover of the wing and an inner bridge.

From the exceedingly complex mechanism for the automaton's movements, we can assume that the intention was to copy the movements of a crowing cock as closely as possible in the mechanical version. The Strasbourg cock—viewed in the broadest sense—stands in the tradition of ancient automata, such as the clepsydrae of Ctesibius[58] or the monumental clock of Gaza described by Procopius,[59] even if a tangible model for the Strasbourg clock cannot be named. And a symbolic connection between the cock and the Three Magi automaton cannot be recognized.

In Christian symbolism, the cock is first of all seen as having the task of awakening people at sunrise; in patristic and medieval texts, the cock's crowing after St. Peter denies Christ three times[60] and Christ's foreseeing this[61] are a frequent topic for interpretations. Besides the meaning of a cock's call in ancient texts, where it had supposedly had the power to avert evil influences, its vigilance and symbolic role as a caller and admonisher in the denial scene are particularly emphasized here.[62] Dasypodius put forward this interpretation of the mechanical cock, which he used for the second Cathedral clock:

> Since, indeed, this poultry cock itself was skillfully made two hundred years ago and placed on the old clock, and since at that time it was customary to commemorate the Passion of Christ in the Christian church, this cock by its crowing warned men of the denial of Peter.[63]

It therefore seems that automata played a role in the liturgy of the Holy Week services. Besides this interpretation, the mechanical cock can be seen as a simple symbol of chronometry, the "winged harbinger of dawn" (*Aurorae volueris*

---

58 On the clepsydrae of Ctesibius see Armstrong 1974; Schürmann 1991, 255–258.

59 Diels 1965, 219ff.

60 Mt 26: 69–75; Mk 14: 66–72; Lk 22: 54–62; Jn 18: 16–27.

61 Mt 26: 34; Mk 14: 30; Lk 22: 34; Jn 13: 38.

62 For these sources see RAC, v. XIII, cols. 863–865, s.v. "Hahn" by Claudia Nauerth; LCI, II, cols. 206–207, s.v. "Hahn" by P. Gerlach; Sauer 1924, 143–145. There is no reason to assume that this interpretation of the cock of the Cathedral clock was unlikely in the late Middle Ages; cf. Ungerer 1922, 9; Ungerer 1924, 128. On the apotropaic meaning of the cock see also Cordonnier 2012, 91–92.

63 "*Siquidem ante 200 annos hic ipse gallus gallinaceus affabre factus fuit: et veteri horologio impositus, atque eo tempore quo Passionem Christi in Ecclesia christiana solitum fuerat commemorare: hic suo cantu abnegationis Petri, homines commonefecit*"; Dasypodius 1580b, fol. HII^r; see also Dasypodius 1580a, 5.

praenuncia), as Nicodemus Frischlin wrote in his poem praising the second Cathedral clock.[64]

There are other mechanical figures in Strasbourg Cathedral: a wooden figure (Samson taming the lion), for example, is located at the bottom tip of the organ case erected in the main nave in 1489; on the left an almost life-size town trumpeter and on the right a bearded man with a cap stand on pedestals on the wall. They were set in motion by means of wires from the organ loft. The organist probably operated the wires via the pedals and accompanied his music with corresponding movements in the figures.[65] Something like this must have existed before 1489, as the canon Peter Schott, who died in 1490, reported to the nuncio Emerich Kemel in a letter that a rustic figure (*rusticana imago*) could be found on the organ, behind which a layperson or cleric was hiding. On Pentecost, at which time extensive festivities and processions took place in Strasbourg so that many people from the surrounding areas came to the city, this so-called *Roraffe* played a role, showering the congregation of believers with shouts, ridicule, and insults as the procession entered the church.[66] Invoices for the 'Frauenwerk' document numerous payments for one of the Cathedral workers who had to shout the calls of the *Roraffe*.[67] In the 21 articles that the Cathedral preacher Geiler von Kaisersberg wrote against the Strasbourg city council in 1501, he complained that the church services were interrupted not only by official town business in the Cathedral, but also by the blasphemous shouting of the *Roraffe*.[68]

The cock now competed for popularity with this obviously very popular figure: Dasypodius included a poem in his description, which he claimed to have found inside the automaton, in which the *Roraffe* complains that he had not been awarded the necessary attention ever since the cock on the clock crows and flaps its wings.[69]

---

64 Frischlin 1575, fol. G^{r}.

65 Winckelmann 1907, 256ff; Laugel 1888, 557–565.

66 The letter is reprinted in Schneegans 1851, 40.

67 Two entries are exemplary: "*ıß ß eime kneht der in den roraffen rufen soll*"; "*xßvii ß orgelknecht und von dem roraffen*"; AMS, 1 OND 43, 1419 (1), fol. 60^{v}; 1431 (1), fol. 50^{r}.

68 Winckelmann 1907, 283–286.

69 *Der Kampff des Roraffen under der Orglen in Münster zu Straßburg mit dem Hanen vor 200. Jharen beschriben daselbst auff dem Uhrwerck* (The battle of the *Roraffe* under the organ in the Strasbourg Cathedral, with the cock on the clockwork described 200 years ago); Dasypodius 1580a, 57–63; although mentioned, it was not repeated in the 1578 edition: Dasypodius 1578a, fol. C1^{r}.

CHAPTER 2

The Clock on the Tower Platform

Besides the clock inside the Cathedral, a chiming tower clock made by the clockmaker Heinrich Halder from Basel was mounted on the southern tower platform of the third story of the facade in 1373, which had been completed in 1365.[1] Koenigshoven (who lists the year as 1372) calls this clock a *Zitglocke* (time bell), compared with *Urlei* for the astronomical clock in the southern transept.[2] Two Cathedral sextons were responsible for lifting the clock weights and for the clock's maintenance. They also had to ring the bells, replace the candles, and keep the Cathedral clean; and they operated the organ bellows for the organist, were responsible for opening and closing the Cathedral, and had to guard the Cathedral at night.[3] Occasionally the sextons were too lazy to climb up to the tower platform just to lift the weights of the clock, so that the tower clock stopped. In an undated directive that was probably from the fifteenth century the proposal was made to give this task to the watchmen, who kept their watch on the platform anyway.[4]

---

1 "*Heinricus Halder horelogifex de Basilea (S.) in presencia Johannis dicti Keszeler p*[res]*b*[ite]*ri. procuratoris, Johannis de Múlnheim militis, Johannis zu dem Trubel et Johannis de Munoltzheim civium Arg. gubernatorum fabrice ecclesie Arg. recognovit, sibi eosdem procuratorem et gubernatores nomine fabrice satisfecisse de horologio campane 24 horas diei et noctis indicantis per ipsum Heinricum in nova turri ecclesie Arg. predicte constructo. actum 16 kal. jul. a. 73*"; AMS, Inventaire des Archives anciennes, no. 8, fol. 36*r*, reprinted in Witte 1900, 459–460, no. 1581. For information on Halder, who built the clock in the Graggen Tower in Lucerne in 1385 and probably the first clock in the Basel Minster, as well, see Marius Fallet-Scheurer 1917, 77–81 and 84.

2 "*Donoch als men zalte noch gotz gebürte 1372 jor, do wart die zitglocke uf dem münster gemaht, die do die stunden sleht* (After which when one counted the year 1372 after God's birth, the time bell was made, which rang the hours there)"; Hegel 1870/71, v. IX, 725.

3 Ordnung der Pfleger, des Schaffners und der Gesinde des Frauenwerkes von 1464 (regulations for the caretakers, supervisors, and common workers of the 'Frauenwerk' from 1464), AMS, Sér. II, 46, no. 15, fol. 17^{v}–18^{v}.

4 "*Uff den Kar fritag alß man den passion In dem münster bredigete do wart ein groß uff wüschen und geloiff In dem volck, deß schellen lütens halp so der wahter uff dem münster det, umb daß die zit glock still stunt und nit slug (nu ist die urle kürtzlich aber stille gestanden daß die glock aber nit geslagen hat, deß halp daß die münster kneht nit allewegen hin uff gon mögent, zu der urle steteß zü sehen alß doch die wahter oder Ir einer so steteß uff dem turn sint wol do zü sehen und fürdrung do zu getün mögent und ob sü eß an dem anfange nit wol kundent, so sint sü eß doch balde zü leren daß ir einer eß so wol kunde alß der münster kneht, dann eß ist nit ein gross kunst Sonder gar balde geleret nuwent* [nur] *von zü sehen und von eim oder zwein slehten zöigen, und were ein notdurfft daß solichs würde geordent und daß den wahtern ouch*

 | DOI:10.1163/9789004423473_004

At the end of the fifteenth century the clock must have been in a defective state because the construction of a new clock was being considered. In 1493 the city of Frankfurt am Main recommended its foreman and clockmaker ("*Werkmeister und Uhrmacher*") Werner Hert (usually called Werner von Buchen), who had already built tower clocks in Worms and Colmar, for building a clock ("*uhrwerck mit eym zeiger in der Höhe*") in Strasbourg.[5] The commission went to the locksmith and clockmaker Hans von Barr in 1494, who received 240 Rhineland guldens for his work.[6] In 1512–1513 the clockmaker

---

geben würde ein slüssel zü der urley und daß doch Ieglich teil bliben solt by dem alten lone und daß die münster kneht nit deste mynner [minder] *zu der urley fürbaß sehen und daß beste tün sollten* (On Good Friday, as the Passion was being preached in the Cathedral, there was considerable unrest and turmoil among the congregation because of the bell-ringing expected from the Cathedral servants; the time bell [clock] stood still and did not strike—the clock had now ceased to run, so that its bell did not chime, because the Cathedral servants could not be bothered to go up to check the clock regularly. As at least one of the watchmen is on the tower in any case, and even if they were originally unable to do this, they can be trained, so that one of them could do it as well as the Cathedral servants. This is not a great feat, rapidly learned by demonstration, and if necessary this should be ordered and the watchmen given a key to the clock. Both groups should be paid their former wages, and the Cathedral servants should do their best to continue to look after the clock.)"; AMS, Sér. II, 46, no. 34. A second record of this case is in Sér. II, 45a (63), no. 22 from 1477.

5 Zülch 1935, 225–226.

6 "*Anno nominato ward das uhrwerckh zu der schlagglocken uff dem münster meister Hansen von Barr dem schlosser und Uhrenmacher, auch burgern zu Straszburg umb zweyhundert und viertzig reinischer guldin, vermög eines uffgerichten kerbzedels, verdingt: ane gemelter summa kam die statt dem werckh umb das halb zu steuer. Und waren hiebey dem verding die strengen, festen, ersamen vnd weisen, herr Wilhelm Böcklin Ritter, herr Jacob Wiszbach, herr Ruly von Druchterszheim und herr Barthlomeus Jörger von wegen der Statt: und von wegen des wercks Junckher Jörg Börer, herr Petter Schott, herr Andreas Habmacher bede altammeister und des wercks pflegere, auch Conradt Hammelburg Schaffner des wercks actum Sambstags nach Jacobi anno 1494* (In the year named the clockwork for the chiming clock of the Cathedral was commissioned to Hansen von Barr, the master locksmith and clockmaker, also citizen of Strasbourg for two hundred and forty Rhineland guldens, by virtue of an established invoice: a reported sum came about for taxing the work by half. And the strict, firm, honorable, and wise Mr. Wilhelm Böcklin Ritter, Mr. Jacob Wiszbach, Mr. Ruly von Druchterszheim and Mr. Bartholomaeus Jörger vouched for this commission: and for the 'Frauenwerk' Junker Jörg Börer, Mr. Peter Schott, Herr Andreas Habmacher, both previous ammeisters and Werk caretakers, also Conradt Hammelburg supervisor of the work, enacted Saturday after St. Jacob's Day anno 1494)"; Dacheux 1887, 17–18. Gérard 1872/73, v. II, 382–382—following the texts of Schadaeus 1617, 21 and Schilter (Hegel 1870/71, 565)—only refers to Johann von Barr as a bell caster who completed the *Nachschlagglocke* (follow-up bell) in 1494. Invoices for the 'Frauenwerk' no longer exist for 1494. In 1510 he made an offer to build a clock for the cathedral chapter in Speyer; Krebs 1968, 292, no. 3091 (June 1, 1510).

Hans Karcher repaired the clock.[7] It disappeared in 1786 and was replaced by a clock built by Johann Philipp Maybaum and Son that is exhibited in the 'Salle d'horlogérie' in the Musée des Arts Décoratifs in Strasbourg.[8]

The tower clock not only regulated public life, its chimes also set the times for contemplation on the Passion of Christ. This is documented by a manuscript on a *Leidensuhr* (Passion Clock) written by a Franciscan monk in Strasbourg in the middle of the fifteenth century:

> It is said, my dear child, that you will be exhorted by the clock, by the hourly chimes of the Strasbourg Cathedral to contemplate the life and suffering of our Lord, as experienced from the evening of Maundy Thursday until the evening of Good Friday. You should therefore know that you must and should take the seventh hour after noon as the first hour of Maundy Thursday evening. That this is so one will recognize as follows: the day and night are just as long as they were for our Lord's suffering, that is, a day has XII hours, or six hours before noon and six hours after noon, before the first hour of the night. The first hour, which is now to be taken as the seventh hour after midday, shall be calculated from the hours struck by the bell of the clock of the Cathedral. The XXIIII hours shall pass and God loving people shall contemplate the life and suffering of our Lord Jesus Christ.[9]

7 AMS, 1 OND 43 (1512–1513), fol. 99r. Transcriptions of the invoice entries for the tower clock in Oestmann 1993/2000, 248.

8 Excerpts from the minutes of the meeting of 1 June 1786 concerning the installation of the tower clock: AMS, Œuvre Notre-Dame, Inventaire des Archives anciennes, No. 15, other documents: Œuvre Notre-Dame, XIXe siècle, no. 588; Musée de l'Œuvre Notre-Dame, box 43; Strasbourg, Archives Départementales, V 348 (1808–1854); as well as AMS, Einundzwanzigerprotokolle, v. 29, 1551, fol. 91r; v. 87, 1606, fol. 266v–267r; v. 262, 1784, 112–113 and 145–147; v. 263, 1785, 83–84; BNUS, Ms. 1645 (Als. 1072), dated 30 April 1784.

9 "*Das ist darumb gesprochen, liebes kint, ob du ermanet wilt werden by dem orley by der stunde glocken des münsters zu Strossburg des lebens und lidens unsers herren, als es sich ergieng von dem grünen donrstag zu nacht bis an den karfritag zu nacht. So soltu wissen, das du die sübende noch mittage must und solt für dich niemen für die erste stunde des grünen donrstages naht. Das dis wor sy, das merck also: wer ingenote tag und naht glich las* [lang] *als es was do unser herre leid, so werent* XII *stund tags also das sehs stunden werent vor mittage und sehs noch mittage biß an die erste stunde der naht. Die erste stunde, das ist nü die sübende stunde noch mittemtage zu nemende und zu rechnende noch dem, als das orley der stund glocken zu dem münster sleht. Der die* XXIIII *stunden darin wil ziehen und darumb so sol der minnende mensche des lebens und des lidens unsers herren Jhesu Christi* [gedenken]"; Oliger 1918, 173.

Following this was a list of the 24 hours associated with the stations of the passion of Christ.[10] The text was clearly directed at the laity,[11] in as much as the contemplation of Christ's suffering was possible for every resident, as the clock's chimes could be heard throughout the city. Even if a similar function cannot be proven for the first Cathedral clock,[12] a contemplative function of the clock with its mechanized mystery play seems not impossible.

---

10 *Ibid.*, 173–175.

11 *Ibid.*, 169–170.

12 Oliger does not distinguish between the Cathedral's tower clock and the astronomical clock, the *Leidensuhr* (Passion Clock) definitely only referred to the "*orley der stund glocken*" (tower clock with hourly chimes), which was called the *Zitglocke* (time bell) in Koenigshoven's chronicle.

CHAPTER 3

The Project for Building a New Clock in the First Half of the Sixteenth Century

After the Reformation was introduced to Strasbourg in the 1520s, efforts were made to replace the astronomical clock with a new one. In 1530 the city asked for permission from the City Council of Frankfurt to inspect the clock in the Cathedral of St. Bartholomew with its Three Magi, stars, planets, and other works of art.[1] That clock, built in 1372–1373, was rebuilt by the clockmaker Johann Schnitzler of Hagenau in 1384,[2] and it is suspected that he used the first Strasbourg astronomical clock as a model.[3] At least that clock had a three tier structure (calendar, astrolabe, moving mechanical figures), just like the Strasbourg one. In 1509–1510 and 1529, the clockmaker Lazarus Kreger von Bare (Barr is located near Strasbourg), who went to Frankfurt in 1508, undertook further repairs and expansion work on the clock.[4] Back in 1493, the builder of the tower clock, Werner Hert, was recommended to Strasbourg. Therefore there were connections between the two free imperial cities, so that the request for technical and artistic information from Frankfurt was certainly not a coincidence. However, it appears that the city council of Strasbourg had false information about the Three Magi automata and the planet hands, as no other source exists to document their presence in the Frankfurt Cathedral clock. If we now consider the accounts of Conrad Dasypodius, which are frequently used for establishing the chronology of events, we must conclude that actual construction on the new clock was not begun until 1547:

> After the works of such an old clock had completely come to an end, an honorable council here recognized in the year of our Lord 1547 approximately to have another clock built, not on this location [i.e. of the first Cathedral clock] but across from it, as now can be seen; had commissioned the honorably learned and understanding Mathematicians

1 Stadtarchiv, Frankfurt am Main, Reichssachen II, 743; cited in Bund 1986, 250.

2 There is no documentation of Schnitzler in Haguenau (information kindly provided by Ms. Pia Wendling, curator, Library and City Archives of Haguenau).

3 Bassermann-Jordan 1929a, 441; for sources on the Frankfurt clock see also Jenzen 1989, 39–49.

4 Lazarus Kreger von Bare received the rights of a citizen of Strasbourg in 1505; AMS, Bürgerbuch I, col. 382 (1505); Wittmer/Meyer 1948/61, v. II, 549, no. 5492; see also Zülch 1935, 242; Bund, 1986, 246.

 | DOI:10.1163/9789004423473_005

> (for which no clockmaker or smith can be found ever again for planning and building such or similar works for Astronomical clocks) namely Doctor Michael Herus, Nicolaus Brucknerus, Christianus Herlinus, and besides them other craftsmen and when the work was so far that the clockmakers had completed numerous gears, and the stand was finished, the stonemason had set up the housing all the way up to the top: By the Mathematicians the Astrolabe, as I later received and still exists, was also planned. After which the work shall be built, and one had determined the spiral staircase to a large extent. But such work was by several deaths and other importunities as occurred then thus prevented and therefore remained unfinished until Anno 1571. [...] At the same time the housing for the twelve signs and the moonlight [dial for the moon] standing so elaborately in front of the Church was also made, the drive mechanism for the new work as can now be seen by means of which the course of the Sun and the Moon and its moonlight could be seen, but everything remaining as planned.[5]

The minutes of the Council of XXI (the committee which was responsible for the construction of the Cathedral) do not document such a decision in 1547, however; the invoices of the 'Frauenwerk' for 1531–1532, which have not been taken into consideration so far, indicate that work was started directly after the inspection of the clock in Frankfurt. Initially an outer dial (diameter 2.85 m) was placed in a square field (side length 3.25 m) on the facade of

5 "*Nach dem nun solches altes Uhrwerck gantz und gar in abgang kommen, so hat ein Ersamer Rhat allhie mein G.G. Herren Anno ungefehr 1547 erkandt, ein ander Uhrwerck aufzurichten, nicht an disem ort, sonderen dargegen uber, wie dann jetzmalen gesehen wirt, und seind auch darzu verordnet gewesen fürnemme gelehrte und verstendige Mathematici (on welche kein uhrenmacher oder schmitt solches und dergleychen Astronomisch uhren werck nimmermehr erfinden angeben anordnen und aufrichten würdt), nemlich Doctor Michael Herus, Nicolaus Brucknerus, Christianus Herlinus und neben ihnen andere Handtwercks leut, und ward das werck so weit gebracht das der Uhrenmacher etliche redder und das gestell verfertigt hat, der Steynmetz das geheuß aufgefuert biß gar nach an den helm, die Mathematici das Astrolabium so ich hernaher bekummen und nach vorhanden ist, auch aufgerissen haben. Nach welchem das werck solte gemacht sein worden, und hat man den Schnecken auch zu gutem theyl außgemacht gehabt. Aber solches werck ist darnach durch etlicher absterben und anderer ungelegenheit, so dazumalen einfielen verhindert und also unaußgemacht verbliben biß Anno 1571.* [...] *Es ist auch zu der selbigen zeit das geheuß mit den zwoelf zeichen und dem Monscheyn so außwendig vor der Kirchen steht gemacht worden, das von dem jetzigen newen werck ein trib gieng, wie dann jetzmalen zu sehen, durch welchen der Sonnen und des Mons lauf und seine Monscheyn zusehen waren, aber wie vorgemelt alles verbliben*"; Dasypodius 1580a, 5–7.

the southern transept towards the *Fronhof*.[6] The dial was located below a late gothic balcony with tracery built in 1493 together with a sundial framed with branches the gnomon of which (no longer existing) is held by half of a figure.[7] The following years were listed on the banderoles held by four putti in the spandrels: 1533 (completion; in Roman numerals), 1572 (transfer to the new clock by Dasypodius and the Habrecht brothers), 1603 and 1669 (repairs or maintenance work; the last three years are given in Arabic numerals). The dial, which was greatly altered in the nineteenth century, used to have Roman numerals for the hours on the outside and twelve gold-plated wrought-copper plates (diameter 430 mm) with the signs of the zodiac.[8] Besides the hour, the positions were indicated for the Sun and Moon in the zodiac (thus there must have been three hands) as well as the length of the day on a compass-rose-like field in the middle.[9] The phases of the Moon were indicated by a rotating sphere above the dial. Hans Haug assumes the design of the copper zodiac plates was influenced by Hans Baldung Grien and dates them at not much later than 1533 after he compared styles extensively.[10] Invoices prove Haug's date to be correct, as the completion of twelve copper signs by the Strasbourg goldsmith Caspar Engel is listed for 1531–1532, whose fellow artisan Stoffel (Christof) Stedlin procured gold leaf for gilding the plates.[11] Thus the plates did not—as Haug postulated[12]—originally have polychrome decorations.

---

6 The claim in Specklin's chronicle that the dial with planets ("*uhrzeyger mit den planetten*") already existed in 1493 is erroneous; Reuss 1890, 474.

7 The original sundial is now located in the Musée de l'Œuvre Notre-Dame, Strasbourg; see Ungerer/Glory 1933, 85–91 and 101–103; Zinner 1964, 17.

8 Figure in Haug 1933, 32–33.

9 Dasypodius 1580a, 53.

10 Haug 1933, 34–42. His assumption that a drawing by Baldung was used as a model for the design of the signs of the zodiac has recently been called into question by Châtelet-Lange 2010, 11.

11 "*Item Meister Caspar Engel dem goldschmid von den zwelff kupferin zeychen zu gebruch des zeygers von yedem iii gulden sampt ii gulden seinen gesellen zu drinckgeldt geschenckt, zu dreyben und zuschlagen geben xvi £ x ß ix ₰; Item kaufft durch Meister Stoffelns Stedlin den goldschmid, anderthalb buch doppel geschlagen gold zu gebruch egemelter zwelff zeichen damit zuvergulden darumb geben v £ v ß ₰* (Likewise goldsmith Master Caspar Engel for twelve copper [zodiacal] signs for the use of the dial, each to 3 guilders including 2 guilders for his journeymen bestowed as gratuity, for driving and beating given 16 pounds [£] 10 shillings [ß] 9 pfennigs [₰]; Likewise purchased by goldsmith Master Stoffel Stedlin one and a half books double-beaten gold for use on the aforementioned twelve [zodiacal] signs, to gild them, for this given 5 pounds [£] 5 shilling pfennigs [ß ₰])"; AMS, 1 OND 43 (1531/32), fol. 57^r.

12 Haug 1933, 29.

Besides, payments were made to the physician Michael Herr (d. 1551) with the construction of an astrolabe dial, who bought paper, glue, and nails for the drawings.[13]

Herr, who was born in Speyer, had studied medicine in Heidelberg in 1508 and enrolled for studies in Vienna in 1526,[14] where he attended a lecture on medicinal astrology by Georg Tannstetter.[15] In 1527–1528 Herr continued his studies in Montpellier, received the rights of citizenship through marriage in Strasbourg in 1528,[16] and became the director of the municipal hospital six years later.[17]

The 'Frauenwerk' commissioned a painter with the name Claus for the *Visierungen* (plans) for two hands as well as for the purchase of various pigments and linseed oil for painting the astrolabe.[18] This painter was Nikolaus

---

13 "*Item kaufft durch Doctor Michel Her iii xxx Bögenkarthen Bapier leym clein und gros negel, Item gleich und anders zu gebruch vorgemelter visierungen darumb geben i £ xiii ß viii ₰; Item demselben uff sein belonung der Visierungen des Astrolabiums so er dem werck gemacht hat geben iiii £ iiii ß ₰* (Likewise purchased by Dr. Michel Herr 33 sheets of paper, glue, small and big nails, Likewise [to] the same and other for the use of the aforementioned drafts, for this given 1 pound 13 shillings 8 pfennigs; Likewise [to] the same [as] remuneration for the drafts of the astrolabe he has executed for the Works [of Strasbourg Cathedral] given 4 pounds 4 shilling pfennigs)"; AMS, 1 OND 43 (1531/32), fol. 57ʳ. Given the clear evidence in the sources, Dengler's reservations about dating the layout of the astrolabe (Dengler 2011, 116) appear completely unfounded.

14 Schrauf 1894/1904, v. III, 167; Gall/Szaivert 1967, v. III, 42; further biographical information in NDBA, XVI, Straßburg 1990, 1543 (Théodore Vetter); ADB, XII, 204/205; Sitzmann 1909/10, v. I, 758/759; Müller-Jahncke 1999. As to his scholarly works—among other things, he was involved, together with Nicolaus Prugner, the woodcarver Johann Weiditz, and numerous pharmacists and herb gatherers, in the publication of the botanical work *Herbarium vivae eicones* initiated by Otto Brunfels—see Chrisman 1982, 175, 179.

15 Herr gave his lecture notes to Otto Brunfels, who published them in Strasbourg under the title *Artificium de applicatione Astrologiae ad Medicinam* [...] in 1531; for information on this see Sudhoff 1902, 45–47.

16 The entry on the rights of citizenship can be found in Bürgerbuch I (1440–1530), col. 634 (1528); in Wittmer/Meyer 1948/61, v. II, 787, no. 8676.

17 Müller-Jahncke 1999, 206.

18 "*Item Meister Caspar dem Urlenmacher uff das verding des wercks der heyligen drey künig Im Münster geben xxvi £ v ß ₰;* [...] *Item Meister Clausen dem Maler von zweyen Visierungen der zweyen zeyger vornen an das Munster und uff dem fronhof zumalen geben v £ xiiii ß ₰* [...] *Item kaufft durch denselben iiii £ Blauwe Lasur yedes £ vor ii gülden zu gebruch des Astrolabiums Im Münster darumb geben iiii £ iiii ß ₰; Item kaufft durch denselben* [Meister Clausen] *Im Martin Hugen gaden vi £ virniß i £ zinober ii £ bly weysz ii £ menig iii £ Silberglett darumb geben ii £ iii ß viii ₰; Item kaufft durch denselben zwo masz leinöle darumb geben v ß ₰* (Likewise given to the clockmaker Master Caspar according to the contract for the work on the Three Magi in the Cathedral 26 pounds 5 shilling pfennigs [...] Likewise to Master Claus the painter for painting two drafts of the two dials in front of the Cathedral and on the 'Fronhof' 5 pounds 13 shilling pfennigs [...] Likewise purchased by the same

Kremer (d. 1553), who was born in Baden-Baden and became a citizen of Strasbourg in 1521. He was a student of Hans Baldung and took over his estate.[19]

Other indications for new construction work are two oil paintings by Hans Baldung Grien. One wooden panel at the Nationalmuseum, Stockholm (196.0 × 63.4 cm, inv. no. NM 1073), displays the planetary god Mercury, standing with a fig in his right hand and the caducaeus in his left, with a lion and a fountain in the background. The back, which was previously divided by a crossbeam, was painted with pictures of Mercury (top) and Saturn (bottom) in grisaille. In the Angermuseum in Erfurt there is a second painting of almost the same size (186.0 × 63.7 cm) in which the creation of man and animals is portrayed.[20] Here too the back, divided in two parts, was covered with grisaille paintings: the sun god Sol (top) and a baluster vase (bottom). Researchers agree that originally there must have been four panels, which resulted in a total of eight fields on the backs for the seven planets. As a result, the bottom field of the Erfurt panel is merely ornamental. Based on the notes of the Baldung researcher Carl Koch, Gert von der Osten suggested a boxlike arrangement of the four panels (like a winged altar) around a dial.[21] But when opened, only the grisaille paintings would have been visible, which hardly seems plausible. Max Hasse therefore pointed out the three-tier structure common in monumental clocks. He postulated that the Mercury panel was located next to the astrolabe on the left, whereas the Erfurt panel with the creation scene was placed on the lower right, next to the calendar disk.[22] His hypothesis that Baldung created the panels for the first Cathedral clock has to be met with skepticism, however. But noticeably frequent sources dated from 1530 on tend to indicate that Baldung and his workshop were involved in the new clock project. This assumption is backed by the stylistic influences on the design of the plates with the signs of the zodiac on the outer dial already mentioned, and possibly also by the fact that he was related to the mathematician Christian Herlin (he was married to

4 pounds blue translucent color, each pound to 2 guilders for the use of the astrolabe in the Cathedral; for this given 4 pounds 4 shilling pfennigs; Likewise purchased by the same [Master Claus] in Martin Hugen's shop 6 pounds varnish, 1 pound vermillion, 2 pounds white lead, 2 pounds minium, 3 pounds lead oxide; for this given 2 pounds 3 shillings 8 pfennigs; Likewise purchased by the same two measures of linseed oil; for this given 5 shilling pfennigs)"; AMS, 1 OND 43 (1531–1532), fol. 56^{v}–57^{r}.

19 Rott 1933/38, v. III (Text), 89–100; v. I (Sources), 225–226.

20 A detailed description of these two paintings and their origins in Osten 1983, 214–216, and fig. pls. 162 (front) and 163 (back).

21 *Ibid.*, 218–221.

22 Hasse 1984.

Margaretha, one of Herlin's sisters).[23] Nonetheless, the manner in which the two-sided panels were attached to the clock housing remains unclear.

It therefore seems reasonable to assume that the city council's decision to build a new clock must have been made earlier (probably around 1530). This, however, cannot be proven, as the minutes of the meetings of the Council of XXI only go back to 1539 (only one volume exists with excerpts from 1523–1533, in which there are no entries on this matter).

Nicolaus Prugner—who was born in Windsheim in central Franconia on 22 October 1488, and enrolled for studies in Vienna in 1516, where he completed his doctorate—most probably participated in the work as early as 1531.[24] He was the prior of an Augustinian monastery in Mulhouse and advocated the introduction of Lutheran reforms. His position in the small free imperial city became increasingly more difficult, and in 1526 he was expelled from the city by the city council. Whether Prugner had compromised himself with his sympathy for radical reforms and the Peasant Uprising has not been clearly proven (Oecolampadius attributed his expulsion to the machinations of Johannes Fabri, vicar general of Constance). He was able to obtain a position as pastor in the town of Benfeld, which the Strasbourg bishop had given to the city as security. There he neglected his clerical duties and intensively pursued mathematics, astronomy, and astrology. In Benfeld, Prugner wrote on 11 February 1531 that the Strasbourg Senate had urged him to complete an

23 Osten 1983, 294–296.

24 The earliest biography by Heinrich Pantaleon (Pantlin, 1522–1596; Burmeister 2015, 435–436), recognizes Prugner not as a theologian, but as a mathematician and astronomer; Pantaleon 1565/66, pt. III, 249. The portrait that was included was not authentic; Pantaleon, lacking a portrait, used the image of a bearded man with a quadrant for several scholars: for example, for the astronomer Georg Peurbach (*ibid.*, pt. II, 442) and Erasmus Oswald Schreckenfuchs (*ibid.*, pt. III, 422). There is a portrait of Prugner (oil on parchment; height 74 cm, width 54 cm) in the Museum für Astronomie und Technikgeschichte in Kassel, inv. no. 1947/65, on which see Mackensen 1995 and Oestmann 1996. Friedrich Hagenauer created a medal in 1541 with a bust of Prugner; Mieg 1957. For the publication of a collection of astrological works by Johannes Herwagen in Basel in 1533—which besides Arabic and 'Chaldean' authors included Ptolemy's *Tetrabiblos*, the didactic poem by Marcus Manilius, *Astronomicon Libri V*, and the hundred aphorisms of astrology, *Centiloquium*, generally attributed to Ptolemy at that time—Prugner submitted a new edition of the *Matheseos* by Firmicus Maternus; Cantamessa 2011, v. I, no. 2633, 908–910; Houzeau/Lancaster 1882/89, v. I.1, 393–394, no. 761. A further edition appeared in 1551. On the biography see: Lutz 1902/12; Röhrich 1855, v. II, 180–201; Mieg 1929; Mieg, 1939; Rau 1963; ADB, v. XXVI, 674–675; Sitzmann 1909/10, v. II, 470–471; NDBA, XXX, Strasbourg 1997, 3056–3057 (Raymond Oberlé); Oestmann 2005, 257–258. Peter Schiller analyzed the two horoscopes supplied by Prugner for Bonifacius and Bruno Amerbach, which unfortunately appear in the publication in a very marred and distorted fashion; Schiller 1997.

"*opus astronomicum*".[25] When the bishop of Strasbourg regained control over Benfeld in 1537, Prugner lost his position and went to the court of the archbishop and elector of Cologne Hermann von Wied (1466–1552), who was more lenient towards Protestantism and therefore excommunicated in 1546, so that Prugner once again had to move. He went to Basel, where he was supported by Count George of Württemberg-Mömpelgard. He was also in close contact with the elector of the Palatinate Ottheinrich (1502–1559). In 1547 Prugner received a pension from the Heidelberg court,[26] and in 1553 he published an astronomical-astrological collection that he dedicated to the elector.[27] By 1548 at the latest, Prugner was residing in Strasbourg and appears to have worked as a physician there.[28]

Prugner's final home was Tübingen, where he enrolled at the university on 26 October 1553. In a letter he is referred to as a professor of mathematics,[29] but there is no proof that he held such a position. When the Tübingen mathematics professor Philipp Imsser (1500–1570) visited Ottheinrich in Baden-Baden in 1554 to gain his support for a new astronomical clock, Prugner accompanied him.[30] Prugner died on 25 February 1557 and was buried in St. George's Collegiate Church in Tübingen. The epitaph on his tombstone has been handed down.[31]

The collaboration of Christian Herlin can only be documented in references from his student and successor Dasypodius. Little is known about the life of the professor of mathematics who died in 1562. He was probably the nephew of the *ammeister* Martin Herlin, studied in Freiburg (1518) and in Heidelberg (1519–1521), went to Tübingen in 1522, and lived in Wittenberg in 1522–1523. In 1526 Herlin enrolled at the University of Vienna. As the first *mathematicus*

25 "*Nam nescius etiam sum, quando senatus ille Argentoratensis me denuo ad complendum, finemque perducendum opus illud astronomicum evocabit*"; Letter to Ulrich Würtenberger, reprinted in Lutz 1902/12, pt. III, 57–58. The original was destroyed in the Franco-Prussian War in 1870.

26 Rott 1905, 71.

27 The first part consisted of an improved version of the Alfonsine Tables of Giovanni Bianchini, including explanations; the second part contained the eclipse tables of Georg von Peuerbach, also with explanations. The municipal library of Strasbourg once owned numerous letters on mathematical and astronomical subjects hand-written by Ottheinrich to Prugner between 1548 and 1554 (Schmidt 1856, IX) which no longer exist. These sources were most probably burned during the bombardment of the library in 1870 during the Franco-Prussian War.

28 Mieg 1939, 44, 48, no. 1.

29 *Ibid.*, p. 48, no. 2.

30 Roth 1877, 168.

31 Westermayer/Wagner/Demmler 1912, 202–203; Rau 1964, 433; Mieg 1970.

publicus in Strasbourg, he taught mathematics as well as astronomy and geography starting around 1530.[32]

Because several invoice books between 1531 and 1569 are missing, we cannot know how far the work was completed beyond the plans for the astrolabe, the construction of the outer dial, and the painting. The paintings could have been intended for the old clock, which was possibly damaged from the iconoclasm following the abolition of the Catholic mass by the city council in 1529–1530.[33] Attention, however, is attracted by two invoices from 1543–1544 and 1548–1549 in which an "*uhren by den dryen künigen*" (clock of the Three Kings) is mentioned,[34] as is a new *Schnecke* (spiral staircase).[35]

The wall opposite the first Cathedral clock in the southern transept was chosen as the location for the stone housing described by Dasypodius, and built in front of a recess with a depth of about 1.5 m[36] as a three-tier structure with a spiral staircase that led to a narrow gallery, intended to provide access to both the outer dial and the upper tiers. The reversed, 'overturned' arrangement of the columns on the spiral staircase is conspicuous. The first three turns of the stairs are supported by Corinthian columns; the four columns of the upper turns (of which the two outer ones have a larger diameter than the two inner ones) have Doric capitals and fluted shafts. The gold leaf on the capitals and the marbling of the column shafts was renovated in the nineteenth century; to what extent this corresponded to the original is not known. On both sides of the clock tower at the level of the astrolabe tier there is a small round medallion, each with the bust of a man carved in stone, with the man facing right (on the left side) and facing left (on the right side). Whether the heads in profile,

32 Sebitz 1641, 279–280; Ficker/Winckelmann 1902/05, v. II, 80; Sitzmann 1909/10, v. I, 755; Kuhn 1971, v. II, 320, no. 1865 (including proof of university matriculation); Schindling 1977, 254–255; NDBA, XVI, Strasbourg 1990, 1534 (Günther Oestmann); Burmeister 2015, 285.

33 Büheler, *Chronique strasbourgeoise*, in Dacheux 1887, 79. Sebald Büheler (1529–1595) was a painter and completed his two-volume chronicles in 1588. He made additions up until 1594; Sitzmann 1909/10, v. I, 258–259; NDBA, V, Strasbourg 1984, 415 (François Joseph Fuchs). The original was burned in 1870. Dacheux compiled the excerpts from Silbermann, Roehrich, Piton, and Schneegans, so that the chronicles—though fragmentary—are once again accessible.

34 AMS, 1 OND 43 (1543–1544), fol. 51r.

35 AMS, 1 OND 43 (1548–1549), fol. 68r. The author's erroneous reading of this entry in the German edition of this book has been rendered correctly in Châtelet-Lange 2010, 13.

36 Before 1348, this was the location of the Holy Sepulcher, and after it was moved to St. Catherine's Chapel, the location of an altar to St. Richard ("*Wan vormols was das heilge grap in dem kemmerlin gegen dem urley, do nu sant Richart alter stet*"); Koenigshoven, *Chronik*, in Hegel 1870/71, 668–669.

which are hardly individualized, were to represent a well-known person—perhaps the builder of the Cathedral—is a moot question.

In the front, the clock tower rests on two pillars with a twined rib vault, the design of which is identical with that of St. Catherine's Chapel (*Figs. 3.1* and *3.2*). St. Catherine's Chapel was built by Master Gerlach in 1340 in the corner between the southern side nave and the southern transept.[37] The original star vault with descending keystones no longer exists and was replaced by Cathedral master builder Bernhard Nonnenmacher in 1542–1546 with the twined rib vault visible today.[38] On its keystone there are two putti holding the stonemason's seal of the 'Frauenwerk' and a banderole around the outside with the inscription: "*GAR DAS HANGEN GEWELB ZERREIS GAR DO MACHT DAS WERCK DIS NEW. AN*[*N*]*O D*[*OMI*]*NI M.D.XLVI.* (Even as the hanging vault was torn down, even as this new work was made in 1546 AD)".[39] An identical keystone (without the putti) is in the clock housing, but the inscription on the banderole reads: "*OPTIMA QVAEQVE DIES MISERIS MORTALIBVS AEVI PRIMA FVGIT, SVBEV*[*N*]*T MORBI TRISTISQVE SENECTVS* (The best days of life are always the first to vanish for miserable mortals, grievous death and old age supersedes)".[40] According to the records of the foreman Johann Georg Heckler (1630–1682), Bernhard Nonnenmacher had been the foreman from 1520 to 1538 and Bernhard von Heidelberg from 1539 to 1551.[41] They were one and the same

---

37 Recht 1974, 54–69.

38 "*Auch in dissem jar 1542, da haben die herren pfleger Unser Lieben Frauen haus alhie zu Strassburg, durch ieren werckmeister disser zeit nemlichen meister Bernhart Lattenmuss* [sic, the name should be Lathomus], *das hangende und künstlich gewölb im Münster in S. Catharinencapell neben der thüren so man bey dem brunnen im Münster hinüss in die steinhütten will gohn, hinweg und ab lassen brechen und ein ganz neu gewölb gemacht das noch stehet und sehr zierlich ist* (Also in this year 1542, the caretakers of Our Lady's house here in Strasbourg, through their foreman Bernhart the Stonemason, have torn down the hanging and artistic vault in St. Catherine's Chapel in the Cathedral next to the doors where one can go out next to the fountain in the Cathedral into the stone cottages and they have made a completely new vault, which still stands and is very elegant)"; Büheler, *Chronique strasbourgeoise*, in Dacheux 1887, 86–87, no. 286. Nonnenmacher's sketch of the vault still exists at the Musée de l'Œuvre Notre-Dame, Strasbourg; see Recht 1989, 373, no. B 29; Bucher 1968, 62–63.

39 Illustration in Haug 1933, 43.

40 Virgil, *Georgics*, III: 66–68; illustration of the keystone in Ungerer 1931, pl. VIII.

41 "*Nach Ihm* [Jacob von Landshut] *ist kommen Hanß Hammerer anno 1510. biß anno 1520 ongefehr nach ihm Bernhard Nonnenmacher, findt ich daß er anno 1538. noch Werkmeister war, aber von anno 1520. bis 1538. Wie lang Hammerer oder Nonnenmacher gewest, einer vor oder nach dem andern kan ich nicht wissen. Anno 1539. ward Bernhard von Heydelberg Werkmeister bis 1551* (After him [Jacob von Landshut], Hanß Hammerer came in from 1510 until 1520 approximately, [and] after him Bernhard Nonnenmacher, I find that he was still foreman in the year 1538, but from 1520 until 1538. How long Hammerer and

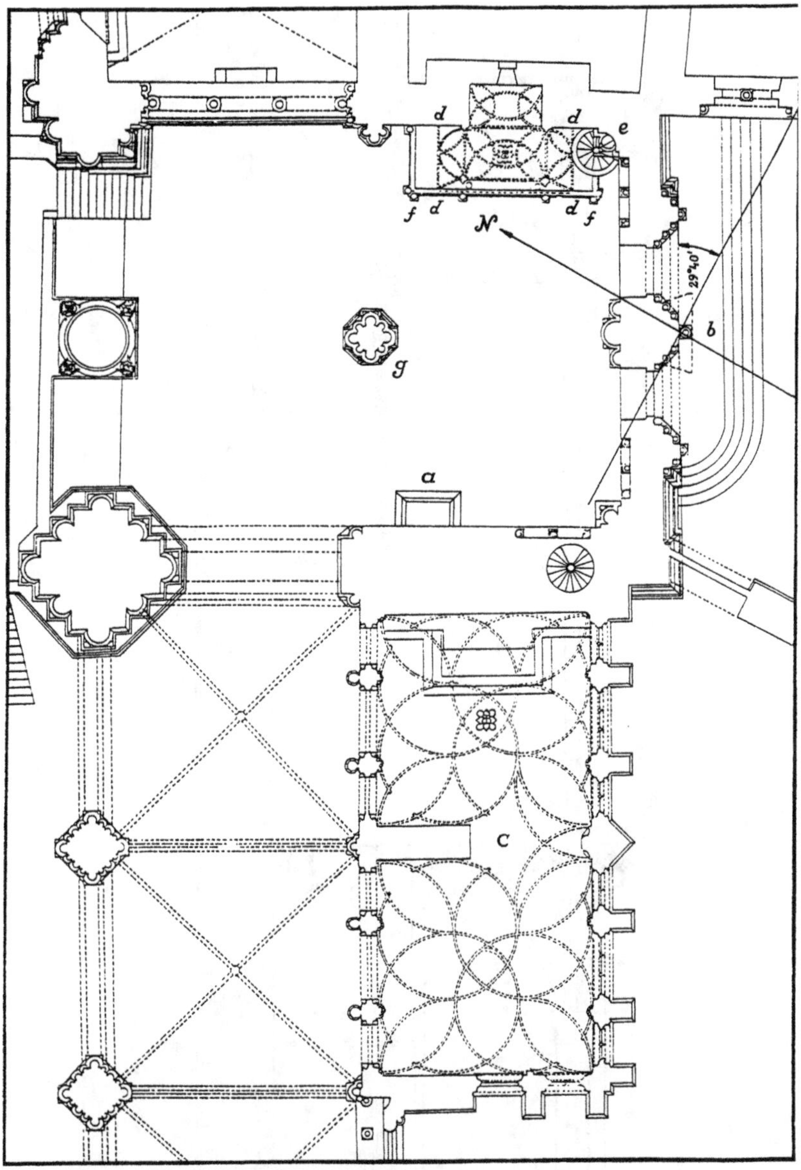

FIGURE 3.1 Floor plan of the Southern Transept
a: Location of the first astronomical clock, 1352–1354
b: Outer dial towards the *Fronhof*, 1533
c: Vault of St. Catherine's Chapel, 1542–1546
d: Original construction of the renewed astronomical clock, 1531–1550
e: Spiral staircase, 1547
f: Wooden case of the clock of 1571–1574
g: Pillar of Angels

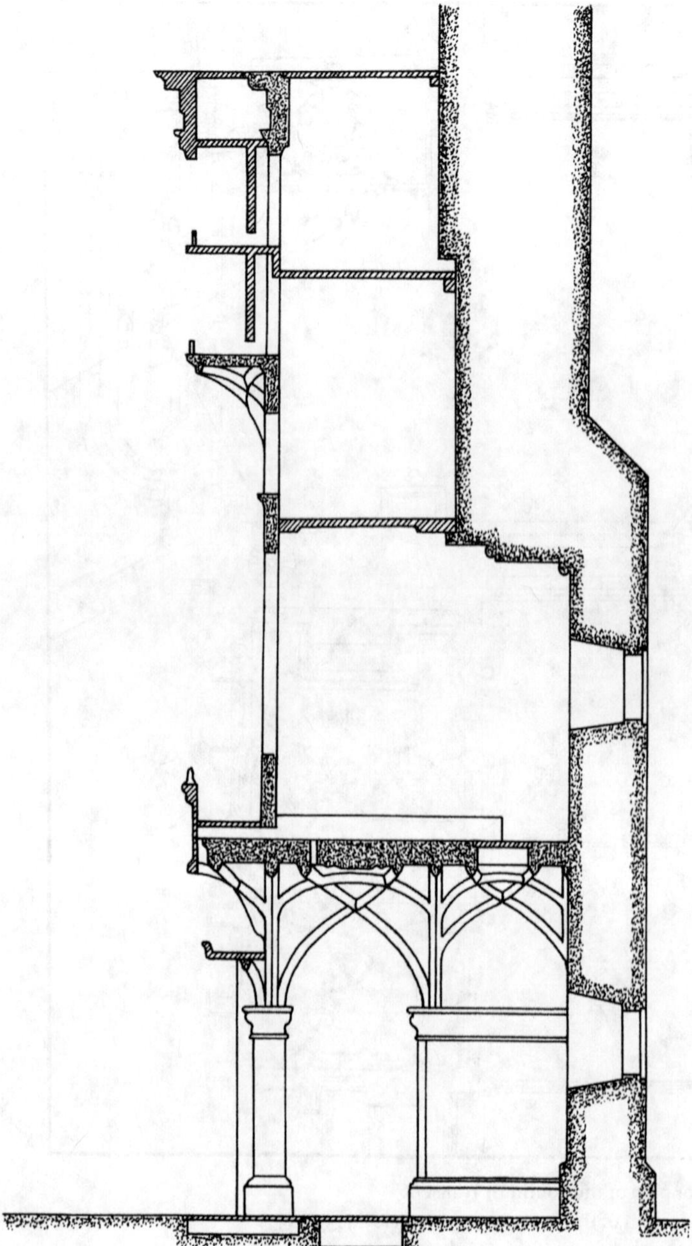

FIGURE 3.2 Cross section of the housing

person, as the latter was mentioned in each of fifteen adjudications between 1521 and 1550,[42] and "*Bernhart Nonnenmacher von Heydelberg*" received the rights of citizenship in 1519.[43] Under his supervision the new clock housing was set up in the southern transept.

The dates originally cited by Dasypodius were therefore correct and the clock housing with the staircase was constructed in 1547, after the vaulting of St. Catherine's Chapel.

In 1550, Catholic masses were celebrated once again in the Cathedral, following the imperial decree of 1548 by Charles V, the Augsburg Interim.[44] This led to a stop in the building activities, in as much as all of the scholars involved were Protestants. When at the end of 1559 the city council finally lifted the Interim, despite protests from the Roman Catholics, and after the Cathedral was temporarily closed, Protestant services were held there again in May 1561.[45] The only scholar who had participated in the work and was still alive was Christian Herlin, and after his death in 1562, the project for building a new astronomical clock came to a complete standstill.

Nonnenmacher were there, one before or after the other, I do not know. From the year 1539 Bernhard von Heidelberg was foreman until 1551)"; BNUS, MS 1373 (Als. 704), fol. 14v. Kraus originally followed Heckler's assumption that there were two successive supervisors (Kraus 1876, 407, 409), but corrected this mistake in the addenda that appeared in 1892 (v. IV, 33). His suggestion—based on Grandidier 1782, 341—that Daniel Specklin could have been the architect for the vault (Kraus 1876, 409) is not tenable, as he was only six years old when the vaulting was carried out. This had already been pointed out by Haug 1933, 30, n. 4. On Nonnenmacher see also Rieger 1996.

42 Schulte 1882, 276–277.

43 "*Item Bernhart Nonnenmacher von Heydelberg der Steinmetze hat das burgrecht von Katherinen meister Hans Hamers Unser Frowen Wergks meister dochter siner elichen Hußfrowen wegen empfangen wil mit den murern dienen. Actum tercia xvij maij* (Likewise Bernhart Nonnenmacher von Heydelberg the stonemason received the rights of citizenship as a result of his marriage with the daughter of St. Catherine's master and foreman of Our Lady's Works, Hans Hammer, his housewife in wedlock, and wants to serve with the masons. The third decreed on May 17th)"; AMS, Bürgerbuch I (1440–1530), col. 472 (1519); in Wittmer/Meyer 1948/61, v. II, 650, no. 6815. Bernhard von Heidelberg is mentioned in the invoices of the 'Frauenwerk' for 1548–1549; AMS, Inventaire des Archives anciennes, no. 43, fol. 68r and 69v.

44 Grandidier 1782, 99–103; Büheler, *Chronique strasbourgeoise*, in Dacheux 1887, 94–95.

45 *Ibid.*, 105–106 and 109.

CHAPTER 4

Completion of the Second Clock in 1571–1574

The second Cathedral clock, the technical execution of which was coordinated by Isaac and Josias Habrecht, was designed by the Strasbourg mathematics professor Conrad Dasypodius. For the decoration of the clock housing the painter Tobias Stimmer from Schaffhausen was commissioned.

4.1 Conrad Dasypodius

Conrad Dasypodius (Hasenfratz)[1] was born in 1529 or 1531 in Frauenfeld in Thurgau, Switzerland, the son of the humanist Petrus Dasypodius, and died in Strasbourg on 21 May 1601.[2] He was educated by Christian Herlin at the

1 There are different variations of the German family name Dasypodius (Rauchfuß, Hase, etc.) in the literature. There is no documentation for this name whatsoever in the town Frauenfeld, where he was born. Büeler could prove that the German name of his father was Hasenfratz (harelip; Greek: *lagócheilos*, Latin: *labium leporinum*). Petrus Hasenfratz was probably unable to form a pleasant sounding name from these translations, so he grecized the first part of his name (Greek: *dasypus* = hare); Büeler 1918. Biographical information on Conrad Dasypodius can be found in Adam 1615, v. I, 441–443, the oldest, frequently quoted biography; Sebitz 1641, 280; Vossius 1650, Ch. XXVI, § 13, 111, and Teissier 1715, v. IV, 371–372, a short biography with information on the works of Dasypodius, based on Adam and Vossius. In Weidler 1741, Ch. XIV, 380–381, there is some brief biographical information besides a list of some of Dasypodius's writings. There is a short biography based on Adam and Witte in Fix 1792, v. I, 282–285; see also Wolf 1845/54 (expanded edition: Wolf 1860); Ficker/Winckelmann 1902/05, v. II, 94; Zedler, VII, col. 225, ADB, IV, 764; Sitzmann 1909/10, v. I, 450; DSB, III, 585; Jud 2008; and Burmeister 2015, 173–174. For a bibliography of his works—the two German descriptions of the Cathedral clock from 1578 and 1580 are missing—see Blumhof 1796, and for Dasypodius's mathematical texts see Kästner 1796/1800, v. I, 332–334. There is no extant portrait.

2 There are contradictions on the biographical dates and place of birth in the literature mentioned above. However, Dasypodius wrote to Johannes Kepler in 1599 that he was in the 70th year of his life; ÖNB, Cod. Vin. 10702, fol. 177; KGW, v. 14, 105. Then he would have been born in 1529. For the date of his death, I have used the information provided in the latest article by Jean Rott, NDBA, VII, 583. After comparing various sources, C.E. Wagner takes 21 April as the date of death; AMS, 90 Z 22, D 21 and D 278. Rudolf Wolf's claim that the clockmaker Jost Bürgi (1552–1632) lived in Strasbourg and was recommended to Landgrave William IV of Hesse-Kassel by Conrad Dasypodius (Wolf 1877, 273–274), cannot be supported by sources. A stay in Milan, Florence or Cremona seems more likely, but absolutely nothing is known about Bürgi's early years before he was called to the court of Kassel in 1579; Kaltenböck 1981, 77–79; Wehrli 2001, 45.

 | DOI:10.1163/9789004423473_006

Gymnasium in Strasbourg and studied in Leuven and Paris after that.[3] He took over the teaching position for mathematics after Herlin died (1562) and became the canon of St. Thomas's Church one year later. In 1567 Dasypodius was awarded the title of magister by the Strasbourg Gymnasium, which had been upgraded to a college by the imperial privilege of Maximilian II in 1566,[4] and held the office of the dean in 1578–1579 and 1588–1589.[5] Directly after completing his work on the Cathedral clock in Strasbourg, Dasypodius, with a letter of recommendation from Maximilian II, applied for another ambitious project which the city council of Olomouc had written up: the astronomical clock on the town hall of the Moravian city was to be renovated and improved.[6] The commission, however, had already been awarded to the imperial mathematician and physician Paulus Fabricius (1519 or 1529–1588) the previous year, on 6 July 1573.[7]

Besides his teaching responsibilities at the Strasbourg academy, Dasypodius was an avid collector, in particular of Greek manuscripts, with the goal of publishing collections of ancient writings on mathematics, mechanics, and music. His ambitious publishing plans—which I will return to later—could only be realized partially.

As was common among humanistic scholars at that time, Conrad Dasypodius corresponded with a wide range of contacts, among which were the Orthodox-Lutheran pastor Matthias Ritter the Younger,[8] Joachim Camerarius, Johannes Kepler, the Heidelberg physician Thomas Erastus,[9] the Viennese

3 According to Adam 1615, 442. However, Dasypodius does not appear in the enrollment lists of the University of Leuven; cf. Reusens/Wils/Schillings 1961/66.

4 Schindling 1977, 63 and 65–66.

5 *Ibid.*, 126–127.

6 Letter of Recommendation for Conrad Dasypodius from Maximilian II, Holy Roman Emperor, from 30 September 1570; Olomouc, Státní okresní archiv [Regional Archives]: Archiv města Olomouc, Zlomky registratur 1426–1786, box 161, inv. no. 4471; Letter to the Olomouc city council written by Conrad Dasyposius from 20 September 1574, inv. no. 4481.

7 The commission has the catalog number 942/2 in the Olomouc Regional Archives; the transcription in Kux/Kreß 1904, 112–113, is flawed and full of mistakes. On Fabricius see: Aschbach 1865/88, v. III, 187–194; Winter 1980; DaCosta Kaufmann 1993; Fröde 2010; Burmeister 2015, 206–208; on the remaining parts of the Olomouc clock, of which the astrolabe rete designed by Fabricius is of particular interest, see Šimková/Horský 1985 and Oestmann 2014a.

8 Matthias Ritter (*c.* 1525–1584) studied in Wittenberg starting in 1542, was *Hofmeister* in Strasbourg and was initially a preacher at St. Catherine's Church and later at the Church of the Discalced (Barfüßerkirche) in Frankfurt am Main; ADB, XXVIII, 666–668. Transcriptions of four letters from Dasypodius to Ritter are extant: Hamburg, Staats- und Universitätsbibliothek, Supellex Epistolica Uffenbachii et Wolfiorum, 2°.

9 Thomas Erastus (1524–1583) studied in Basel and Bologna, and became the personal physician of the elector of the Palatinate Ottheinrich in 1558, who appointed him professor of

court librarian Hugo Blotius, and Heinrich Rantzau. Johannes Praetorius (1537–1616), who was appointed professor of higher mathematics at the University of Wittenberg in 1571 and taught astronomy in Altdorf starting in 1576, was probably also in contact with Dasypodius (at least he sent presentation copies of two of his books to Praetorius).[10]

While preparing his *Harmonice mundi*, Johannes Kepler expressed the wish to see the manuscript of Ptolemy's *Harmonics* that was in Dasypodius's possession as well as his commentary on that work, and contacted Dasypodius to this effect.[11] Kepler's requests, however, were not fulfilled, as the elderly Dasypodius was seriously ill and Kepler did not receive a response until 1615, when Matthias Bernegger, who had taken over Dasypodius's estate, provided him with detailed information on the content and scope of the manuscripts.[12]

Heinrich Rantzau (1526–1598), the Danish governor of Schleswig-Holstein, and Dasypodius had in common an intensive occupation with astrology and its practical applications.[13] Rantzau had studied in Wittenberg, had a profound knowledge of the subject, and had a band of professional writers compile numerous astrological works. Dasypodius dedicated his commentary of Ptolemy's *Tetrabiblos* to him in 1578 and prefaced that work with the horoscopes of Rantzau and his wife, together with a particularly favorable interpretation.[14] This homage from the Strasbourg mathematics professor must have flattered Rantzau extraordinarily, and he printed the horoscope in all of his astrological books. The Hamburg goldsmith Jacob Mores the Older (d. 1609) presented his patron with a golden casket with inlaid silver in which the horoscope was

medicine in Heidelberg. He polemicized against astrology in his *Astrologia confutata* (1557); ADB, VI, 180–182; NDB, IV, 560; Gunnoe 2010; Burmeister 2015, 199–200.

10 They were *Isaaci Monachi Scholia in Euclidis Elementorum Geometriae sex priores libros* published in Strasbourg in 1579 (Schweinfurt, Stadtbibliothek: AvS 6500) with the inscription: *Clariss*[imo] *Viro D*[omino] *Joanni Praetorio amicitiae ergo Cunradus Dasypodius d*[ono] *d*[edit] *1579* on the title page and the publication from the same year *Oratio Cunradi Dasypodii De Disciplinis Mathematicis* [...] with an identical inscription (*Ibid.*, AvS 6497). On the two works see Schäfer 1993, 158–161; a description of the *Oratio* is in Martin 1854, 238–240.

11 Letter from Dasypodius to Johannes Kepler from 31 December 1599, ÖNB, Cod. Vin. 10702, fol. 177; KGW, XIV, 105. On 12 July 1600 Kepler informed the Bavarian Chancellor Herwart von Hohenburg that: "*Scripsi ad Dasypodium. Habere se respondit, nihilque praeterea. At nisi ego sperassem illum possidere hunc librum* [Ptolemy's *Harmonics*], *non scripsissem nec petiissem copiam.* [...] *Ita naso ducor*", KGW, XIV, 131; see also Klein 1971, 52; Balmer 1971, 123–124.

12 KGW, XVII, 131, 136 and 139.

13 Oestmann 2004.

14 *Ibid.*, 104–107, 124–125 and 222–230. The horoscope is not directly attributable to Dasypodius, but possibly was cast by Heinrich Rantzau himself.

engraved.[15] It was a very personal gift, as he presented the governor his destiny, secured in a treasure.

4.2 Isaac and Josias Habrecht

The brothers Isaac Habrecht (1544–1620) and Josias Habrecht (1552–1575?)[16] were born in Schaffhausen, where their father Joachim received the rights of citizenship on 15 October 1540,[17] and made a name for himself in his profession by completing the Solothurn tower clock (1545)—started by the Winterthur clockmaker Laurentius Liechti[18]—and by building the astronomical clock on the Fronwag Tower in Schaffhausen (1562–1564).[19] The two brothers learned their craft from their father, and the older one, Isaac, completed his qualifications as master craftsman in Schaffhausen.

Upon completion of the Cathedral clock, Isaac settled in Strasbourg, as a twelve-year guaranty for the repair of eventual problems was included in the commission for the Cathedral clock. In this way Isaac Habrecht received the rights of citizenship in 1574[20] and, after delivering a table clock with an astrolabe and displays for various calendar dates as well as for the phases of the moon four years later, was admitted to the guild of smiths in Strasbourg.[21]

Still extant are two of his other astronomical clocks, which are miniature copies of the Cathedral clock. One of these was built in 1589 and is now

15 *Ibid.*, 107 and 112.

16 The most important source on the genealogy of the Habrecht family is a booklet written by Isaac Habrecht II (1589–1633), a son of Isaac Habrecht, in 1620, "*Genealogia, das ist Geburt Büchlin und Register unserer Voreltern*"; BNUS, MS 1566 (Alsat. 1001). The information in Edel 1777, 193–195 and Edel 1843, 23–26 is based on this manuscript. Primary publication: Ungerer 1925; followed by Weitenmühlen 1950.

17 Schaffhausen, Stadtarchiv, city invoices AII 05/208, 1540/1541, fol. 3^r.

18 Details on this in Kauffmann 1930 and Ungerer 1931, 400–406.

19 First description: Spleiss 1702; see also Ungerer 1931, 396–398; Frauenfelder 1951, 231–232.

20 "*Isaak Haberecht der uhrenmacher von Schaffhausen hatt das burgkrecht kaufft und dient zu den schmiden.* [Juravit] *den 21ten Decembris 1574* (Isaak Haberecht, the clockmaker from Schaffhausen, purchased the rights of citizenship and served among the smiths. [He swore on] 21 December 1574)"; AMS, Bürgerbuch III (1559–1730), col. 237.

21 Toulouse, Musée Paul Dupuy, inv. no. 18.069 (height 38 cm, base 26 × 18.5 cm, diameter of the astrolabe dial 12.7 cm); Ungerer 1927a, 43–50; Gélis 1949, 177–183; Maurice 1976, v. II, 33, no. 187; *Musée Dupuy* 1978, 23–26; Hayard 2004, 26–39. For a detailed account of the of the astrolabe of this clock, together with a list of the fixed stars included on the rete, see D'Hollander 1993, 113–125.

in the British Museum,[22] the other, built in 1594, is in Rosenborg Castle in Copenhagen.[23]

Isaac Habrecht installed an artistic clock in the tower of Donaueschingen castle in 1577 for 87 guldens.[24] In 1580, together with his journeyman Michael Müller, he carried out important restorations and enlarged the clock in the Heilbronn town hall built by Hans Paulus in 1525,[25] and immediately after that (1580–1581) the new astronomical clock with an astrolabe dial was built on the town hall in Ulm.[26] Another clock, probably commissioned by Hans Fugger in 1583, is treated in detail in *Appendix* 2.

On 29 October 1611 the Duke of Lorraine, Henry II 'the Good' (1563–1624) paid a total of 1375 francs to Habrecht for building a clock similar to that in Strasbourg Cathedral, which was to be installed in the Ducal Palace in Nancy. Of this sum, 125 francs were allotted to transporting the clock from Strasbourg to Nancy.[27] In 1627 the painting gallery was destroyed by fire and the clock was probably destroyed, as well.[28] It was probably similar to the clocks in London and Copenhagen.

Isaac Habrecht's son Abraham (1578–1650?) was also employed by the Duke of Lorraine. On 14 July 1618 he received 81 guldens for servicing (or adjusting?) a *grand horloge* he had built. The amount was also to cover the travel costs from Strasbourg to Paris, which is where the Duke had had the clock sent.[29]

Isaac Habrecht's brother Josias left Strasbourg one year before the completion of the Cathedral clock and was initially in Zürich for a short time, where he married. After that he lived in Kaiserswerth near Düsseldorf. There Josias Habrecht was to build a tower clock for the palace of Count Salentin of Isenburg, the archbishop of Cologne, but before work was completed he died, probably in 1575.[30]

---

22 Good 1985; Norman 1929.

23 Maurice 1976, v. II, 40–41, no. 231; Lühning 1975, 82–83.

24 Rott 1933/38, v. I, 156.

25 Heilbronn 1865, 179–180; Rauch 1929; Ungerer 1928, 79–82 (for a project sketch of the Heilbronn clock); Helfrich-Dörner 1957; Ungerer 1931, 224–228; Leichtle 1940, 87–90.

26 Arlt 1888 (description of the dial); Ungerer 1931, 290–293; Leichtle 1940, 90–91.

27 Nancy, Archives Départementales de Meurthe et Moselle: B 1332.

28 Information kindly provided by the directorate of the Archives Meurthe et Moselle on 13 October 1995.

29 Payment instructions for the treasurer Nicolas de Pallenoy; Nancy, Archives Départementales de Meurthe et Moselle: B 1398.

30 Besides the information in the genealogy of the Habrecht family (BNUS, MS 1566 (Als. 1001), fols. 3^r–4^r) some notes that Isaac Habrecht wrote on his copy of the commission for the construction of the Cathedral clock are the only sources on the life of his younger brother: "*Des nacht geschreyben uhrwerck hab ich Isaac Habrecht uhrenmacher*

Members of the widespread Habrecht family, in which the title of *Münsteruhrmacher* (Cathedral clockmaker) was inherited over generations, were active in Strasbourg until the first half of the eighteenth century; the dynasty ended when Abraham Habrecht IV (1693–1732) died.

4.3 Tobias Stimmer

Tobias Stimmer, born in 1539, received his first artistic training from his father, who was a school master in Schaffhausen from 1532 on and dedicated himself to calligraphy, bookbinding, and glass painting. Nothing is known of his teachers, and researchers have come to differing opinions, without finding satisfying conclusions.[31]

The situation surrounding Stimmer's frequently discussed trip to Italy is similar. Stolberg assumed that the soft contours and vibrant colors of the paintings on the Cathedral clock as well as the sketches with architectural plans pointed to a stay in Venice.[32] Whether, after completing his apprenticeship in 1557–1559, the artist from Schaffhausen worked under Giambattista Zelotti on the frescos in the Palazzo Trevisan in Murano[33] or was even influenced by Tintoretto in Venice in 1568[34] cannot be proven beyond any doubt.

---

von schaffhusen nachvolgend verdeing werck an meicht genomen dyeweil mein Bruder nocht ein ledyger gesell war. gab ime ein wochen ein Daller vür seinen wochenlonn ob das werck gar verfertygt war war er Nach kaeiserswerd vomm Beischof von Kollen erforderth macht ime ein gross uhrwerck ins Schloss im Rein und starb da selbst (The completion of the clockwork described herewith I, Isaac Habrecht, clockmaker from Schaffhausen, took upon myself according to this contract while my brother was still an unwed fellow. Gave him one taler as his weekly pay. Before the work was completed, he was called to Kaiserswerth by the Bishop of Cologne to make for him a great clock in the palace on the Rhine and there he himself died)"; BNUS, MS 1563 (Als. 998), fol. 1^{v}.

31 According to the art historian Berthold Haendcke (1862–1951), Stolberg assumed that Stimmer worked for Hans Asper temporarily; Stolberg 1901, 2. Felix Lindtmeyer from Schaffhausen has also been suggested as his teacher; Bendel 1940, 13–14. Thöne on the other hand points out upper German influence, in particular by the painter Philipp Memberger the Older in Constance and by a portrait painter from Antwerp who worked there and whose name is not known; Thöne 1936, 20–21.

32 Stolberg 1898, 14; Stolberg 1901, 3. Based on architectural compositions with views into high-lying halls with side rooms, influence from Paolo Veronese (1528-1588) was also assumed; Brun 1905/17, v. III, 256.

33 Bendel 1925/26, 132–135; Bendel 1940, 15–17 and 48–49.

34 Barnass 1930, 168–170; Barnass 1932, 28–33. For an opposing view see Thöne 1936, 19–20. His interpretation that the Italian and Dutch influences were secondary "to his lively powerful, near-natural, down-to-earth spirit" (*ibid.*, 84–85) have to be treated with just as much skepticism as the exaggerated speculations of Bendel and Barnass. An explanation

These assumptions are based entirely on stylistic arguments that cannot be confirmed by documents.[35] The only thing that is certain is that Stimmer traveled to Como probably in 1569–1570, commissioned by the Basel publisher Pietro Perna to make sketches of portraits in the collection of Paolo Giovio.[36] As early as 1568 Stimmer provided plans for woodcuts for Strasbourg printers and publishers. After completing the painting on the facade of the *Haus zum Ritter* (Knight's House) in Schaffhausen, he settled in Strasbourg in 1570, but did not acquire the rights of citizenship until 22 October 1582.[37]

4.4 The Commission of 1571 and the Completion of the Second Cathedral Clock

In 1571 the Habrecht brothers traveled to Strasbourg in order to offer to complete the unfinished work for the city council. To demonstrate their qualifications they brought an astrolabe and an a celestial globe with them.[38] The two clockmakers contacted Conrad Dasypodius through several Schaffhausen students and asked him to support their application with the city council and 'Frauenwerk' caretakers. Dasypodius—as his father before him—was the overseer for the Schaffhausen students, that is, he was in charge of supervising the scholarship recipients of the city, who were frequently sent to the Strasbourg

that goes beyond a critical comparison of styles does not seem to be possible with available sources.

35 See the summarizing discussion of the different hypotheses by Fritz T. Schulz; Thieme/Becker, XXXII, 61.

36 Rave 1959, 150–151.

37 "*Thobias Stimer von Schaffhausen der maler hat das Burgrecht kauft und will zuer steltzen dienen actum den 22. VIIIbr 1582* (Thobias Stimer, the painter from Schaffhausen, purchased the rights of citizenship and wants to serve in the Stilt, enacted 22 October 1582)"; AMS, Bürgerbuch III, col. 393. In the *Zunft zur Stelzen* (Guild of the Stilt) artistic professions—such as painters and sculptors, goldsmiths, printers, bookbinders, engravers etc.—were incorporated.

38 "*Sphaeram materialem wie die Mathematici solche Instrumenta nennen, welche die stunden und Planeten stunden, auch Son und Mons lauff und die Monschein durch redder und derselbigen tryb zeigten* (Sphaera materialis, as the mathematicians name such instruments which show the hours and the planet hours and the paths of the Sun and Moon and the moonlight by means of gears and similar drives)"; Dasypodius 1580a, 7. It is possible that the clock-driven armillary sphere made by Josias Habrecht in Rosenborg Castle in Copenhagen, dated 1572, is identical with this instrument; Ungerer 1925, 101, 146. See also Bertele 1961, 34–35; Ernst Schlee 1965, 93–94 (on its provenance); Maurice 1976, v. II, 43, no. 250; Maurice/Mayr 1980, 294–295; Kejlbo 1995, 72–75 and 190–191.

academy to prepare them for university studies.[39] He entreated the city council and the caretakers in their behalf, even though he was apparently not initially convinced of the qualifications of the Habrecht brothers:

> On this matter I soon appealed to the caretakers and after that to my grace commanding Lords, that these two are so far to be trusted, accepted and recognized, so far as I, Cunradus Dasypodius, wish to be good to these two clockmakers, that they have the knowledge and craft to create such a work (the recognition of which was difficult enough for me), but I have risked it with them.[40]

Dasypodius was commissioned with planning the astronomical clock. At first he supplied a *modulus* (construction pattern), and shortly after that the ground plan and elevation design:

> And on me, Conrad Dasypodius, ordinary professor of mathematics of our Academy, they imposed and entrusted this task, namely that I earnestly make plans for constructing such a magnificent, splendid, and artistic work to that end which I have stated, and first of all I set before the *Aediles*[41] and master craftsmen a model, in accordance with which such a machine would be skilfully formed and constructed by the workers. And so a space of some days having passed, I set before the *Aediles* a ground plan and elevation of the whole machine which is now seen by the eyes.[42]

---

39 Dasypodius 1580a, 8; Lang 1932, 45 and 71–74; Lang 1903, 151–152.

40 "*Darauf hab ich als bald bey den herren Pflegeren auch hernacher meynen G.G.H. angesucht, ist ihnen zweyen so fern zu vertrawen bewilligt und erkandt, so fern ich Cunradus Dasypodius gut für dise zwen Uhrenmacher sein woelle, das sie solches werck zu verrichten ein wissens und kunst haben, welche erkandtnuß mir schwer genug fiel, aber habs mit ihnen gewagt*"; Dasypodius 1580a, 8.

41 The members of the Council of XXI (*Einundzwanziger*), which, inter alia, were responsible for the construction of the Cathedral.

42 "*Ac mihi Cunrado Dasypodio Academiae nostrae ordinario Mathematωn professori, hoc imposuerunt et mandarunt negotium: ut sollicite cogitarem de tam magnifico, et splendido, ac artificioso opere, in eum quem iam dixi finem extruendo: modulumque imprimis Aedilibus, et officinatoribus, proponerem: iuxta quem machina talis affabre ab opificibus formaretur et extrueretur. Spatio itaque aliquot dierum sumpto: ichnographicam, et orthographicam, totius, quae nunc oculis conspicitur machinae, delineationem*"; Dasypodius 1580b, fols. FII^{v}–FIII^{r}.

Dasypodius had Tobias Stimmer paint this plan, which is referred to as a "*Visierung*" in the German description.[43] He would have known Stimmer from 1567, who was then residing in Strasbourg, because, in return for an edition of Euclid that he had dedicated to the city council of Schaffhausen, he received a silver goblet for which Stimmer had produced the drawings.[44] In addition, a younger brother of Stimmer studied at the academy and was therefore under Dasypodius's supervision.[45]

Besides the specifications for and probably the technical components of the clock, the plan also included its sculptural ornamentation and the topics of the illustrated tables.[46] This extremely important document remained in Dasypodius's possession[47] and has been lost.[48] In the German description, mention is made of a first concept that had been placed by Dasypodius in his books.[49] This concept will be dealt with in more detail in *Part 4*.

43 Dasypodius 1580a, 9.

44 "*Item xß. gaben wir Tobias Stymmer dem maler, von ainem Muster zemalen zu ainem sylberin becher, so unser herren, dem Dasipodio zu Strassburg vereeren wellen, um dz dediciert buch. sambstag vor palmarum* [22 March 1567] (Likewise 10 shillings we gave Tobias Stymmer the painter of a model painted for a silver goblet to honor our lord Dasypodius of Strasbourg, because of the dedicated book. Saturday before Palm Sunday)"; Schaffhausen, Stadtarchiv, City invoices from 1566–1567, A II 05/247, 183. "At that time there was very active contact between Schaffhausen and Strasbourg. Some young men from Schaffhausen were almost always in Strasbourg, studying at the academy, and their diligence and conduct were the topic of lively correspondence between the Strasbourg professors and the Schaffhausen school board, called the Scholarchenrathe. This contact was surely the reason why Conrad Dasyposius, instructor of mathematics in Strasbourg, dedicated his lovely new edition of Euclid's Geometry to the Senate of the Republic of Schaffhausen in the summer of 1566. The dedication copy still exists and is kept in the Ministerialbibliothek and, with its richly ornamented cover, is a valuable memorial to the artistic sensibilities of that time. The courier who brought the book to Schaffhausen was provided food and lodging at the city's expense and he was given two talers; the author of the book was sent a silver goblet with a value of 40 guldens by the mayor and the council, for which Tobias Stimmer had provided the drawing"; Baeschlin 1880, v. II, 5–6. A sheet kept in the Museum zu Allerheiligen in Schaffhausen (inv. no. B 1453) was possibly the drawing for the goblet; *Kunstmuseum Basel* 1984, 332, no. 204. A figure is in Bendel 1940, 217, who erroneously interpreted the drawing as the plan for a plate.

45 Thöne 1936, 16, n. 39.

46 Dasypodius 1580a, 15–16.

47 *Ibid.*, Preface, fol. B IV^r.

48 Attempts by the author to find the plan in Strasbourg and Schaffhausen have been fruitless. Should the plan have survived into modern times, it was probably destroyed when the municipal library of Strasbourg was burned on 24 August 1870, during the Franco-Prussian War.

49 Dasypodius 1580a, Preface, fol. B IV^r.

Dasypodius presented his plans to his colleagues Michael Beuther[50] and Laurentius Tuppius,[51] who found no fault in them. To ensure the quality of his plans, he called upon the Freiburg mathematician Erasmus Oswald Schreckenfuchs,[52] who together with other university scholars expressed his approval. Finally the plans were shown to the Habrecht brothers, and the caretakers asked them if the work could be completed in the proposed form.

After these preliminary negotiations, on 23 July 1571, the city council decided to have the clock built.[53] Three days after that a contract was written up between the caretakers Diebold Johann von Mundoltzheim (*stettmeister*), Michael Liechtensteiger (former *ammeister*), Friedrich von Gottesheim (member of the XIII), the supervisor (*procurator*) of the 'Frauenwerk', Aristarchus Mornhinweg, as well as the Habrecht brothers. Conrad Dasypodius was named inspector of the work on the Cathedral clock.[54]

The text of the contract precisely lists the required specifications for the clock, their arrangement, and the division of the work.[55] Noteworthy is the brief, summarizing mention of the artists involved. Other than Isaac and

50 Michael Beuther (1522–1587) was a history professor at the Strasbourg academy starting in 1565, and was its first dean; Burmeister 2015, 100–103.

51 The lawyer Laurentius Tuppius (1528–1614) also taught at the Strasbourg academy and concerned himself with astrological matters; *ibid.*, 598–599.

52 *Ibid.*, 10. Erasmus Oswald Schreckenfuchs (1511–1575 or 1579) was born in Merkenstein, studied in Ingolstadt, Leipzig, and Basel. Besides mathematics, he taught Hebrew, initially in Memmingen and Tübingen, then in Freiburg, and published all of Ptolemy's works but *Geography*. He also wrote a commentary of Johannes de Sacrobosco's *Sphaera* and another on the *Almagest*. Among the extant manuscripts is one dealing with the construction and use of the astrolabe: "*De Astrolabii compositone compendium*", Salzburg, Universitätsbibliothek M I 127, 601–654; in Zinner 1925, 294, no. 9485 under the catalogue no. V1 JS 2 260/1. Biographical information in Pantaleon 1565/66, v. I, 299–300; Zedler XXV, col. 2355, ADB, XXXII, 467–468, Schreiber 1868, v. II, 253–262. On the writings of Schreckenfuchs see: Khautz 1755, 184–203; Gericke 1955, 29–34; Burmeister 2015, 540–541. A portrait is in Reusner 1587, fol. Y VIII^v.

53 AMS, minutes of the XXI (Einundzwanzigerprotokolle), v. 51, 1571, fol. 663^r–664^v.

54 Dasypodius 1580a, 10. The *ammeister* was elected by the guilds for one year; he headed the city administration of Strasbourg together with four *stettmeisters* chosen from the nobility, each of whom held office for a quarter of a year, and the 30-member city council. The 'XIII' were responsible for general political tasks and foreign policy; this 'privy council' was made up of four members each of the patricians, the guild members, former *ammeisters*, and *stettmeisters*. They were elected for life by the city council and later by the city council together with the 'XXI'. The main accountant and supervisor of the mason's lodge was the procurator. For detailed information on the administration of Strasbourg see Winckelmann, 1903, 493–537 and 600–639.

55 AMS, 1 OND 15. The second copy of the treaty mentioned on fol. 6^v, which was formerly in the possession of Isaac Habrecht, is also preserved; BNUS, MS 1563 (Als. 998).

Josias Habrecht, the artisans who worked on the housing remain anonymous. According to the contract, the technical components were to be built in Schaffhausen and transported to Strasbourg within one year.[56]

On top of the responsibility for the entire project, which Dasypodius felt was oppressive, "worry and other misfortune" soon put him temporarily sick in bed.[57] Dasypodius was probably referring to personal and confessional differences of opinion which had existed since 1569 between the rector of the Strasbourg academy, Johannes Sturm (1507–1589), and the dean and president of the church convention Johannes Marbach (1521–1581), which escalated into personal denunciation starting in 1571. Sturm accused Marbach of dismantling the humanistic curricula of the school.[58] In 1578 the conflict between the orthodox Lutheran camp and humanistic anti-Lutheran camp sympathizing with Reformed theology reignited around the anti-Calvinist theses of the teacher Johannes Pappus (1549–1610), which finally led to Sturm's dismissal.[59] During these conflicts, Conrad Dasypodius supported Sturm.[60] Dasypodius's teaching responsibilities and his extensive publication plans as well as the continuing conflicts in the academy were probably the reason why he sought the support of his friend David Wolckenstein (1534–1592).[61] The mathematician came to Strasbourg in 1572 and was appointed the second professor of

56 The author was unable to find the monogram of Isaac Habrecht with the date 1572, which according to Ungerer was on one of the pillars of the clock. There are no records of the construction of the Cathedral clock in the archives of Schaffhausen; no mention is made of such in the council minutes, missives, or duplicates *'Copeyenbücher'* of the city during 1571–1574.

57 Dasypodius 1580a, 10–11.

58 Schindling 1977, 130ff.

59 *Ibid.*, 138–143.

60 *Ibid.*, 136, n. 38; Chrisman 1982, 250.

61 David Wolckenstein (Nephelitus) was born in Breslau (today's Wrocław). He had studied at the gymnasium in Breslau, in Frankfurt am Oder in 1553, and finally in Wittenberg in 1555, where Melanchthon was his teacher. Dasypodius asked him to come from Augsburg, where he worked as a professor of mathematics, to assist with the work on the clock; Sebitz 1641, 280; Zedler, LVIII, col. 401; Sitzmann 1909/10, v. II, 1011; ADB, XLIV, 136–137. In 1585 Wolckenstein received the rights of citizenship: *"Der Wollgelert Magister Davit Wolckenstein von Breßla in der Schleßingen der allhiesigen Academie Profeßorem hat daß Burckhrechtt kauft und will zu der Steltzen dienen. actum den 23t*[en] *december 1585* (The well educated Master David Wolckenstein from Breslau in Silesia, professor of the local academy, purchased the rights of citizenship and wants to serve in the [Guild of the] Stilt. Enacted 23 December 1585)"; AMS, Bürgerbuch III, col. 448. Extensive material on Wolckenstein's biography and his musical activities is in the Estate of Charles Eugène Wagner (AMS), 90 Z 15, 1576–1589, and 90 Z 18, 1195–1316. On the astronomical clock designed by him in 1583 see *Appendix* 2.

mathematics and professor of music at the Strasbourg academy two years later.[62] Dasypodius entrusted him with his plan and Wolckenstein promised to carry out everything in accordance with his instructions and, in addition, to take over the inspection of the work.[63]

On 24 June 1574, after three years of construction, the clock was finally set in motion; however, the 700 guldens originally budgeted by the 'Frauenwerk' proved to be insufficient, which led Isaac Habrecht to apply for additional funds.[64]

The cooperation between Dasypodius and the Habrecht brothers was apparently very tense as Dasypodius did not feel that his accomplishments surrounding the plans and construction were acknowledged sufficiently and filed a petition to the city council in his and Wolckenstein's name on this matter on 5 June 1577.[65] The council decided to hand over the supplication of the two mathematicians to the caretakers, and council members were assigned to arbitrate.[66] The caretakers submitted a report in which they pointed out that Dasypodius had promised to assist the clockmakers with his astronomical and mathematical knowledge without pay other than a present from the 'Frauenwerk' to honor him.[67] After completion of his work, he demanded 400 pound pfennigs, after which an agreement was made with him and Wolckenstein in which they received 300 guldens, corresponding to 150 pound pfennigs,[68] and a silver goblet (*geschürrlin*). They also claimed that the clockmakers played a bigger role in the invention and disposition of the work than Dasypodius,[69] and additionally that Wolckenstein carried out more work on the clock than he did, so that Dasypodius should be satisfied with the remuneration that he received. The city council followed the arguments of the caretakers and exhorted the two parties to come to a settlement.[70]

---

62 Schindling 1977, 260.

63 Dasypodius 1580a, 12.

64 For invoices for multiple objects completed for the Cathedral clock later, see BNUS, MS 1565 (Als. 1000).

65 AMS, Sér. II, 46, no. 35.

66 Council decision of 5 June 1577, AMS, minutes of the XXI (Einundzwanzigerprotokolle), v. 58, 1577, fols. 327^{v}–328^{v}.

67 AMS, Sér. II, 46, No. 35, fol. 2^{v}–3^{r}, 4^{v}; Regest in Kraus 1876, 412.

68 AMS, Sér. II, 46, fol. 5^{r}. On the units of currency: 1 pound (£) corresponds to 20 shillings (ß) to 12 pfennigs (₰) each; 1 gulden corresponds to ½ pound; Hanauer 1876/78, v. I, 19–21.

69 "*Inventio und dispositio*"; AMS, Sér. II, 46, fol. 6^{r}.

70 City council decision of 19 October 1577, AMS, minutes of the XXI (Einundzwanzigerprotokolle), v. 58, 1577, fols. 638^{v}–639^{r}. The dedication of the two editions of the *Warhafftigen Außlegung* to the city council of Schaffhausen was possibly an answer to the rejective

From their report, it is clear that the caretakers considered the Habrecht brothers the real creators of the Cathedral clock and that Dasypodius was only attributed with having advised them on the project. Presumably the Habrecht brothers—in particular, Isaac, who remained in Strasbourg and took care of the maintenance of the new clock after his brother left—claimed to have not only constructed the clock's technical components, but also to have come up with the conceptual plan themselves.

In the spring of the following year the conflicts broke out again. On 5 April 1578 Habrecht accused Dasypodius of publishing his description of the Cathedral clock shortly before that, in which he "attributed the work to himself alone and belittled him, the supplicant [Habrecht], to the greatest degree, and placed so much glory for himself in it that it stank".[71] Habrecht demanded that the published copies be turned over to the chancellery, or that the book be attributed to him. A few days later, on 12 April, the conflict came to a head again, as Habrecht felt that he was grievously offended ("*schmechlich angetastet*"). The council exhorted the opposing parties once again to come to a peaceful settlement.[72]

In 1608 Johann Jakob Ebelmann's engraving of Isaac Habrecht (*Fig. 4.1*) appeared with the inscription: "*Argyropum automati, inventor, fabricator, et autor, Imortale Habrecht nomen Isacus habet*".[73]

Habrecht possibly commissioned the copper engraving himself. Publicly he was considered the "*Inventor, fabricator et autor*" of the new Cathedral clock. At the same time, Dasypodius in his *Heron mechanicus* attributed the invention (*inventio*) as well as the arrangement of the entire work expressly to himself and credited Wolckenstein and Stimmer with completing certain components, whereas the work of all the other artisans was merely assigned to them:

> I care nothing for the judgments of the others who, while they either do not understand these arts and disciplines and rashly wish to bring judgment on these machines of ours, or, stirred by animosity and malevolence, are eager to draw off the praise which is justly due to us by their

---

position of the Strasbourg council. The dedication can be found in fol. A II^r of the preface in both editions.

71 "[Daß Dasypodius sich] *dieses Werck allein zumesse und ihn, den Supplicanten* [Habrecht] *zum höchsten vercleinert, seines eignen rhums sovil darin stecke, das sie darvon stinck*"; AMS, minutes of the XXI [Einundzwanzigerprotokolle], v. 58, 1577, fol. 190^v.

72 *Ibid.*, fols. 209^v–210^r.

73 Original size 210 × 144 mm; Strasbourg, Musée des Arts Décoratifs, Cabinet des Estampes. A smaller engraved copy, 155 × 100 mm, without the ornamental border is in Edel 1777 between pp. 194 and 195, and in Friese 1792, v. I, between pp. 222 and 223.

FIGURE 4.1 Johann Jakob Ebelmann, portrait of Isaac Habrecht; copperplate engraving, 1608, 144 × 210 mm

> own depraved judgment, corrupted by their desires. Let no one arrogate to himself the first invention of this work because it is mine, nor the first arrangement of the particular parts of the whole work because it is also mine. But the other items which were accomplished with effort, zeal, care, and solicitude in constructing or polishing off the work are both mine, my colleague's [David Wolckenstein] and Tobias Stimmer's. It was we who laid out for the master craftsmen and handicraftsmen all that had been approved by the Tresviri Aediles.[74]

The claim of intellectual authorship is emphasized numerous times by him in the German descriptions of the Cathedral clock of 1578 and 1580.[75] In his insistence on this matter—independent of the extent to which the Habrecht brothers actually claimed authorship—Dasypodius's concept of himself as a scholar, the significance that he placed on the *artes liberales*, and his perception of the state of development of clock making have to be taken into consideration, which will be dealt with in more detail in *Part 4*. Apparently criticism was voiced on the arrangement of the clock and its completion, against which Dasypodius took a stand by publishing his descriptions of the clock.[76]

---

74 "*Caeterorum nihil facio iuditia qui dum vel non intelligunt has artes et disciplinas: temere iuditium ferre de his nostris machinis volunt: vel odio et malevolentia excitati laudem quae nobis merito debetur: suo pravo, et affectibus corrupto iuditio detrahere student. Primam totius operis, inventionem quia mea est, nemo sibi arroget: nec primam totius operis praecipuarum partium dispositionem: quae etiam mea est: caetera vero quae venustatis, commoditatis, ac magnificentiae gratia addita sunt: quaevel labore, studio, cura, ac sollicitudine, in extruendo expoliendo vel opere, facta sunt: et mea, et collegae mei, et Thobiae Stimmeri sunt: qui omnia quae a Triumviris Aedilibus approbata fuerunt: officinatoribus, et manuariis opificibus proposuimus*"; Dasypodius 1580b, fols. H IV^{r}–IV^{v}.

75 Dasypodius 1580a, Preface, fols. BIV^{r}–BIV^{v}, C^{r}, 17.

76 *Ibid.*, Preface, fols. BII^{v}–BIII^{v}.

PART 2

The Artistic Embellishments of the Second Clock

CHAPTER 5

The Artistic Decoration of the Clock Housing

5.1 Pictorial Representations of the Clock

Even if the housing of the second Cathedral clock still exists, substantial changes were carried out when Schwilgué installed the new clock in it in the nineteenth century, which forces us to look at pre-nineteenth century pictures of the clock. Fortunately Tobias Stimmer depicted the clock's exterior precisely in two woodcuts. The larger of the two woodcuts, 519 × 290 mm, was produced in 1574, the year in which the clock was completed, and was printed from two wooden blocks (*Fig. 5.1*). The view is from the front—except for the side view of the weight tower—and resembles an elevation. The entire clock in the southern transept, however, can only be viewed at an angle, so that we must assume that Stimmer used Dasypodius's ground plan and elevation design mentioned in *Part 1*. A smaller version of the woodcut, 385 × 223 mm, also appeared with a poem by Johann Fischart (1546/1547–1590) explaining the functions displayed by the clock and its artistic embellishments (*Fig. 5.2*).[1] Furthermore smaller woodcuts exist, including those preceding Dasypodius's two German descriptions of the Cathedral clock and his *Heron mechanicus* (*Figs. 5.3–5.5*), and numerous copies of Stimmer's woodcut.[2] As a result of their detailed representation of the clock, Isaak Brunn's copper engraving of 1609 (page size 385 × 265 mm; *Fig. 5.6*),[3] and the pen-and-ink drawing by Johann Jakob Arhardt (1613–1674) of 1673 (size 265 × 492 mm; *Fig. 5.7*)[4] are particularly important.

1 On Fischart's poem see Hauffen 1896.

2 Kraus 1876, 480; *Kunstmuseum Basel* 1984, 108–109; Harms/Kemp 1987, 430. Prints in different states of Stimmer's original woodcut as well as numerous copies are in the copper engraving department of the Musée des Arts Décoratifs in Strasbourg, case VII: Cathédrale, Horloge astronomique. Until the nineteenth century, 39 depictions of the Cathedral clock were produced; Stolberg 1898, 5, n. 4.

3 Wolfenbüttel, Herzog-August-Bibliothek, 36.16 Geom. 2° (6).

4 Vienna, Albertina, inv. no. AZ 9654, dated 23 June 1673. On Arhardt (1613–1674), who served as an engineer under the margrave of Baden-Durlach and later as Strasbourg's municipal master builder, see Rott 1917, 82–106 and 163–164.

 | DOI:10.1163/9789004423473_007

FIGURE 5.1 Tobias Stimmer, The Strasbourg Cathedral Clock; woodcut, 1574, larger version

FIGURE 5.2 The Strasbourg Cathedral Clock; woodcut by Tobias Stimmer, 1574, smaller version

CVNRADI DASYPODII
Heron Mechanicus:
Seu
De Mechanicis artibus, atq; disciplinis.
Eiusdem
Horologij astronomici, Argentorati in summo Templo erecti, descriptio.

ARGENT. Excudebat Nicolaus VVyriot, 1580.

FIGURE 5.3 Conrad Dasypodius, *Heron mechanicus*, 1580, title page

Warhafftige
Außlegung des Astronomischen
Vhrwercks zu Straßburg/beschriben
Durch
M. Cunradum Dasypodium/der solches Astronomische
Vhrwerck anfenglichs erfunden/vnd angeben.

Getruckt zu Straßburg bey Nyclauß Wyriot.
M. D. LXXVIII.

FIGURE 5.4 Conrad Dasypodius, *Warhafftige Außlegung*, 1578, title page

FIGURE 5.5 Conrad Dasypodius, *Warhafftige Außlegung*, 1580, title page

FIGURE 5.6 Isaak Brunn, The Strasbourg Cathedral Clock; copper engraving, 1609

FIGURE 5.7 Johann Jakob Arhardt, The Strasbourg Cathedral Clock; pen and ink drawing, 23 June 1673

5.2 General Description of the Clock Housing

During the construction of the clock, the stone housing built and almost completed in the second third of the sixteenth century was used unchanged,[5] but was almost completely hidden under the wooden covering with imitation marble, gilt, or multicolored columns, consoles, cornices, and numerous fields for pictures. The vault of the lowest tier was completely hidden by an expansive three-axis foundation that was 7.32 m wide and 4.26 m tall, and the original clock ensemble with a spiral staircase was complemented with a three-layered wooden tower for the clock weights, resulting in a more symmetrical arrangement. As in the first Cathedral clock, the central tower with a height of 18.06 m had a three-tier arrangement for the calendar disk, astrolabe and automata.

In order to facilitate orientation in the following description of the clock housing, each object is given a number that can be found on a copy of Stimmer's woodcut in *Fig. 5.8.*

5.3 The Lowest Tier

The calendar disk, with paintings in the spandrels, is located on the front of the wide bottom tier, together with tables for the solar and lunar eclipses, six painted panels, and an opening in which the regent of the day appears.

The wooden calendar disk (*Fig. 5.8, no. 1*), which will be described in more detail in the section on the technical equipment of the clock, is flanked on the right by Diana (*Luna*) and on the left by Apollo (*Sol*), whose arrow functioned as a pointer for the outer ring for the day on the calendar (*Fig. 5.8, nos. 2* and *3*). With the index finger of her right hand, Diana points to the "day that makes the exact opposite for the other half year".[6] The possible astrological meaning of this juxtaposition will be dealt with later. The two statues are not in their original state; their arms were sawn off, exchanged, and reattached at different angles, which becomes obvious when compared with the existing illustrations by Stimmer.[7] On today's clock, Apollo holds a bow with his left

5 Dasypodius 1580a, 6–7.

6 "[...] *tag so stracks gegen dem anderen deß halb Jhar machet*"; *ibid.*, 23. Just as blazons in heraldry use the point of view of the shield carrier, Dasypodius's description places Diana on the left and Apollo on the right. See also the 1578 edition of the *Warhafftigen Außlegung*, fol. E1^r. However, in *Heron mechanicus* he does not indicate the locations in fols. GIII^r–GIII^v.

7 Strasbourg, Musée des Arts Décoratifs, fig. in Pariset 1932, 127. Stimmer's draft shows Apollo's right hand resting on a shield, on which allegedly the names of Dasypodius, Wolckenstein, Stimmer and the brothers Habrecht were written; Bach/Rieb 1992, 32.

FIGURE 5.8 Tobias Stimmer, The Strasbourg Cathedral Clock, with legend; woodcut, 1574

FIGURE 5.9 Center of the calendar plate

hand, which originally was an attribute for Diana. Apollo's shield and the dog at Diana's feet have also disappeared; in addition, Diana's left breast, which originally was partially exposed, has been covered by a crudely applied fold in her gown.

At the center of the calendar disk was a round plate (*Fig. 5.8, no. 4*), on which a map of Germany with the Rhine River and the bordering countries was painted, together with a bird's-eye view of Strasbourg and the following inscription on a field in the lower left-hand corner (*Fig. 5.9*): "*Cunrad Dasypodius et Dauid Wolkenstein Vratisl. designabant: Thobias Stimer pingebat. An*[n]*o* MDLXXIIII".

The round plate has a red border with a carved edge, on which the four times of the day (points of the compass) are indicated: top "*Auff gang.*" (sunrise); right "*Mittag.*" (noon); bottom "*Under*[?]*gang.*" (sunset); and left "*Mittnacht.*"

FIGURE 5.10 Tobias Stimmer, The Four Empires; corner spandrel paintings for the calendar plate

(midnight).[8] In the spandrels of the square field for the calendar disk are personifications of the four world empires (*Fig. 5.8, nos. 5–8*; *Fig. 5.10*); on their shields, the four great beasts from Daniel's dream (Dan. 7: 3–8) are depicted:

Upper left: a man with a mustache, wearing a turban, holding a scepter in his right hand, and standing behind an oval shield with an upright winged lion on it, with its paws raised, and with the inscription "*ASSYRIA*".

Upper right: an entire figure with a crown and a full beard in magnificent garments with a cape, holding a scepter in his right hand; the shield in front of him has the inscription "*PERSIA*" and a bear-like figure painted on it.

Lower left: a personification of Greece with a scepter; Stimmer probably depicted Alexander the Great. The shield has a four-head winged leopard (an allusion to the division of Alexander's empire into the four kingdoms of the Diadochi) and the inscription "*GRAECIA*".

---

8 Considering the image in the center of the calendar disk as a representation of the geocentric world system, as Dengler does, seems very far-fetched: "Possibly one saw in this arrangement of peripheral numbers [the calendar disk] and the central image an abbreviated form of the geocentric system of the world on which the 'astrolabium' was based. Dasypodius did not, however, comment on this"; Dengler 2011, 142.

Lower right: a representation of the Roman empire as a semi-inclined three-quarter figure with a helmet, scepter, and an inclined shield, with a goat-like beast with ten horns and the inscription "*ROMA*" on it.

All four figures are depicted in front of clouds of smoke which come from the heads of the winds in the four corners.

The tables with the eclipses (*Fig. 5.8, no. 9*): Dasypodius calculated the beginning, type, and duration of the solar and lunar eclipses for the years 1573–1605 using the Prutenic and Alfonsine Tables.[9] This information was provided on tables to the right and left of the calendar disk,[10] the left of which was decorated with two putti sitting on fruit, four busts, lion heads, and masks. The right table is somewhat less ornate and only has three busts in the top row, followed by three animal heads with garlands and three herms further down. Below the numbers every eclipse is shown in a dark field surrounded by clouds; either the Moon is shown crossing the Sun or the shadow of the Earth is shown crossing the Moon at the beginning, middle, and end of the eclipse. Each field is divided into two halves by the ecliptic; a vertical line marks the middle of the eclipse.

Around the border of the left table, inscriptions indicate the four points of the compass: top "*SEPTENTRIO*", right "*OCCIDENS*", bottom "*MERIDIES*", left "*ORIENS*". The right table bears the names of the chief winds on its border: top "*BOREAS*", right "*ZEPHYRUS*", bottom "*AVSTER*", left "*SVBSOLANVS*". The times listed on the tables are not in their original state, as the Strasbourg mathematician Matthias Bernegger recalculated the dates for the years 1613–1649 in 1609, which were then written on the fields after they were repainted.[11] However, as Dasypodius included his results in the second edition of his German description of 1580 and in *Heron mechanicus*, the original dates have been handed down.[12]

Regent of the day (*Fig. 5.8, no. 10*): There is an opening above the calendar disk in which the respective regent for the day appears on a chariot pulled by animals.[13] Stimmer's seven designs for the planetary chariots (grisaille on canvas) are in the Musée des Arts Décoratifs in Strasbourg. Of the sculptures

9 Dasypodius 1580a, Preface, fol. B IV[r]; Dasypodius 1580b, fol. G III[v].

10 Strasbourg, Musée des Arts Décoratifs. Figs. in Bendel 1940, 202, 204; Ungerer 1922, pls. XVII–XVIII.

11 Bünger 1893, 62.

12 Dasypodius 1580a, 33–43; Dasypodius 1580b, fols. I[r]–K[r]. The Strasbourg mathematician and astronomer Julius Reichelt (1637–1717)—on which see: Sitzmann 1909/10, v. II, 524—came up with a new set of data for the years 1670–1699, which were then painted on a pair of newly created tables by Johann Mock.

13 For information on the development and spread of the planetary week in antiquity see: RE, VII, cols. 2556–2578, s.v. "Hebdomas" by Franz Boll; XX, cols. 2143–2147, s.v. "Planeten" by Wilhelm and Hans Georg Gundel; Colson 1926; Zerubavel 1985, 5–20.

made according to Stimmer's designs, Apollo, Diana, and Jupiter were already missing in 1836. Three chariots were rebuilt in 1838–1842, all the draft animals were replaced, and the original solid wheels with the signs of the zodiac were replaced with spoked wheels.[14] The following brief descriptions are therefore based on Stimmer's grisailles:

1. *Dies Solis* (Sunday): A crowned male figure with a scepter and a solar disk on a chariot pulled by two horses. The sign on the wheel was Leo.
2. *Dies Lunae* (Monday): Diana is depicted with a bow and arrow, as well as a crescent Moon in her hands. Her chariot was pulled by two deer and had the sign Cancer on its wheel.
3. *Dies Martis* (Tuesday): A warrior with a helmet and sword; his chariot is pulled by two panthers. Scorpio is on the chariot wheel and a ram (Aries) lies at the feet of the god of war.
4. *Dies Mercurii* (Wednesday): Mercury with a winged helmet, caduceus, and armillary sphere. Two cocks pull his chariot. The sign on the wheel is Gemini; a personification of Virgo is sitting in the front of the chariot.
5. *Dies Iovis* (Thursday): A crowned male figure with a sword and scepter. In this chariot, which is pulled by two peacocks, an archer (Sagittarius) is kneeling who has just shot an arrow. The sign on the wheel is Pisces.
6. *Dies Veneris* (Friday): Venus is holding a heart out towards the viewer, which is pierced by an arrow from Amor, who is sitting in the front of the chariot. Two doves are harnessed to the chariot; Taurus is on the wheel. Scales (Libra) are attached to the back of the chariot.
7. *Dies Saturni* (Saturday): Cronos/Saturn holds a scythe in his right hand; the bearded old man is devouring one of his children. A goat (Capricorn) lies in the front of the chariot, which is pulled by two dragon-like creatures. Aquarius is on the wheel.

The planetary chariots of the Strasbourg clock follow the representational tradition of the Children of the Planets,[15] and the sequence of woodcuts that Georg Pencz made in 1531[16] probably inspired Stimmer (*Fig. 5.11*). Pencz on the

---

14 Schwilgué, Description, v. II, 245 (OND).

15 Contrary to older research, which tried to contribute representations of the Children of the Planets to classical traditions for imagery that evolved from the Arab-Islamic cultures (Panofsky/Saxl 1923, 121–136; Saxl 1918/19), the Jena art historian Dieter Blume places their development in southern Germany during the Middle Ages (Blume 2000, 158–191; Blume 2004), just as Hauber 1916, 264–268 had assumed. On the origins and spread of the images of the Children of the Planets see also Breanne 2012 and Shamos 2013.

16 Lippmann 1895, 13 and Pauli 1901, 371–374, nos. 904–910, attributed the sequence of the Children of the Planets to Sebald Beham. Roettinger, however, was able to convincingly prove that the sequence came from the works of Pencz; Roettinger 1914a, 20ff, 36, nos. 4–11.

FIGURE 5.11 Georg Pencz, Children of Mercury; wood engraving, 1531

other hand most probably based his series of the Children of the Planets on a sequence that has been attributed to the Florentine goldsmith Baccio Baldini[17] and divided his representation into two parts, just as his model did: on the top are the personifications of the individual planets, riding their triumphal chariots across the clouds, and below them are scenes with the people under their regency, depicted with their professions and activities. Here the same arrangement of the signs of the zodiac can be found as in the Strasbourg Cathedral clock.

This corresponds to the astrological placement in houses based on the Egyptian '*Thema Mundi*' (Horoscope of the World),[18] which shows the positions of the seven visible planets at the beginning of the universe (*Fig. 5.12*). The signs of the zodiac are actually the 'houses' of the moving stars. A planet has the greatest influence when it enters its domicile and rules the house (*oikodespotism*, an essential dignity in astrology). In order to assign the twelve signs of the zodiac to the seven planets, the Sun and Moon were only assigned one sign (a diurnal house or a nocturnal house), the other planets had both a diurnal and a nocturnal house each (*Tab. 5.1*).[19]

17 Attribution to Baccio Baldini is based, however, solely on a brief note from Vasari, according to which Baldini was one of the first to use copper engraving and had worked from drawings by Botticelli. Lippmann justifiably questioned this; Lippmann 1895, 1ff, 13. Adolf Rosenberg was the first to recognize that the sequence, which he also attributed to Sebald Beham, used the Florentine images of the Children of the Planets as a model; Rosenberg 1875, 76, 127. An extremely popular and widespread astrological work published by Cammerlander in Strasbourg in 1541 reproduced the planetary gods in their chariots in a considerably coarser mirror image: "*Das Planeten Buch von natur eygenthumb und wirckung der siben Planeten und zwoelf zeychen deß himels, auch von den 28. Mansionibus das ist Stellungen deß Monds wie und was sie wircken in den menschen darauß ein jeder sein eygen natur gut oder boeß leichtlich erlernen mag* (The Book of the Planets: On the nature, characteristics and effects of the seven planets and twelve celestial signs, also on the 28 Mansions, that is stations of the Moon, how and what they effect in people, from which the very nature of each person, good or bad, easily can be learned)"; Zinner 1941, 198, no. 1781a.

18 Firmicus Maternus ascribed the Horoscope of the World to the legendary Egyptian King Nechepso and his high priest Petosiris, who supposedly received their knowledge from the gods Aesculapius and Anubis. They, in turn, were said to have received this information from Hermes (the Egyptian Thoth); Prugner 1551, 45–46; Firmicus Maternus 1897/1913, V. I, III.1, 91; Firmicus Maternus 1975, 61; Bouché-Leclercq 1899, 187. Ptolemy did not base his placement in houses on Egyptian revelations, but rather on physical considerations; Ptolemy 1980, I.17, 79–83.

19 Uhlemann 1857, 62 and 64; Bouché-Leclerq 1899, 188.

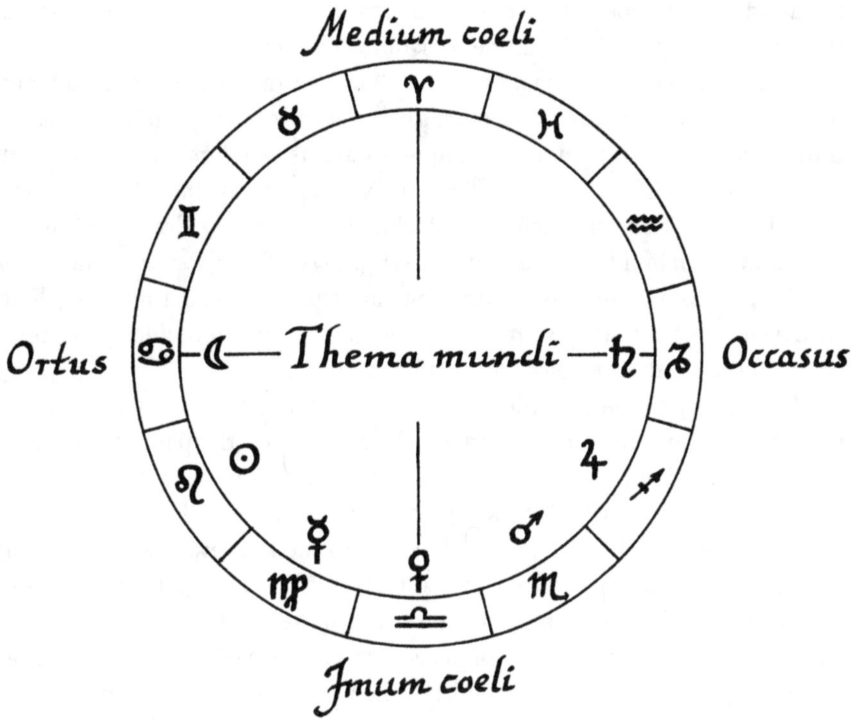

FIGURE 5.12 '*Thema mundi*'. All the planets are in the middle of the sign in the lower half of the heavens.

TABLE 5.1 Diurnal and nocturnal houses of the seven planets

Planet	Day house	Night house
Sun	♌ Leo	–
Moon	–	♋ Cancer
Mars	♏ Scorpio	♈ Aries
Mercury	♍ Virgo	♊ Gemini
Jupiter	♐ Sagittarius	♓ Pisces
Venus	♎ Libra	♉ Taurus
Saturn	♑ Capricorn	♒ Aquarius

These astrological views are illustrated in the representation of the day regents with their signs of the zodiac in the Strasbourg Cathedral clock. Stimmer's *Dies Veneris* differs somewhat from the other figures: Amor sits in the chariot (an allusion to the goddess of love, Venus), the symbol for the day house of this planet, scales (Libra), hangs at the back of the chariot.

The tradition of the imagery of the Children of the Planets is also alluded to by the armillary sphere on Mercury's chariot; according to ancient thinking, those born under this planet were frequently artists, scholars, and craftsmen.[20]

The pen-and-ink drawing of 'Mercury's children' by the Master of the Housebook, for example, includes a clockmaker using a quadrant at his workbench, an organ builder, and a goldsmith (*Fig. 5.13*).[21] Georg Pencz pictured two scholars, one of whom is measuring angular distances on an armillary sphere with a compass, and they were accompanied by an organ builder, goldsmith, accountant, woodcarver, painter, and physician.

The painting in the upper left part of the bottom tier (360 × 1740 mm) depicts the creation story (*Figs. 5.14* and *5.8, no. 11*). Above the painting is the inscription "*IN PRINCIPIO CREAVIT DEVS COELVM ET TERRAM Gen. 1*".

Eve rises from the rib of the sleeping Adam; God the Father is depicted in proper Lutheran fashion as a halo of light with the Hebrew, Greek, and Latin names for God. The Sun and Moon are shown in the two upper corners of the painting; Adam lies on plants and a bird sits on the left. On the right side the horizon disappears in the mists of heaven. The scene clearly depicts the separation of the firmament of heaven from the waters on the second day of creation (Gen. 1: 6–8). The panel therefore provides a summary of the six creation days. The two hares at Adam's feet as symbols of intemperance and unchastity point to the Fall of Man.[22]

---

20 "For if Mercury governs action, to speak generally, he makes his subjects scribes, men of business, calculators, teachers, merchants, bankers, sooth-sayers, astrologers, sacrificers, and in general those who perform their functions by means of documents, interpretation, and giving and taking"; Ptolemy 1980, IV.4, 383. "Again, when two planets are found to rule action, if Mercury and Venus take the rulership, they bring about action expressed by the arts of the Muses, musical instruments, melodies, or poems, and rhythm, particularly when they have exchanged places [their houses]. For they produce workers in the theater, actors, dealers in slaves, makers of musical instruments, members of the chorus, makers of strings, painters, dancers, weavers, and wax-molders. [...] If Mercury and Mars together assume the lordship of action, they produce sculptors, armorers, makers of sacred monuments, modelers, wrestlers, physicians, surgeons, accusers, adulterers, evil-doers, forgers"; *ibid.*, 387. "Mercury, by himself taking the domination of the soul, in an honorable position makes those who are born under him wise, shrewd, thoughtful, learned, inventive, experienced, good calculators, inquirers into nature, speculative, gifted, emulous, beneficent, prudent, good at conjecture, mathematicians, partakers in mysteries, successful in attaining their ends"; *ibid.*, III.13, 359/361. See also Hauber 1916, 137–139; Linnard/Robey 2019a, 2019b.

21 Hauber 1916, 121, pl. XXXI.

22 LCI, II, cols. 221–222, s.v. "Hase" by Wolfgang Kemp.

FIGURE 5.13 Children of Mercury; drypoint by the Master of the Housebook, second half of the fifteenth century

FIGURE 5.14 Tobias Stimmer, Creation of the World and Resurrection of the Dead

Directly under this panel and allocated to it typologically is a painting (550 × 1510 mm) depicting the resurrection of the dead (*Fig. 5.8, no. 12*). The inscriptions on the ledge below the upper panel read "*ECCE EGO CREO COELOS NOVOS ET TERRAM NOVAM—ISAIAE 65*",[23] and "*EXPERGISCIMI ET LAETAMINI Q*[UI] *HABITATIS IN PVLVERE—ISAIAE 26*".[24]

Besides corpses, Stimmer places numerous nudes in various poses in this painting.[25] The painter skillfully solves the artistic problems created by the extremely rectangular shape of the panels—particularly of the upper panel—by connecting the composition of the two paintings with the beams of the Sun (in the right-hand corner of the upper panel): the light from the creation scene dimly illuminates the resurrection of the dead.

The panel in the upper right part of the bottom tier (350 × 1440 mm) has the inscription "*ASCENDISTI IN ALTVM CEPISTI CAPTIVITATEM Psalm 68*" (Thou

23 Vulgata: "*Ecce enim ego creo caelos novos et terram novam*". KJV: "For, behold, I create new heavens, and a new earth"; Isaiah 65: 17.

24 Vulgata: "*Expergiscimini et laudate, qui habitatis in pulvere*". KJV: "Awake and sing, ye that dwell in the dust"; Isaiah 26: 19.

25 This scene also appears later in the margin of his illustrated Bible, *Neue Künstliche Figuren Biblischer Historien grüntlich von Tobia Stimmer gerissen*, Basel 1576.

FIGURE 5.15 Tobias Stimmer, Last Judgment

hast ascended on high, thou hast led captivity captive).[26] Christ is depicted here as Judge at the Last Judgment with his arms spread and surrounded by an aureole (*Fig. 5.15* and *5.8, no. 13*). In his hands he holds a scepter, a palm branch, a whip, and a sword, as symbols of his office, and he is throned on four glowing red and yellow wheels that are interwoven, forming two pairs of wheels,[27] and is accompanied by three winged animals—an ox, a lion, and an eagle—as well as a man. On the ground at the sides lie Death as a skeleton and the Devil in the shape of a black human-like creature with claws and horns, with their hands bound. Two angels blast their trumpets announcing the Last Judgment; the seven-horned lamb from the Apocalypse (Rev. 5: 6) lies between the wheels of the throne. The figure of Christ the Judge with his gloriole is surrounded in a semicircle by the 144,000 chosen (Rev. 14: 1–3); at their front are King David with his harp on the left, and a prophet (Ezekiel?) with a flaming brazier on the right. Stimmer's painting of the throned Christ with the fiery wheels combines Ezekiel's vision with the Revelation of St. John in the Apocalypse:

26 The text is cited from the Vulgata, in which the verse is Psalm 67: 19. The citation "68" is based on Luther's translation; Psalm 68: 19. In the KJV this is, however, Psalm 68: 18.

27 This depiction is also used in the margin of Stimmer's illustrated Bible.

> As for the likeness of their faces, they four had the face of a man, and the face of a lion, on the right side: and they four had the face of an ox on the left side; they four also had the face of an eagle. […] Now as I beheld the living creatures, behold one wheel upon the earth by the living creatures with four faces. The appearance of the wheels and their work was like unto the colour of a beryl: and they four had one likeness: and their appearance and their work was as it were a wheel in the middle of a wheel.[28]
>
> And immediately I was in the spirit: and, behold, a throne was set in heaven, and one sat on the throne. […] And before the throne there was a sea of glass like unto crystal: and in the midst of the throne, and round about the throne, were four beasts full of eyes before and behind. And the first beast was like a lion, and the second beast like a calf, and the third beast had a face as a man, and the fourth beast was like a flying eagle.[29]

In the Revelation of St. John the four beasts are described as individual creatures, unlike Ezekiel's description of them as a tetramorph, and from early Christian times they were interpreted as the four Evangelists by Irenaeus and Hippolyte. However, Hieronymus and Gregor the Great were the first to assign the creatures in the fashion common today: man or angel (Matthew); ox or bull (Luke); lion (Mark); eagle (John). The throned figure provides another typological connection between the Old and New Testaments: Ezekiel's vision has been interpreted as a revelation of the preexistence; the description cited from Revelations as a demonstration of the elevation of Christ and thus his divinity.[30]

On the lower panel (530 × 1170 mm) with the inscription "*VENITE BENEDICTI P*[AT]*RIS MEI POSSIDETE REGNV*[M] *VOB*[IS] *PARATVM. DISCEDITE A ME MALEDICTI IN IGNEM AETERNV*[M] *Math. 25*"[31] on the bottom side of the

28 "*Similitudo autem vultus eorum, facies hominis et facies leonis a dextris ipsorum quattuor, facies autem bovis a sinistris ipsorum quattuor, et facies aquilae desuper ipsorum quattuor.* […] *Cumque adspicerem animalia, apparuit rota una super terram iuxta animalia habens quattuor facies. Et adspectus rotarum et opus earum quasi visio maris; et una similitudo ipsarum quattuor; et adspectus earum et opera quasi sit rota in medio rotae*"; Ezek. 1: 10; 15–16.

29 "*Et statim fui in spiritu; et ecce sedes posita erat in caelo, et supra sedem sedens.* […] *Et in conspectu sedis tanquam mare vitreum simile crystallo; et in medio sedis et in circuitu sedis quattuor animalia plena oculis ante et retro. Et animal primum simile leoni, et secundum animal simile vitulo, et tertium animal habens faciem quasi hominis et quartum animal simile aquilae volante*"; Rev. 4: 2, 6–7.

30 Schiller 1971, v. III, 234.

31 Vulgata: "*Venite, benedicti Patris mei, possidete paratum vobis regnum a constitutione mundi*"; KJV: "Come, ye blessed of my Father, inherit the kingdom prepared for you from the foundation of the world"; Mt. 25: 34. Vulgata: "*Discedite a me, maledicti, in ignem aeternum*"; KJV: "Depart from me, ye cursed, into everlasting fire"; Mt. 25: 41.

ledge below the upper panel, the Last Judgment scene described above is specifically conceived to apply to individual people (*Fig. 5.8, no. 14*). Two people lie dying at the bottom: the one on the left is a believer; the sinner on the right faces Death, who is depicted as a skeleton with a shovel, throned on a black block in the middle. The believer has his hand folded over his heart and is surrounded by three women, allegories for faith, hope, and charity (love), while the sinner's right hand has already been shackled by the devil. Before he dies he is being offered a cup of gall from a richly dressed 'Lady World', crowned with a glass armillary sphere. In the background the fires of purgatory are visible. As with the panels on the left side of the lowest tier, Stimmer combined these two paintings using formal devices, as Stolberg pointed out:[32]

1. The triangle constructed above the base line of the lower panel has its apex in Christ's head and is approximately equilateral.
2. The allegories for hope and sin (*Frau Welt*) are enclosed in this triangle.
3. The figures lying at Christ's feet, Death, and the Devil are grouped within an arc the center of which is on the triangle's perpendicular where the perpendicular intersects the baseline of the two paintings. If lines are drawn from the two upper corners of the Last Judgment to the intersection of the perpendicular with the baseline of the upper panel, the result is the line formed by the heads of the chosen.

These two pairs of paintings are connected by the calendar disk with the representations of the four empires in the spandrels; the ensemble thus obviously illustrated the beginning and end of the history of salvation.[33]

The two figures lying down on the panels above the opening for the regents of the day (*Fig. 5.8, nos. 15* and *16*) reemphasize the contrast between sinful and Christian living (*Fig. 5.16*). On the left is a full female figure with bare breasts and a glass of wine in her hand. The serpent of Paradise winds its way towards her with an apple; Death emerges in the corner behind her. She steps on a book with her foot, which clearly represents the Bible. On the right, the allegory of Christian living is represented with the Eucharistic chalice and the Bible in her hands; light shines on her from an aureole, from which a dove representing the Holy Spirit flies to her. A stone tablet with commandments is in the background.

---

32 Stolberg 1898, preface, VII–VIII; 11–12.

33 A painting similarly summarizing Creation, the Apocalyptic vision, and the Last Judgment was created by Wendel Dietterlin a few years later, in 1590–1592, on the ceiling of the *Lusthaus* of Duke Louis III of Württemberg in Stuttgart. The duke's senior *Oberrat* (councilor), Georg Gadner, and the leading Protestant theologian and court chaplain Lucas Osiander conceptualized the painting.

FIGURE 5.16 Tobias Stimmer, Sinful Life and Godly Life; Corner spandrel paintings in the bottom tier

The inscription above the panels is a verse from the Epistle to the Romans split in half: "*PECCATI STIPENDIVM MORS EST Rom: 6*" on the left, and "*DONVM VITA AETERNA PER CHR*[ISTU]*M Rom: 6*" on the right, which is different from the Vulgata text.[34]

Between the two paintings is the quarter-hour dial, with two cherubs sitting next to it (*Fig. 5.8, nos. 17* and *18*). The right cherub, next to a skull, turns its hourglass at the fourth quarter chimes of the clock, while the left cherub is set in motion by the striking of the full hour, whose beats he indicates with its sceptre.[35] This automaton reminds the viewer of the Fall of Man every hour, while the figure with the hourglass and skull reminds of the transience of all life on Earth.

On the two posts for the low platform on the bottom tier, two semi-reclining putti (*Fig. 5.8, nos. 19* and *20*) hold up scrolls with the inscriptions "*OMNIS CARO FOENUM*" (All flesh is grass; Is. 40: 6) and "*ALTERAE VITAE SPES*" (Hope of another life) (*Fig. 5.17*). The right putto rests his arm on a skull, the left one has an apple, a snake, and an hourglass as attributes.[36]

Finally, two lions sit on the front corners of the platform on top of the bottom tier (*Fig. 5.8, nos. 21* and *22*); one holds a helmet with crest and covers with its paw, the other a shield with the coat of arms of Strasbourg. According to

---

34 Vulgata: "*Stipendia enim peccati mors gratia autem Dei vita aeterna in Christo Iesu Domino nostro*". KJV: "For the wages of sin is death; but the gift of God is eternal life through Jesus Christ our Lord"; Rom. 6: 23.

35 The latter is in the Musée des Arts Décoratifs in Strasbourg. According to Bach/Rieb/Wilhelm 1992, 55, the other cherub was returned to a platform of the bottom tier, after its arms and the hourglass were changed.

36 Neither figure still exists. Stimmer's plans are in the Musée des Arts Décoratifs in Strasbourg, illustrated in Bendel 1940, 201. The inscriptions can be seen on the engraving by Isaak Brunn from 1609 (Fig. 5.6).

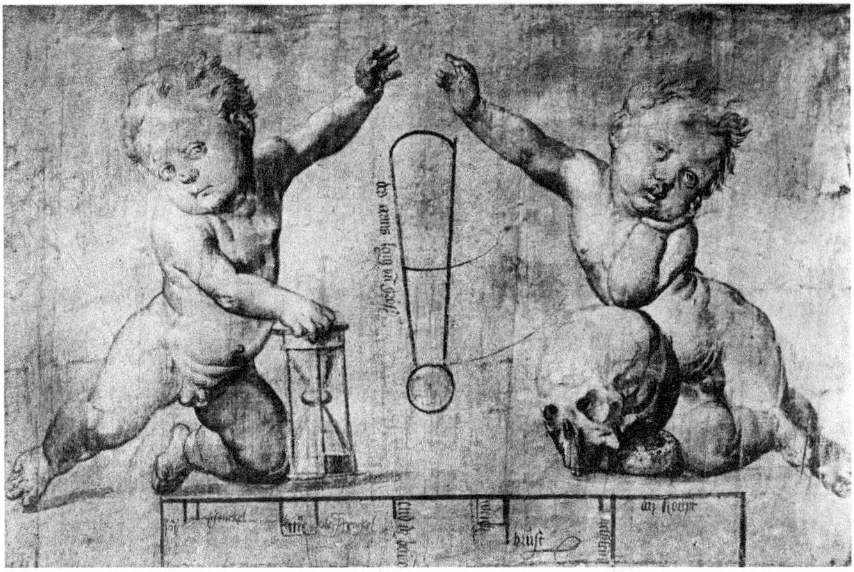

FIGURE 5.17 Tobias Stimmer, conceptual sketch of two putti

Dasypodius something else was originally to have been placed here, but that never happened.[37]

On the staircase, at the same level as the platform above the bottom tier a putto with an hourglass can be seen (*Fig. 5.8, no. 23*), whose left hand rests on a skull (*Fig. 5.18*). Next to this figure, which was carved during the first phase of construction of the new clock, are three inscriptions in Hebrew, Greek, and Latin. The Hebrew inscription is the verse "*Tempus nascendi et tempus moriendi*" (A time to be born, and a time to die; Ecclesiastes 3: 2). Under it is a quote in Greek from Pindar's ode *Man is but a dream of a shadow*, followed by the Latin words "*HOMO BVLLA*". The metaphor "*Homo bulla*" was formulated by Marcus Terentius Varro in his *De re rustica*;[38] Erasmus of Rotterdam dedicated a lengthy section of his *Adagia* (1500) to it, thus introducing it to Renaissance literature.[39]

The scholar and medalist Giovanni Boldù created the motif of the putto with the skull and used it for the first time on a medal in 1458.[40] At the close

37 Dasypodius 1580a, 48. Grandidier erroneously claims that one lion holds the coat of arms of the city, while the other holds the coat of arms of the master builder, and that neither was placed on the clock until 1669; Grandidier 1782, 293.

38 Marcus Terentius Varro, *De re rustica*, 1.1.1.

39 Stechow 1938, 227.

40 Janson 1937, 428–429.

FIGURE 5.18 Sleeping putto with a skull and an hourglass on the spiral staircase

of the fifteenth century the motif developed into an independent pictorial formula and probably become known in northern Europe via an Italian woodcut with the inscription "*L'hora passa*", which depicted a putto with an hourglass, resting its arm on a skull.[41] Here, however, the melancholy reflective putto (*memento mori*) was interpreted as sleeping, thus allowing for a moralizing explanation of the image (a representation of listlessness).[42] This image on the staircase of the Cathedral clock is particularly important, as this is the earliest known example of the motif together with the metaphor '*homo bulla*', thus converting the motif into a symbol for vanity.[43]

5.4 The Central Clock Tower

The four seasons are illustrated in the spandrels around the astrolabe.

On the upper left (*Fig. 5.8, no. 24*), spring is depicted as a young man with a wreath on his head, looking to his left and smelling a flower held in his left hand. A falcon sits on his right hand, which is protected by a leather glove.

---

41 *Ibid.*, 432, fig. 426.

42 *Ibid.*, 436.

43 A putto blowing bubbles is shown for the first time on the back of a portrait of Cornelius Ketel in 1574 (round portrait, diameter 430 mm, Amsterdam, Rijksmuseum); Stechow 1938, 227. In 1594 Hendrick Goltzius created a putto who was blowing bubbles while sitting on a skull, a motif that became popular in genre paintings in the Netherlands in the seventeenth century; Janson 1937, 446–447.

On the upper right (*Fig. 5.8, no. 25*), summer is personified as a farmer with a straw hat, holding a pitchfork in his hand; a sickle rests on his shoulder. In the background lightning shines in a dark sky.

Bacchus personifies autumn in the lower right spandrel (*Fig. 5.8, no. 26*), sitting on a wine barrel while wine is flowing out of the tube held in his hands. A pelican and two ducks can be seen on the left.

In the lower left spandrel (*Fig. 5.8, no. 27*), winter is depicted as an old man in a fur coat, crouching close to a fire to warm his feet. A dog lies curled up on the right. In the background Death emerges out of the dark with an hourglass.

In a few places the figures in the spandrels overlap the painted border of the astrolabe: the cap of the old man, the bill of the pelican, Bacchus's head. In this manner, and particularly by means of the shadow cast on the border of the astrolabe by the left foot of the young man, which appears to be projecting out of the picture, Stimmer creates the illusion of three-dimensionality.

These scenes are framed on the right and left by two marbled wooden columns with foliated Corinthian-like capitals.

The four seasons, the iconography of which the medieval artists copied directly from late antiquity, was a popular topic, especially in book illuminations and church sculptures, where figures with attributes represented rural work, preferably figures that were carrying out tasks—comparable to the monthly labors—associated with the seasonal activities.[44] Stimmer's paintings of the four seasons also follow the traditions for representing the monthly labors. The young man with the floral wreath and falcon corresponds to the motif of a mounted falconer, which was particularly popular in France as an image for May;[45] corresponding representations can be found elsewhere in the Cathedral: the figures of the wise and foolish virgins (southern portal of the west facade) stand on cubes set at an angle, the faces of which depict the twelve signs of the zodiac with the respective monthly labors in quatrefoils. A youth with flowers in both hands represents April (Taurus) and the image for May is a nobly attired rider with the sign Gemini. The hand is missing on his left arm, which is held at an angle, and on which a falcon probably sat. A man sitting at a fire personified February, shown with Pisces. This image first appeared in the Middle Ages and was usually placed as the second month in

44 For details, see Hanfmann 1951, v. 1, 210ff and 262ff. The development of the images of the monthly labors from the predominantly abstract ancient personifications for the individual months to the 'active images' of the Middle Ages, which gradually developed into genre scenes with landscapes in the background, can only be briefly mentioned here; see Riegl 1889, 36ff, 49–51 and 71; Webster 1938, passim (esp. 97ff); Brandt 1927/28, v. 1, 148ff.

45 Koseleff 1934, 33–37.

the year.[46] At the same time, the spandrel paintings of the astrolabe illustrate the four temperaments or humors. The dark skin color of the old man[47] with his dog—an animal that was thought to be Saturnian[48]—indicate that he was a melancholic; spring corresponded to a sanguine humor, while summer was choleric and autumn phlegmatic. The same parallels between the seasons and the humors can be found in the woodcut illustrations in the *Quattuor Libri Amorum* of Conrad Celtis;[49] in ancient times, however, the phlegmatic humor was allotted to winter and the melancholy temperament to autumn.[50] This shift in Celtis can be attributed to the text, but also to the different climatic conditions: autumn in southern Europe is cold and dry, and is associated with the north wind and black gall in ancient sources; in northern Europe, however, such weather corresponds closer to winter. On the other hand, the characteristics attributed to winter (cold and damp / west wind / phlegm) in ancient times fit more closely for autumn at northern latitudes.[51] Additionally, the spandrel paintings illustrate the four elements: spring = air (alluded to by the falcon); summer = fire (lightning in the sky); autumn = water (Bacchus with wine flowing, water fowl are at his feet); winter = earth as the fourth element, as the old man will become dust after his imminent death.

The parallels between the seasons and the temperaments were recognized by the clock's contemporaries, as can be seen from Johann Fischart's poem describing the clock in the small edition of the woodcut: "*Erstlich ain Astrolabium* [...] *Ummalet mit vier Jareszeiten Und vier Complexion zur seiten*" (To start, an Astrolabe [...] Surrounded by four Seasonal tides and four Humors at their sides).

Assuming the seasons symbolize the four ages of man—which finds parallels among the Pythagoreans and in Ovid,[52] among others—is suggested by the ages of the personifications and the appearance of Death in the winter scene as well as by the fact that Dasypodius expressly mentioned this: "Besides this

46 On this, see the synoptic tables in Webster 1938, 175–179, which, however, only cover the months' cycles of the twelfth century. Comparable studies for later centuries have not been carried out, so far. This image possibly resulted from a misinterpretation of the sacrifice scene in the Chronograph of Filocalus (354) as the warming of hands; Riegl 1889, 39 and 45.

47 The faces of both Saturn's children as well as melancholics were black or black-brown according to ancient authors, a view that was passed on in the medieval treatises on medicine and astrology as well as in the widespread *Complexbüchlein*; documented in Klibansky/Panofsky/Saxl 1964, 290.

48 *Ibid.*, 322–323.

49 Chojecka, 1967, 85, 91.

50 Boll/Bezold 1926, 54.

51 Klibansky/Panofsky/Saxl 1964, 280.

52 *Metamorphoses*, XV, 199–213; on this see Boll 1913, 13–14; Hanfmann 1951, v. I, 124–125.

there are the four tides of the year, Spring, Summer, Autumn, Winter painted with the four noble ages in the corners next to the Astrolabe, so that nothing shall be wanting in the description of time".[53]

In addition to these quaternities, there is also an astrological level of meaning, which will be dealt with when the astrolabe is described. The eclipse tables and spandrel paintings next to the astrolabe, as well as the automata figures of the four ages of man (see below) point to numerous quaternities (four winds = four points of the compass = four seasons = four humors = four elements = four ages of man) and allude to the cosmos as the model for the astronomical clock.

Tetraktys (quaternities) had already played a decisive role for the Pythagoreans; the world and mankind were ruled by four principles and they were the first to come up with a correlation between the ages of man and the seasons. Around 400 BC, a parallel then was formed between the four seasons, humoral fluids, qualities, and ages of man in *De natura hominis,* which was part of the *Corpus Hippocraticum*.[54] Departing from a depiction of melancholy in the pseudo-Aristotelian *Problemata*, the school of Aristotle then developed the quadruple scheme of the temperaments and assigned the four qualities to the four elements of Empedocles.[55] Speculations on tetrads were also carried out by astrologers and led to expanded systems, such as that of Antiochos of Athens (first century BC).[56] From these quadruple combinations, Galen finally set up a comprehensive system[57] for the micro- and macro-cosmos, which was expanded to include the history of salvation in the early Christian era (the fourfold exegesis, four Evangelists, four rivers of paradise, four arms of the cross, etc.).[58] The creation of such parallels can be found in the encyclopedias and commentaries of the Middle Ages, such as those of Isidore of Seville or Vincent of Beauvais, frequently in the form of geometrical diagrams (circle diagrams).[59] In the Renaissance, cosmological quaternities were incorporated into all fields of study and became a common, popular topic in humanist circles, as well as in the visual arts.[60] The illustrations in the *Quattuor Libri Amorum* (1502) of Conrad Celtis and the synoptic depiction of quartanes in Book II, Chapter 7,

---

53 *"Neben disem allem seindt die vier zeit des Jahrs Früling, Sommer, Herbst, Winter mit den vier fürnemmen alteren abgemahlet in den neben ecken des Astrolabii, damit an beschreibung der zeyt nichts manglete"*; Dasypodius 1580a, 47; Dasypodius 1580b, fol. G III^{v}.

54 Schöner 1964, 17ff.

55 *Ibid.*, 66–67 and 71.

56 Boll/Bezold 1926, 54 (also in Schöner 1964, 98); Gundel/Gundel 1966, 115–117.

57 Schöner 1968, 87ff.

58 Meyer 1975, 123–127; Maurmann 1976, 188–200; Meyer/Suntrup 1987, cols. 332–337.

59 On this, see Wickersheimer 1914, 157–164; Beer 1952; Euw 1975, 90–95; Murdoch 1984, 52–53 and 356. 368; North 2004; Müller 2008.

60 Heninger 1977, 102ff.

of *Occulta Philosophia* (1531–1533) by Agrippa von Nettesheim form remarkable examples.[61]

In the widely read 'standard work' on astrology, the *Tetrabiblos* by Claudius Ptolemy, parallels are also set up between the seasons, regions of the world, winds, and ages of man.[62] As this work—as will be shown later—played an important role in the design of the Cathedral clock, it is quite probable that Dasypodius was motivated by it to emphasize combinations of quaternities in his concepts of images and figures for the clock.

Above the astrolabe is the inscription: "*D*[OMI]*N*[U]*S LUX MEA QVEM TIMEBO*".[63] Centered above the two half panels with this inscription was a dial with an indicator for the phases and age of the Moon and a star-studded sky is painted on the arch above it. On the border above the semicircular opening for the disk representing the Moon is the inscription: "*QVAE EST HAEC TAM ILLVSTRIS SIMILIS AVRORAE PVLCRA* [sic] *VT LVNA PVRA VT SOL*" (Who is she that looketh forth as the morning, fair as the Moon, clear as the Sun).[64]

Arched buttresses support a projecting platform on which the four ages of man appear. On the left and right, the Apocalyptic woman clothed with the Sun and the Antichrist are depicted in the arched fields of the buttresses (*Fig. 5.8, nos. 28* and *29*). The Apocalyptic woman is interpreted in the patristic literature of the early Christian theologians as Ecclesia,[65] and was painted by Stimmer wearing a crown, with her arms outstretched, and standing on a crescent moon; the inscription above her reads "*ECCLESIA CHRISTI EXVLANS*" (*Fig. 5.19*). The Antichrist, with the inscription "*SERPENS ANTIQVVS ANTICHRISTVS*", is depicted as a winged, lion-like creature with seven heads. The two paintings refer to descriptions in the Book of Revelation:

> And there appeared a great wonder in heaven; a woman clothed with the sun, and the moon under her feet, and upon her head a crown of twelve stars: […] And there appeared another wonder in heaven; and behold a great red dragon, having seven heads and ten horns, and seven crowns upon his heads.[66]

---

61 *Ibid.*, 116–117.

62 Ptolemy 1980, I.10, 60–65.

63 Vulgata: "*Dominus illuminatio mea et salus mea: quem timebo?*"; Ps. 26, 1. KJV: "The Lord is my light and my salvation; whom shall I fear?"; Ps. 27, 1.

64 Vulgata: "*Quae est ista, quae progreditur quasi aurora consurgens, pulchra ut luna, electa ut sol*"; Song 6: 10. However, on Brunn's engraving (Fig. 5.6) the inscription reads "*QVAE EST ILLVD, QVOD PROSPECTAT, SICVT AVRORA, FORMOSUS VT LUNA, ELECTU UT SOL*".

65 LCI, I, col. 146, s.v. "Apokalyptisches Weib" by J. Fronrobert.

66 "*Et signum magnum apparuit in caelo; mulier amicta sole, et luna sub pedibus eius, et in capite eius corona stellarum duodecim.* […] *Et visum est aliud signum in caelo: et ecce*

FIGURE 5.19 Tobias Stimmer, Ecclesia and the Antichrist; corner spandrel paintings in the top tier

Stimmer probably used Dürer's woodcut the *Apokalypse* as a model for designing these two fields (*Fig. 5.20*).

On the platform, the wooden statues of the four ages of man (*Fig. 5.8, no. 30*) were mounted on a rotating disk, which rang out the quarter hours:[67]

1. Childhood (h. 610 mm): a child taking a step and holding a ball in its raised right hand.
2. Youth (h. 710 mm): a youth with a laurel wreath and cape, holding a bow and arrow. Around his loins he wears a wide wreath of victory.
3. Manhood (h. 690 mm): a warrior with a full beard in a ceremonial suit of armor, his torso bent back to the left, holding a club in his right hand.
4. Old age (h. 710 mm): an old man with a beard, bent over, wearing a fur coat with a cape, and holding a cane with both hands.

The bell for the hours is above these figures on a higher level which is concluded with arches adorned with tracery. There is another rotating disk on this level, on which Christ and Death stand opposite each other.

Christ (*Fig. 5.8, no. 31*, h. 780 mm) wears a robe which hangs from his shoulders and is held together by a strap near his clavicle. The figure, taking a step and making a gesture of blessing with his right hand, holds a palm branch and a cross in his left hand.

Death (*Fig. 5.8, no. 32*, h. 780 mm) is depicted drastically as a skeleton covered with skin, holding a long bone horizontally in front of him. Snakes are wrapped around his left leg, climbing towards his upper torso.

---

draco magnus rufus habens capita septem et cornua decem, et in capitibus suis septem diademata"; Rev. 12: 1 and 3.

67 The figures and the extant plans for the statues of childhood, youth, manhood, and Death by Stimmer are in the Musée des Arts Décoratifs in Strasbourg.

FIGURE 5.20 Albrecht Dürer, The Woman of the Apocalypse and the Seven-Headed Dragon; woodcut, 1497

Finally, the wooden figures of a lion and a griffin must be mentioned, which sit on the capitals on the same level as the disk with the ages of man (*Fig. 5.8, nos. 33* and *34*). The shield held by the lion carries the symbol of the masons of the 'Frauenwerk' and that of the griffin has the coat of arms of the master builder Uhlberger: a horizontally split shield in the lower field of which the base is formed by a trimount, while in the upper field is a half-figure of a griffin holding a compass and a sphere in its claws.

The seal impression in Archives Municipales de Strasbourg (*Fig. 5.21*) shows the entire coat of arms with a frog-mouth helmet, cover, and a griffin on the right as a crest, holding a compass and sphere in its claws.[68]

The openwork crown of the central clock tower is made up of complicated arches with fleur-de-lis-shaped crockets—similar to the openwork superstructures of winged altarpieces.

The design for this was probably created by the master builder of the Cathedral Hans Thoman Uhlberger, who also built the wooden cover for the stone clock housing. Uhlberger is mentioned as the *Stadtbaumeister* (municipal master builder) of Sélestat (Schlettstadt) in 1563.[69] Two years later he was chosen to be the master builder of the Cathedral in Strasbourg and thus the supervisor of the German builders' lodge, succeeding Marx Schan. He held this position for 43 years, until he died in 1608.[70]

---

68 Strasbourg, AMS: Fastinger Collection, casts by Charles Haudot, no. 199[4].

69 "Ordnungen und Artikel der Steinmetzen-Brüderschaft (Regulations and articles of the society of masons)" from 29 Sept. 1563, reprinted in *Jahrbuch der kunsthistorischen Sammlungen des Allerhöchsten Kaiserhauses*, 15, 1894, v. II, 58–61; on page 61 Uhlberger's signature as "*statmeister zu Schletstatt*".

70 Reinhardt 1972, 28—following the thinking of Kraus 1876, 411–412, who based his information on notes from the Cathedral's master builder Heckler—assumed that the father of Hans Thoman Uhlberger created the clock housing. Heckler wrote: "*Anno 1565 auf Micheli ist Hans Uhlberger Werkmeister worden bis 1576. von 1576 bis 1608. war Hanß Thoman Ulberger Werkmeister; von Pauli Bekerung 1608. bis den 13ten Julii 1610. war Hanß Carln Uhlberger des Hanß Thoman Uhlberger Sohn Werkmeister* (on Michaelmas, in the year 1565, Hans Uhlberger became master builder until 1576; from 1576 until 1608 Hanß Thoman Ulberger was master builder; von Pauli Bekerung from 1608 until 13 July 1610. Hanß Carln Uhlberger was the son of the master builder Hanß Thoman Uhlberger)"; BNUS, MS 1373 (Als. 704), fol. 15[r]. The list of master builders compiled by the Cathedral Office of Construction in 1905 also claims that Hans Uhlberger worked as master builder from 1565 to 1576, and his son Hans Thoman from 1576 to 1608; *Straßburger Münsterblatt: Organ des Straßburger Münster-Vereins* 2, 1905, 31–32. This assumption is, however, erroneous, as Otto Winckelmann showed; Winckelmann/Schmitz 1894, 266. According to the minutes of the XXI, v. 81 for 1600, fol. 13[v], H. Th. Uhlberger had—as he personally claimed—served in this position continuously for 35 years, that is, since 1565. In 1608 Uhlberger died at an old age and on 22 January 1608 an inventory was compiled of all the figures, materials

FIGURE 5.21 Seal of Hans Thoman Uhlberger, Master Builder of the Strasbourg Cathedral

In the uppermost part of the tower, statues of the three Evangelists Mark, Luke, John with their respective symbols (*Fig. 5.8, nos. 35, 36* and *37*), and two angels playing musical instruments (*Fig. 5.8, nos. 38* and *39*) can be seen in the woodcut. There are shields with the coats of arms of several Cathedral caretakers and *ammeisters* on the pedestals for these figures. Between the two angels there was an iron, multicolored ogee arch on which the coat of arms of Strasbourg—today at the Musée des Arts Décoratifs—was mounted.

---

and tools left behind by him in the 'Frauenhaus'; AMS, Sér. II, 46, no. 36. Additionally, it should be observed that his father received the rights of citizenship in 1507: "*Item Hans Ulberg der murer hat das burgrecht entpfangen von Magdalenen Jacob von Barre des murers seligen witwen siner hussfrowen wegen secunda ipsa Conversionis pauli*"; Bürgerbuch I, col. 391, in Wittmer/Meyer 1948/61, v. II, 561, no. 5641. Uhlberger worked in the masons' court as municipal works master. His name appears in two documents: "*Johannis Vlberger lapicide*", AMS, Kontraktstube, v. XVII, 1524–1526, fol. 283^{v} [1526]; "*Hanns Ulberger murer*", AMS, Œuvre Notre-Dame, Inventaire des Archives anciennes, no. 4, fol. 35^{r}, Spruchbrief von 1527.

Nothing is known of the whereabouts of the statues; they are neither in the Musée de l'Œuvre Notre-Dame nor in the Musée des Arts Décoratifs in Strasbourg.[71] The statues that are now on the clock all stem from the nineteenth century.[72] However, the small statue of the master builder Uhlberger on the finial dates from the building period of the second clock. The gaunt man holds a compass in his right hand, while resting his left hand on a shield with the mason's mark of Uhlberger.[73] This figure is cut off by the border of the two woodcuts by Stimmer; Arhardt's drawing from 1673, however, shows it quite clearly.

5.5 The Tower for the Weights

The tower for the weights, with the iron cock from the first Cathedral clock mounted on its top, has a total of seven panels with paintings on three levels. These panels are framed by Doric columns on the bottom level, Ionic columns in the middle, and by Corinthian columns on the top level.

On the front, at the very bottom of the tower, is the portrait of Nicolaus Copernicus (*Fig. 5.8, no. 40*), which will be dealt with in more detail in the next chapter. Above it is the 'Great Image' or colossus from Nebuchadnezzar's dream (*Fig. 5.8, no. 41*), and at the top is a painting of the celestial goddess Urania (*Fig. 5.8, no. 42*).

Urania is a winged full-length female figure leaning against a wall, with her right knee bent and her right foot resting on a sphere. With a compass in her right hand, she measures a celestial globe with lines signifying tropics, colures, and the ecliptic, which rests on her right thigh and is held in place by her left hand. Basic geometrical shapes (circle, triangle, square) are painted on her right wing together with the label "*GEOMETRIA*". On her left wing are the numbers 1, 2, 3, 4 and the label "*ARITHMETICA*". Geometry and arithmetic are represented here as 'subsidiary sciences' or, rather, as the preconditions for astronomy. They are the "wings for the human spirit", which permit man to rise to the heavens. This is Melanchthon's interpretation of the wings:

---

71 Information kindly provided by Dr. Jean-Daniel Ludmann (Musée des Arts Décoratifs) and the Cathedral architect Jean-Richard Haeusser (Musée de l'Œuvre Notre-Dame).

72 Alfred and Théodore Ungerer erroneously took these sculptures to be the originals; Ungerer 1922, 22 and 46.

73 Winckelmann/Schmitz 1894, 267, besides the plaster cast of the statue on the finial, mention a second figure in the 'Frauenhaus' which also has a shield with a mason's mark similar to that of H. Th. Uhlberger. Whether this figure was of the father, Hans Uhlberger, or possibly the original figure, cannot be determined.

> In the *Phaedrus* Plato invents two kinds of soul, one of which he says is winged, the other has lost its wings. Then he says that those that have wings fly up to heaven and delight in meeting and conversing with God [...]. And, flying throughout all of heaven, these souls are charmed by the beauty of divine things and by the sweetness of the knowledge and of the virtue of that admirable order, and they wish to enjoy this one pleasure forever. [...] Even if Plato thinks of the wings as heroic impulses of the minds, it is nevertheless not these impulses alone that lift up the minds, but the arts are also needed by which these impulses are raised up. Consequently, the wings of the human mind are arithmetic and geometry. If someone endowed with a not sordid intellect attached these [wings] to himself, he would easily enter heaven and would wander freely in the heavenly company, and enjoy that light and wisdom.[74]

The 'Great Image' or colossus from Nebuchadnezzar's dream (Dan. 2: 31–32) is shown almost head-on in contrapposto. The painting on the upright rectangular panel (1840 × 635 mm; Strasbourg, Musée des Arts Décoratifs) ends in a semicircle at the top, which was the result of the original frame of the tower and gave the impression of a niche for the figure. The figure wears a golden crown on his head and holds a scepter in his right hand. Above it is the horizontal inscription "*Daniel = II CA*". Corresponding to the description in Daniel, his breastplate and his arms were of silver, such that the breastplate armor did not cover the lower rib cage or the stomach.

There are three painted panels on the left side of the weights tower, which Stimmer adorned with monochrome reddish-brown paintings of the three *Moirai* (Fates or, in Latin, *Parcae*):

Lachesis, at the top (*Fig. 5.8, no. 43*), appears as an airily draped girl with a floral crown, taking a step with a slight twist of her upper body towards the viewer. In her right hand she holds a distaff mounted on a long rod, from which she spins the thread of life with her left hand.

74 *In arithmeticen praefatio Georgii Ioachimi Rhetici* (1536), in Melanchthon 1999, 93. "*Plato in Phaedro fingit duplices animas, quarum alteras ait alatas esse, alteris, inquit, decidisse alas. Porro alatas ait volitare in coelum, fruentes congressu colloquioque Dei* [...]. *Atque hae animae toto volitantes coelo captae pulchritudine divinarum rerum, et illius admirandi ordinis et suavitate doctrinae et virtutis, hac una voluptate perpetuo frui cupiunt* [...]. *Etsi autem Plato alas intelligit heroicos impetus ingeniorum, tamen ne illi quidem impetus soli animos subvehunt, sed opus est etiam artibus, quibus illi ipsi impetus attollantur. Sunt igitur alae mentis humanae, Arithmetica et Geometria. Has si alligaverit sibi aliquis praeditus ingenio non sordido, facilime penetrabit in coelum ac libere in coelestium coetu vagabitur, et illa luce ac sapientia fruetur*"; CR, XI, col. 288. On how the tradition of the motif was transmitted further see Remmert 2009, 45–47.

Clotho, in the middle (*Fig. 5.8, no. 44*), holds the end of the thread of life in her raised left hand, which is wrapped around the front of her body and which she picks up with her right hand.[75] This figure seems to form a spiraling curve around her firmly placed left leg as a result of her bent free leg, which barely touches the ground, her extended right arm, which is twisted towards the left front part of the panel, the motion of which is picked up by billowing folds of the robe over her left shoulder, and the simultaneous twist of the figure's torso slightly to her right.[76]

Atropos, at the bottom (*Fig. 5.8, no. 45*), is shown as the frontal view of an old woman with her eyes lowered, wrapped in a flowing robe, and holding large shears in her raised right hand. She is about to cut the thread of life, which is wound on a spool.

At the same time, the ages of the three women present the stages of human life:

> So that no spot remains empty, but rather at every one not something of no avail, but food for thought is to be seen, so we have painted three different women […] on the back of this chest towards the choir, who are called the Parcae by the poets, each of whom has a special name, the one named Lachesis, she holds the distaff; Chlotho, she spins; Atropos, she cuts the thread the time of human life off; in such manner the poets wanted to show how we come to this world, and strive as well as court, at last comes Death, who of our life, the thread takes and cuts off.[77]

---

75 Clotho is not holding the tip of her robe, as Stolberg interpreted the painting; Stolberg 1898, 22.

76 Stimmer used a similar pose around 1578 in his "Versuchung eines Reiters" (Temptation of a Knight); Berlin, Kupferstichkabinett, inv. no. 901. See also Thöne's catalog of Stimmer's drawings; Thöne 1936, 89, no. 27, pl. 33, fig. 78. According to Thöne, the drawing served as a pattern for a field in a ceiling painting in the Baden-Baden castle; *ibid.*, 77. A copy of the drawing is in Kupferstichkabinett (Department of Prints and Drawings) in Basel, inv. no. U4, 63; *ibid.*, 95, no. 108.

77 "*Damit aber kein ort laehr stünde, sonder an einem jeden etwas nicht vergeblich, sonder zu einer gedechtnuß zusehen were, so haben wir zu ruck gegen dem Chor an disem kasten drey underschydliche Frawenbildt* […] *gemahlet, welche bey den Poeten Parcae genennet werden, deren ein jede ein sonderen nammen hat, als namlich Lachesis die haltet die kunckel, Chlotho die spinnet, Atropos die schneidt den faden ab, dardurch die Poeten die zeit deß Menschlich laebens haben wollen anzeygen, wie wir auff diße welt kommen und streben auch werben, zu letst kompt der Todt, der unserem leben den faden nimpt und abschneydt*"; Dasypodius 1580a, 51–52. See also, for example, the woodcut of the three Fates by Hans Baldung Grien (1513).

The representations of the three Fates and the ages of man around the astrolabe as well as the figures of the automata that chime the quarter hours can possibly be brought into correlation with the 'inverted' arrangement of the columns on the spiral staircase. In his *Theatrum Vitae Humanae*, Vredeman de Vries (1527–1604 or 1623) set up a relationship between the arrangement of columns and the ages of man.[78] Dasypodius can hardly have been familiar with this 'column philosophy' from the Netherlands at the time he was designing the Cathedral clock, as de Vries's work did not appear in Amsterdam until 1577. Nonetheless the arrangement of the panels with the Fates is an example of Mannerist column exegesis: as the personification of their ages, Atropos was assigned Doric, Clotho Ionic, and the youthful Lachesis Corinthian columns.

On the right side of the weights tower there is a panel with emblems (*Fig. 5.8, no. 46*), which will be dealt with in more detail in *Chapter 11*.

---

78 Forssman 1956, 156–159.

CHAPTER 6

The Portrait of Copernicus: Did Conrad Dasypodius Adhere to Heliocentrism?

Copernicus is depicted on the bottom front panel (1680 × 580 mm) on the front of the weights tower as a standing figure, turned slightly to the left (*Fig. 6.1*). The astronomer is wearing a *Schaube* (broad tunic) with a fur collar at his neckline and a wide red vest. Essentially the painting is only a bust, as an upright rectangular tablet, on which Copernicus's right hand rests, conceals the entire lower half of the figure, such that only his right leg is visible.[1] The tablet carries the inscription: "*NICOLAI COPERNICI VE=RA EFIGIES EX IPSIVS AVTOGRAPHO DEPI=CTA*".

The signature "*TS*", for Tobias Stimmer, is located at the very bottom. In his left hand Copernicus holds a sprig of lily of the valley (*Convallaria majalis* Linné, 1753). This plant was recommended as a tonic for the heart and brain by various medieval authors and alludes to Copernicus's occupation as a physician.[2]

After his studies of canon law and astronomy in Bologna (1496–1500), Copernicus received permission from the cathedral chapter of Frombork (German: Frauenburg) to study medicine for two years in Padua. He did not, however, complete an academic degree in that discipline, but received his doctorate in canon law in Ferrara in 1503 and returned to Ermland in the same year, where, besides administrative duties and astronomical studies, Copernicus worked as a physician.[3]

Although the attributes on the ledge in the painting's foreground—compass with open legs, cone, quadrant and celestial globe—identify Copernicus as an astronomer, he is not depicted as the discoverer of a new order for the universe. In fact, it was not until 1591 that a portrait of Copernicus with a greatly

1 The painting is now in the middle field of the tower. The painting of the "Great Image" or colossus is displayed in the Musée des Arts Décoratifs in Strasbourg. The present clock now has the 1843 portrait of Schwilgué by Gabriel Christophe Guérin (1790–1846) at the bottom.

2 Behling 1957, 78–79. On portraits of physicians with lilies of the valley as attributes see Brunn 1937, 27ff—who also discusses the Strasbourg portrait; *ibid.*, 32–36—and Flik-Fizek 1992, 168–169. In the picture of the municipal physician of Zürich and polymath Conrad Gesner in Nicolaus Reusner's *Icones*, fol. RVII^r, besides two roses, lily of the valley is also shown.

3 For details on the medical activities of Copernicus see Berg 1943, 172–201; Schmauch 1940/43; Eis 1952; Rosenberg 1973; Buczkowski 1989.

 | DOI:10.1163/9789004423473_008

FIGURE 6.1
Tobias Stimmer: Portrait of Nicolaus Copernicus; oil painting on the weight drive tower of the Strasbourg Cathedral Clock

simplified version of the heliocentric system appeared on a table clock by Jost Bürgi and Hans Jakob Emck.[4]

On the two Stimmer woodcuts, Copernicus does appear turned partially to the left, but the tablet is on the left and the figure's left arm is completely visible. Here one must assume that Stimmer's rendering of the clock was inaccurate, as the painting is—as an investigation proved in 1923—in good condition and only has minor amounts of overpainting.[5] Both Isaak Brunn's engraving and Arhardt's drawing, that were created directly in front of the clock, have the same arrangement as can be seen today in the Cathedral.

As Dasypodius wrote, he had a pattern for the portrait of Copernicus sent to him from Danzig (now Gdańsk):

> Following this [colossus from Nebuchadnezzar's dream] on the bottom layer is a truthfully rendered portrayal of the exquisite and learned Mathematician Nic. Copernicus […] as was sent to me from Danzig by the honorable and greatly learned Doctor Tiedemann Giese and copied from the original in the most diligent and most excellently honed manner by Tobias Stimmer, the painting which we wanted to have placed here as a special memorial, as no other has such meaning and importance, as good then as can be found at any location on the entire Astronomical Clock.[6]

Tiedemann Giese was a son of the Danzig merchant Georg Giese, whose portrait was painted by Hans Holbein the Younger, and a cousin of Copernicus's friend Bishop Tiedemann Giese (1480–1550).[7]

---

4 Mackensen 1978b, 132–138; Metze 2004, 101–102.

5 Batowski 1942, 29, n. 23 (1). Zygmunt Batowski (1876–1944) was an art historian and lectured as a professor at the University of Warsaw from 1919 on. The German translation of his book *Wizerunki Kopernika* (Toruń, 1933), was published in 1942 by the Publikationsstelle Berlin-Dahlem. This was an office of the Prussian State Archives which continued to exist until 1945, in which research of the *Ostforschung* was carried out with a National Socialist orientation. In contrast to Batowski's assessment, Józef Flik claimed there were massive amounts of overpainting after he x-rayed the panel in the 1970s. However his findings cannot be unequivocally verified; Kühne/Kirschner 2004, 333; Metze 2004, 30.

6 "*Nach disem auf dem understen ist des herrlichen und Gelehrten Mathematici, Nic. Copernici warhafftige abconterfet* […] *so mir auß Danzig durch den Ehrenfesten und hochgelehrten herren Doctor Tideman Gyse zukommen und auß dem original auf das allerfleysigst und scharpfest durch Tobiam Stimmer abgemahlet worden, welche gemaeldt wir zu einer sonderen gedechtnus hieher haben setzen wollen, dan keines ist, das nicht sein bedeutung und anzeygung hette, wie dan solches an allen orten des gantzen Astronomischen Uhrwercks zu finden ist*"; Dasypodius 1580a, 51.

7 Brachvogel 1919, 596.

The pattern, as the portrait and the description in the *Warhafftigen Außlegung* substantiate, was supposed to have been an '*autographon*' of Copernicus. His biographer Pierre Gassendi (1592–1655) claimed that the Frombork Cathedral canon was engaged in artistic activities, basing his narration on an "old written record":

> To be sure, he [Copernicus] was occupied with all the fields of mathematics; in particular, however, he concentrated on perspective, and by means of this opportunity he learned to paint and was proficient at doing so up to such a level that it was said that he painted excellently from his mirror image. He, however, came up with the plan to paint because he was thinking about traveling—preferably to Italy—not only in order to draw, but also to express in drawing, to the extent to which he was able, what he considered to be worthy.[8]

Even if one doubts the value of the source of these frequently cited comments,[9] the possibility cannot be completely excluded that Copernicus was able to draw a self-portrait. Nothing else, however, is known about the type of art used for the '*autographon*' or what it possibly looked like—Stimmer probably used a drawing of the bust of the astronomer, which he enlarged into a full-length figure—and it can be assumed that the portrait was returned to Danzig from Strasbourg.[10] The '*autographon*' was possibly created in Heilsberg between 1503 and 1510, where Copernicus was occupied as a companion, secretary, and

---

8 "*Cum parteis vero omneis Matheseos curaret, tum Perspectivae speciatim incubuit, eiusque occasione Picturam tum addidicit, tum eo usque calluit, ut perhibeatur etiam se ad speculum eximie pinxisse. Consilium autem pingendi ex eo cepit, quod peregrinationem, ac potissimum in Italiam cogitans, in animo haberet, non modo adumbrare, sed graphice etiam, quantum posset, exprimere quicquid occurreret observatu dignum*"; Gassendi 1658, 499. If this was true, Copernicus would have been left-handed, as Batowski concluded; Batowski 1942, 6.

9 Schwarz considered Gassendi's account unreliable and assumed that the '*autographon*' was merely a drawing that he himself had signed or designated for use; Schwarz 1943a, 145–146; Schwarz 1943b, 112–113. Kühne also came to a skeptical conclusion: "Copernicus as an artist and practicing 'perspectivist', as long as no new sources become known, which is hardly to be expected, has to be placed in the realm of biographical fables"; Kühne 2001, 241–242.

10 The speculative attempt to date the creation of the '*autographon*' to 1499, based on a 1509 portrait of an 'unknown person' by Lucas Cranach the Elder—on which see Schenk zu Schweinsberg 1943, 279–281—, has proven to be incorrect. The portrait in question is of Georg Spalatin, the court chaplain and librarian of Frederick the Wise; Sander 1950, 45–48; Hamel 1995; Höss 1996. It was formerly part of the collection of the Berlin publishing couple Franz and Frieda Lipperheide, and had been missing since the Second World War.

physician for his uncle, Bishop Lukas Watzenrode, and acquired numerous works on medicine for the library of the bishopric.

A portrait of Copernicus was mentioned in 1580 and 1598 as being in the possession of the chapter of the Frombork Cathedral.[11] Whether it was the '*autographon*' itself or an oil copy can no longer be determined, as the portrait has disappeared.[12]

Besides pictures of other important astronomers, the Danish astronomer Tycho Brahe received a portrait of Copernicus which inspired his short poem: "*In D. Nicolai Copernici Toronensis effigiem, quam ipsemet sua manu e speculo depinxisse dicitur*".[13] One might assume that such a portrait was the '*autographon*'.

Elias Olsen Morsing (d. 1590), a student of Tycho Brahe, was sent to Frombork by Brahe in 1584 in order to determine the latitude of the location precisely.[14] On that occasion the Frombork canon Johannes Hannov gave Olsen a triquetrum (or parallactic ruler) that had belonged to Copernicus as a present for Brahe. Olsen possibly received the '*autographon*' or a copy of it at that time. Unfortunately, the picture has also been lost, so that nothing more exact can be said about what was passed down to Brahe.[15]

The woodcut attributed to Christoph Murer (1558–1614) in Reusner's work on portraits (*Fig. 6.2*),[16] is directly connected to the '*autographon*'—possibly

In 1993 it reappeared on the art market and was purchased by the Museum der Bildenden Künste in Leipzig.

11 Hipler 1875, 86–88.

12 On the portraits that descended from the Frombork portrait, see Schwarz 1943a, 154–162, stemma on 171.

13 TBOO, VI (Epistolarum astronomicum), 270–271. See also Schwarz 1943a, 153.

14 Dreyer 1890, 123–125; Christianson 2000, 323–325.

15 Hipler claims—without naming a source—that the '*autographon*' in Tycho's possession was burned in 1597; Hipler 1875, 91. This cannot be true; in March of that year Brahe left the island of Hven after he fell out of favor with the Danish king, Christian IV; Dreyer 1890, 238. The portrait was probably used as a pattern for another painting, which Brahe had mounted on the walls of his study in Stjerneborg, built in 1584, together with the portraits of seven other astronomers. It carried the inscription: "*NICOLAVS COPERNICVS TORONENSIS VIXIT CIRCA ANNUM 1520. POST CHRISTVM*"; TBOO, VI, 275.

16 Reusner 1587, fol. GVI^v. According to Reusner, preface, fol. IV^v, Tobias Stimmer was the primary painter of the portrait. The publisher Berhard Jobin, in his German edition *Contrafacturbuch, Ware und lebendige Bildnussen etlicher weit berhümbten und hochgelerten Männer in Teutschland*, Strasbourg, 1587, makes the more specific claim that Stimmer and his pupil Christoph Murer (Maurer) had worked together, the latter by diligently imitating and copying Stimmer's style ("*fleissige Nachraisung und verfolgung seiner Art*"). In a letter from Reusner to Heinrich Rantzau from 31 December 1586; Vienna, ÖNB, Cod. Vin. 9737 1, a painter from Zürich is mentioned who drew the portrait on woodblocks for the *Icones* and was most probably the same Christoph Murer. Stimmer cannot

FIGURE 6.2 Christoph Murer: Portrait of Nicolaus Copernicus; woodcut

via the intermediary step of a copy of the original[17]—and was copied in numerous later portraits. An unsigned, undated copper engraving[18] in the Muzeum Narodowe Zbiory Czartoryskich, Kraków, showing the bust of Copernicus turned partially to the right with lily of the valley and holly in his right hand was probably also based either directly on the '*autographon*' or on an existing copy (*Fig. 6.3*). The portrait in Reusner's book also has the head turned partially to the right and therefore cannot have been the model used for the Kraków engraving. The painting on the Strasbourg clock therefore represents the oldest and most authentic portrait of Copernicus, which has given rise to many erroneous interpretations.[19]

Angelo Rocca's absurd claims that Copernicus was the builder of the second Cathedral clock and was blinded upon its completion[20] were already rejected by Grandidier[21] and need not be further discussed here. The question still remains, however, why Dasypodius had the portrait placed on the clock and why he made the effort of obtaining an authentic pattern.

Alfred and Théodore Ungerer came up with the theory that Dasypodius had to have been an adherent of the Copernican teachings, which seemed obvious at first. As he already had the 'Ptolemaic'—that is, geocentric—astrolabe designed by Herlin, Prugner, and Herr, he supposedly decided to use it in order not to delay the completion of the clock. The portrait of Copernicus was to be

possibly have been the painter, as he died in 1584; on the claim that this was Murer, see Thöne 1935, 29; Zinner 1988, 462. The assumption that the woodcut used a copper engraving of considerably poorer artistic quality as a model, which shows Copernicus's bust with a book in his right hand, is surely inaccurate; *ibid.*, 456–457 and 461–462. Zinner exaggerates the importance of the picture, basing his claim on the fact that it appears both in the manuscript and in a copy of the first edition of *De revolutionibus*; Schwarz 1943a, 159–160; Schenk zu Schweinsberg 1943, 268–269. Erich Sommerfeld even postulates the existence of a second self-portrait of Copernicus that he supposedly drew shortly before he died, which was used as a model for the woodcut; Sommerfeld 1981, 5–6, bases his arguments on a hypothetical source for which there is not the slightest documentation.

17 Schwarz 1943a, 148–149.

18 Inv. no. MNK-XV-R.802, size: 145 × 98 mm (trimmed, no plate margin). The marginal note in another handwriting at the bottom right reads: "*sc. Jenischinus*" (Balthasar Jenichen?). Information kindly provided by Ms. Ewa Czepielowa (Muzeum Narodowe). On this extremely rare print—of which only one other copy has been documented in the Bibliothèque Nationale in Paris—see Batowski 1942, 14–15; Kühne/Kirschner 2004, 358–359; Metze 2004, 47–51. Metze rejects the attribution to Jenichen, however.

19 See, for example, the appalling nonsense in Anon. 1840.

20 Rocca 1612, 86; later repeated in his *Opera omnia*, Rome, 1719, v. I, 170.

21 Grandidier 1782, 286–287.

FIGURE 6.3 Balthasar Jenichen (?): Portrait of Nicolaus Copernicus; copper engraving

mounted as an 'apology' as it were, in order to demonstrate his real stance.[22] The view is also expressed in the literature that Dasypodius—besides mounting the portrait of Copernicus—also had the heliocentric system represented in the form of the celestial globe, in order to prove his adherence.[23] On the other hand, this view is rendered less plausible, first of all by the fact that an astrolabe with the stereographic projection of the celestial sphere for the respective latitude requires a geocentric perspective *ex constructione*, and, secondly, because installing a celestial globe does not prove adherence to the Copernican system, as the constellations are represented on it such that the viewer sees the fixed star sphere from the outside (that is, one has to imagine the Earth at the center).[24] If Dasypodius had really wanted to provide proof for the new heliocentric system, an armillary sphere with the Sun at the center would have been necessary, around which the Earth and its Moon would have orbited on a radial arm.[25]

For both astrolabes and celestial globes, simple design aspects require a geocentric starting point, making an interpretation which sees a demonstration of cosmological views here incorrect.

As further proof of their theory that Dasypodius was a follower of Copernicus, the Ungerers used a book that was allegedly written by him with the title *Hypotyposes* [brief description, outline] *orbium Coelestium, quas appellant Theoricas Planetarum: congruentes cum Tabulis Alphonsinis et Copernici, seu etiam tabulis Prutenicis: in usum Scholarum publicatae* (Strasbourg, 1568).[26]

---

22 Ungerer 1922, 16–17; Ungerer 1925, 109. Since the publications of the Ungerers, this opinion has been repeated many times uncritically and has been expanded with further fantasy, e.g. by Bach 1978, 41–43.

23 Beyer 1973, 410.

24 The celestial globe will be dealt with in detail in *Chapter 8*.

25 Such instruments already existed in the second half of the sixteenth century, as Tycho Brahe was confronted in 1568–1569 by an instrument built by Schreckenfuchs which was intended to demonstrate planetary motion using both the Ptolemaic as well as the Copernican systems, but was not reliable in demonstrating the latter according to Brahe; TBOO, VII, 79. Whether the armillary sphere built by Johann Wagner in 1540 (Nuremberg, Germanisches Nationalmuseum, inv. no. WI 1167) had possibly already been rebuilt in the sixteenth century in order to display the Earth and Moon orbiting around the Sun, as proposed by Zinner 1979, 580, appears dubious. Connecting geocentric reference circles (meridians, celestial equator, colures) with a heliocentric system in the middle is inconsistent, however.

26 Cantamessa 2011, v. I, 713–714, no. 2066. In Houzeau/Lancaster 1882/89, v. I.1, 592, no. 2639, the year of publication is erroneously listed as 1565 and the book is attributed to Erasmus Reinhold. Dasypodius's dedicatory letter for William IV is also in the Cologne reprint: Peucer 1573, fols. AII^{r}–AV^{r}. It is identical with the Strasbourg edition: Peucer 1568, fols. AII^{r}–AVII^{v}.

In the dedication to William IV, Landgrave of Hesse-Kassel—which the Ungerers apparently did not consult—Dasypodius wrote that he had received the manuscript for the work which he published from his relative, the Strasbourg printer Theodor Rihel, and claimed that he did not know the author.[27] Dasypodius asserted that he would never tear something away from someone else and claim it as his own.[28] As a possible author he named the astronomer known as '*alterum Ptolemaeum*', Erasmus Reinhold.[29] In 1571 his *Hypotheses Astronomicae, seu Theoriae Planetarum*[30] appeared in Wittenberg under the name of Caspar Peucer (1525–1602).[31] He also dedicated his book to Landgrave William IV; in the preface he wrote that Dasypodius had had the manuscript published without permission and added with indignation that he should have published his own books.[32] He declared that the teachings of Copernicus should by no means be introduced in colleges.[33] Peucer presented the Ptolemaic system, but used the figures and explanatory models of Copernicus; however, he did not deal with the three motions of the Earth.[34] This position seems contradictory at first, but can be explained by the reception of the Copernican teachings in Wittenberg.

In 1543 Copernicus's main work was published in Nuremberg with a preface by the Lutheran theologian Andreas Osiander—unauthorized by Copernicus—in which the new system was viewed as a working hypothesis and its technical-astronomical character was emphasized.[35] With this opinion,

---

27 *Ibid.*, Preface, fol. AIIII^r–AV^r.

28 *Ibid.*, fol. AVI^r–AVI^v.

29 *Ibid.*, fol. AIIII^v. It is in fact possible that the text was partially written by Reinhold; Barker 2009, 94–99.

30 Zinner 1941, 250, no. 2552.

31 Peucer was the son-in-law of Melanchthon. Since 1545 he was appointed as professor of philosophy in Wittenberg, then as professor of mathematics (1554), and finally (1560) as professor of medicine; see Friedensburg 1917, 274–304, Brosseder 2004, 22–25, and Weichenhan 2005.

32 Peucer 1571, preface, fol. 6^r.

33 *Ibid.*, fol. 5^v. In 1554, in an announcement for a lecture on Ptolemy, Peucer asked his listeners to provide themselves with a book on Copernican doctrines; Friedensburg 1917, 292.

34 For a brief overview see Delambre 1821, v. I, 146–147. In the text, Peucer notes explicitly: "*insistemus vestigiis Ptolemaei et veterum aliorum, omissis recentibus Copernici hypothesibus, quas Aristarchum Samium et quosdam alios veteres sequutus, suo quodam consilio usurpavit*"; Peucer 1568, 34.

35 Johannes Kepler discovered from notes by the Nuremberg patrician Hieronymus Schreiber in his copy of *De revolutionibus* that Osiander was the author of the anonymous preface; *Apologia Tychonis contra Ursum*, 1600–01, in KGW, XX.1, 27–28. In 1584 Giordano Bruno had decidedly declared in his *La cena de le ceneri* (Ash Wednesday Supper) that the preface had not been written by Copernicus.

Osiander set the stage for a separation of the astronomical content of the work from the fundamental cosmological reflections upon which it was based, as they were described in Book I of Copernicus's *De revolutionibus*.[36]

This position, referred to as the 'Wittenberg interpretation',[37] was characteristic for a group of scholars around Melanchthon, including Erasmus Reinhold (1511–1553) and Caspar Peucer. Reinhold, who had been a professor of mathematics in Wittenberg since 1554,[38] considered Copernicus to be the reformer of astronomy, who set the planetary theory on solid ground once again by adhering to strict uniform orbital motion, renouncing the use of equants favoring epicyclical systems.[39] Reinhold pragmatically accepted the Copernican teachings in order to produce more exact planetary tables;[40] however, he did not pay much attention to the cosmological explanations in Book I of the *De revolutionibus*.[41]

36 Characteristic for this type of reception is also the assessment of Gemma Frisius in a letter to Bishop Johannes Dantiscus on 20 July: "*Neque ego nunc disputo de hypothesibus, quibus ille utitur pro sua demonstratione, quales sint, aut quantum veritatis habeant. Mea enim non refert terramne dicat circumvolvi, an immotam consistere; modo syderum motus temporumque intervalla habeamus ad amussim discreta et in exactissimum calculum redacta* (Now I will not discuss the hypotheses either, which he [Copernicus] uses for his demonstration, which they are, how true they are: It does not matter to me whether he says that the Earth rolls around or remains motionless, provided that we have the movements of the stars and the periods of time that are exactly distinguished and handled by the most exact calculation)"; Curtze 1874b, 319–320; Original in the University Library Uppsala.

37 Westman 1975a, 166–167; Westman 1975b, 286–287; Moran 1973, 13–17; Christianson 1973, 3ff. Thorndike speaks tellingly of a 'Circle of Melanchthon'; Thorndike 1923/41, v. 5, 378–405. Georg Joachim Rheticus (1514–1574) is therefore unique when he returned from his sojourn in Ermland (1539–1541), as an enthusiastic supporter of Copernicus. Even if researchers have assigned different protagonists to the 'Circle of Melanchthon', the disqualification of Thorndike's interpretation as a myth by Brosseder 2004, 11–17, seems exaggerated.

38 Reinhold was born in Saalfeld (Thuringia), matriculated in winter semester 1530–1531 in Wittenberg and graduated as a magister in 1535. He was dean of the Faculty of Philosophy in 1540–1541, and finally became rector in 1549–1550; Friedensburg 1917, 232–234; Koch 1908.

39 Westman 1975a, 175–177.

40 Reinhold's *Prutenicae tabulae coelestium motuum* appeared in Tübingen in 1551; Gingerich 1973, 44–45. On the post-Copernican Reception of Ptolemy, see also Omodeo/Tupikova 2013.

41 This can be concluded from notes in the margins of Reinhold's personal copy of *De revolutionibus*. There are hardly any comments on Book I, unlike the other parts of the work; Gingerich 1973, 56–58; Gingerich 2002, 217 and 269. That Reinhold can by no means be considered one of the early supporters of the Copernican teachings *in cosmologicis* was made clear by Aleksander Birkenmajer with the aid of a handwritten comment on the *De revolutionibus* by Reinhold; Birkenmajer 1960.

For Philipp Melanchthon, three motions of the Earth contradicted the Bible and the postulate of Aristotle according to which only one motion could be attributed to a body. Probably influenced by the *Narratio prima*, published by Georg Joachim Rheticus in 1540, one of the first short expositions on the Copernican system, Melanchthon assessed the new system pejoratively and commented in a letter to Burkhard Mithobius (1504–1565) on 16 October 1541: "*sicut ille Sarmaticus Astronomus, qui movet terram et figit Solem. Profecto sapientes gubernatores deberent ingeniorum petulantiam cohercere*".[42]

In his *Initia doctrina physicae*, published in 1549, Melanchthon felt that the teachings of Copernicus were merely a revival of the absurd and erroneous views of Aristarchus, motivated by an obsession with innovation.[43] On the other hand, he appreciated Copernicus's theory on the Moon,[44] and used data from his work.[45]

A considerable weakening of Melanchthon's generally rejective stance can be found in the second edition of the *Initia*, published in 1550, which led the Galileo Galilei researcher Emil Wohlwill to point out that all of the dismissive comments on the teachings of Copernicus are missing in the revised text, which remained that way in the later editions of this widely used textbook.[46]

---

42 CR, IV, col. 679. In his letter, Melanchthon lamented the signs of cultural decay and draws parallels between *De revolutionibus* and a text in which the Melsungen pastor Johannes Lening defends the bigamy of the landgrave of Hesse. He felt both authors should have been reined in because, by using exaggerated arguments, they had carried what they thought was a good line of thinking *ad absurdum*. In this context, Wilhelm Maurer correctly indicated and stated that this comparison "clearly did not flatter Copernicus, but was not a denunciatory call for the constable, either"; Maurer 1964, 61.

43 CR, XIII, col. 216. Walther Ludwig translated Melanchthon's *Initia* into German; Melanchthon 2008.

44 "*Sequar usitatam doctrinam a Ptolemaeo traditam, quam hactenus plurimi secuti sunt, etsi enim recens Copernici fabricatio orbium Lunae admodum concinna est, tamen nos haec recensemus, ut aliquomodo studiosos invitemus ad communem doctrinam in scholis receptam* (Even if the orbits of the Moon recently fabricated by Copernicus are very agreeable, we will nonetheless review the common doctrine in order to invite the students in some manner to follow the teachings generally used in the schools)"; CR, XIII, col. 244.

45 CR, XIII, col. 225, 241 (apogee of the Sun); 262 (apogee values for the upper planets). In the same year Melanchthon wrote the following in the lecture given by Erasmus Reinhold in memory of Caspar Cruciger: "*His et similibus observationibus moti, Copernicum magis admirari et amare coepimus. Eumque secuti, nostras observationes cum omnibus praecedentibus in hoc genere contulimus. In qua consideratione animadvertimus anni veram seu adparentem magnitudinem eius, quam Copernicus temporalem, Graeci τσπιχσν, Latini vertentem dicunt, nunc iterum maiorem esse, quam vel Albategnii vel Alphonsi temporibus*"; CR, XI, col. 839.

46 Wohlwill 1904, 260–267. Corresponding passages in the editions from 1549 and 1550 are compared there.

Even if Melanchthon did not advocate Copernicus's cosmological ideas, he seems to have arrived at a more moderate position after an in-depth study of the *De revolutionibus*.[47]

Like his teacher and predecessor, Caspar Peucer saw the Copernican teachings as a means for calculating more exact values for the motion of the planets. The lack of precision in the Alfonsine Tables and the 'absurd' postulates of Copernicus were equally unsatisfactory for him; the solution in his opinion lay in adjusting new observations and calculations based on the teachings of Copernicus to fit with Ptolemy's model.[48] This combination of Ptolemaic hypotheses with Copernican calculations was criticized by Tycho Brahe in a lecture series in Copenhagen in 1574–1575.[49]

The publication of the *Hypotyposes* by Dasypodius can therefore not be seen as him simply taking sides with the Copernican teachings, but indicates their partial reception, while rejecting the fundamental heliocentric system with the three motions of the Earth. In his public lectures, Dasypodius favored the Ptolemaic point of view,[50] and his two publications on the 1577 comet confirm that he completely supported the traditional Aristotelian-Ptolemaic

47 Müller 1963/64, 24. In this connection, some remarks on Luther's attitude to Copernicanism, which he knew at best by hearsay, may be given. Luther casually commented on Copernicus in a table talk only once, and his words have been handed down in two versions, which has given rise to diverging interpretations; Bornkamm 1943, 173; Norlind 1953, 274–276 (critically on this Meyer 1954); Wardeska 1973, 164–168; Barker 2005. Incidentally, Luther's alleged energetic opposition to Copernican doctrine is an excellent case study for the repeated copying and extrapolation of unchecked interpretations of sources, as Andreas Kleinert has aptly pointed out; Kleinert 2003.

48 Peucer 1571, fol. 3^{r}.

49 "*Indicavi nihilominus generali quadam expositione, quomodo in reliquis Planetis haec ipsa etiam essent intelligenda, et qua ratione illorum apparentiae ad terrae stabilitatem, manentibus Copernici numeris, adaptari possent, insinuavi, idque aliter quam a Peucero et Dasipodio hoc frustra factitantibus in libello quodam nuper de ea re edito praestitum esset. Illi enim Copernici calculum hypothesibus Ptolemaicis atque Alphonsinis inconvenienter applicarunt*"; TBOO, I, 173.

50 "*Quid de terrae motu sentis? Authoritate sacrarum literarum immotam eam statuo, quemadmodum Ptolemaeus, et ante eum plures alii Mathematici nullum terrae tribuerunt motum, existimantes, quod illa motione partes eius collaberentur, neque amplius nascentibus rebus, atque animantibus commodum habitandi domicilium praeberet. Quod vero Copernicus triplicem terrae motum ascribat, ex hypothesi hoc docendi gratia, atque ad coeli φαινόμενα totius salvanda facit: quemadmodum ipse fatetur in libris suis Revolutionum. Hactenus hypotheseis fuerunt, quibus singula insequentia superstruuntur, tanquam semper firmis atque indubitatis fundamentis. Astronomorum siquidem proprium est, ex hypothesibus φαινόμενα coelestia tanquam corollaria concludere, eaque vel per circulos in primo mobili postea vel per orbes in secundis mobilibus demonstrare: His addunt duo alia principia, postulata scil: et axiomata*"; Malleolus 1628, 24–25.

doctrine.[51] In conformity with the Aristotelian postulate of the invariability of the heavens, comets were basically interpreted as phenomena of the sphere of fire in the sublunary world,[52] and Dasypodius held to this explanation completely.[53] The astrological importance of comets (in particular the 1577 comet) played an extensive role in his book,[54] which Tycho Brahe criticized.[55]

Therefore Conrad Dasypodius could hardly have been a supporter of Copernican cosmology.[56] This follows unequivocally from the fact that he was fully convinced by the astrology of Ptolemy's *Tetrabiblos* (for convincing proof see *Part 4*). The astrological premises of the *Tetrabiblos* presuppose Ptolemy's planetary theory; however, one can by no means infer from this that accepting the Copernican teachings automatically implies a rejection of astrology.

---

51 Dasypodius 1578b, 1578c. The 1577 comet resulted in a flood of publications which are discussed in detail by Hellman 1944. On Dasypodius see *ibid.*, 236–243.

52 Aristotle 1955, v. III, Book I, Ch. 7, 344 a l. 7–344 b l. 18; [Pseudo-Aristotle], *De mundo*, 392 a l. 33–392 b l. 5.

53 "*Hinc factum est, ut variae sint variorum philosophorum sententiae, de Cometis, eorumque naturis, proprietatibus, causis et effectibus: nihilominus tamen inter omnes omnium opiniones: proxime ad veritatem accedere uidetur, Aristotelis sententia: qui docet Cometam esse exhalationem terrestrem, calidam siccamque, quae ad supremam aeris regionem,* (*quam ignem elementarem vocant*) *virtute stellarum elata: in ea aliquanto temporis spatio continuata successione eiusdem materiae, colligitur: tandem cum iam ad ignem concipiendum praeparata est: accedit vis illa stellarum: atque etiam virtus motus corporum coelestium: quae incendit, materiam viscosam, terrestrem: diuturno tempore, et continua serie in suprema aeris regione collectam, qua inflammatione facta: lumen illud apparet in aere pendulum: diversa tamen forma: facta et habita consideratione materiae collectae, et iam incensae, ac inflammatae*"; Dasypodius 1578c, Ch. I "De materia, forma, et causis efficientibus Cometarum", fol. BII^r^; see also the preface fol. AII^r^–AIII^v^. Being consistent, Giovanni Battista Riccioli puts Dasypodius in the 'first class' in the chapter "De loco Cometarum, seu Distantia a Terra" of his *Almagestum novum*. "*Nunc nuda ac mera nomina recensebo Authorum in tres classes distributa* [...]. *Prima Classis est eorum qui omnes Cometas aut indefinite infra Lunam fuisse asseruerunt, cuiusmodi fuere Aristoteles cum Peripateticis* [...] *Conradus Dasypodius* [...]. *Secunda classis point Cometas omnes supra Lunam: in qua classe fuerunt* [...] *Tycho* [...]. *Tertia Classis point vel aliquos Cometas supra aliquos infra, vel eosdem modo infra et modo supra Lunam*"; Riccioli 1651, v. I.2, 28.

54 *Ibid.*, Ch. III "De effectibus Cometarum", fols. DI^r^–DIII^v^; Ch. IV "Exemplis facta horum effectuum demonstration", fols. DIII^v^–EI^r^; Ch. V "De Cometa qui anno 1577 apparuit", fols. EI^v^–EIV^r^. See also Hellman 1944, 240–243.

55 "*M. Cunradus Dasypodius Argentoratensis Libellum de Cometis edidit, in quo totius circa Astrologicas praedictiones occupatur, et in his Dogmaticus Ptolomaei ipsiusque interpretum vestigiis praecipue insistit*"; TBOO, IV (*De Mundi aetherei recentioribus phaenomenis*), 361–362. However, Brahe by no means rejected astrology, on which see Oestmann 2003. In the aforesaid work on the comet of 1577, which he considered to be an object in the fixed star sphere because the parallax could not be determined, Tycho forewent astrological explanations; Hellman 1944, 122.

56 Schindling; 1977, 259–260, also came to this conclusion.

The teachings of the influence of the stars are founded on Aristotelian cosmology with the basic assumption of a separation of the sublunary and superlunary spheres,[57] which is why the effects of the macrocosm on the Earth in the heliocentric system needed new or expanded explanations. However, the assumption that all planets, including the Earth, revolve around the Sun presented no major problems, as—regardless of whether a geocentric or heliocentric system was used—the important thing was to determine the configuration of the heavenly bodies for a specific location at a specific time, as Christopher Heydon (1561–1623) clearly asserted in his *A Defence of Iudiciall Astrologie* (1603):

> For whether [...] (as Copernicus saith) the Sun be the center of the world, and the earth be in the Sunnes place, betweene the sphere of Mars and Venus, the Astrologer careth not. For so by any of these Hypotheses, he may come to the true place and motion of the Starres, this varietie of opinions, whether such things be indeede, and in what order they be, is no impeachment to the principles of Arte.[58]

Thus the growing acceptance of the teachings of Copernicus did not directly effect a decline in astrology, as occurred during the seventeenth century. In the sixteenth century, it was taken for granted that astrology was one of the sciences, and only a few scholars categorically rejected the influence of the stars on the Earth and its inhabitants. Criticism was at most directed at the excesses of judicial astrology and the adherents of extreme fatalism.

Copernicus himself apparently gave some thought to astrological questions, as one must conclude from some handwritten comments by him in a work by Abī-l-Rijāl (Haly Abenragel).[59] Certainly this interest must be seen in connection with his work as a physician. Georg Joachim Rheticus based his speculations about the rise and fall of the world empires on the movement of the lines of the apsides discovered by Copernicus,[60] and Johannes Kepler was also both

57 *Locus classicus* for the connection between astrology and Aristotelian physics is Book II, Ch. X of the *De generatione et corruptione*.

58 Heydon 1977, 371.

59 *Albohazen Haly filius Abenragel Praeclarissimus Liber Completus in iudiciis astrorum*, Venice, 1485; Curtze 1875, 244–246; Prowe 1883/84, v. I.2, 401, v. II, 219. Whether, however, the very far-reaching and controversial conclusions of Robert S. Westman on Copernicus's stance on astrology are correct (Westman 2012) must remain undecided in this context.

60 Rosen 1971, 121–122; Kraai 2003, 91–100. Characteristic for the perspective of the Age of Enlightenment is the assessment of this passage by the Göttingen professor of mathematics Abraham Gotthelf Kästner (1719–1800): "*Man sollte doch denken, wer die Sonne für einen Fixstern annimt, und die Erde um sie gehen läßt, wie andre Planeten, könne von*

an adherent of Copernican teachings and an avid supporter of astrology, who tried to give it a new scientific basis. His reformed astrology, however, has little to do with the 'classical' astrological doctrines of Ptolemy.

Why then did Nicolaus Copernicus get a prominent place on the Strasbourg Cathedral clock instead of the astronomer from Alexandria? The answer is provided by the mechanism which Dasypodius designed for representing the motion of the planets on the astrolabe. In this mechanism the gears for the planets are driven by a central gear for the Sun, and for this design Dasypodius was able to use the relationship between the sidereal and the synodic periods for planetary orbits demonstrated by Copernicus. His mechanical representation of the universe was decidedly inspired by Copernicus (for particulars see *Chapter 9*). Dasypodius was—despite his fundamental Ptolemaic belief *in astronomicis*—willing to take up and to make use of Copernicus's ideas *in mechanicis*. The portrait of Copernicus on the clock can be seen as an expression of the extent to which he felt indebted to his great predecessor for the design of his clock. The publication of Caspar Peucer's *Hypotyposes* makes it plausible that the Strasbourg professor of mathematics favored the 'Wittenberg interpretation', the followers of which were willing pragmatically to use the data and calculations of Copernicus, but rejected his cosmological views.[61] In this respect, the portrait of Copernicus appears in a new light, because, even if one considers the tradition of displaying the images of astronomers on monumental clocks—for example, the clocks in Lund, Doberan, Stralsund, and Lübeck, which besides Ptolemy and King Alfonso x, 'the Wise', show Arabian scholars in the spandrel fields around their astrolabes[62]—the Strasbourg portrait remains singular. The mathematics professor Dasypodius, who was completely caught up in the traditional understanding of the universe, did not display a picture

---

Astrologie nicht viel halten, für welche die Sonne ein Planet seyn muß, und alles um die Erde gehn: Gleichwohl fand Rhäticus in dem Wege, den der eccentrischen Erdbahn Mittelpunct nach dem Copernicus beschrieben, das wahre astrologische Glücksrad (One would assume that someone who accepts that the Sun is a fixed star and lets the Earth orbit it like other planets would not think much of astrology, in which the Sun has to be a planet and everything revolves around the Earth. Nonetheless Rheticus considered the path that the central point of the Earth's eccentric orbit described according to Copernicus to be the true astrological wheel of fortune)"; Kästner 1796/1800, v. II, 368.

61 Chrisman's interpretation, according to which Dasypodius ignored in a sense the Copernican view of the universe, is therefore completely inaccurate; Chrisman 1982, 257–258.

62 More on this in Boer/Strohmaier 1979; Trinkert 2014. On the clock in Lübeck see Hass 1927, v. 6, 275–276. For illustrations of the spandrel figures of the astronomical clock in the Lund Cathedral see Wåhlin 1923, 120–123.

of a venerable astronomer from the past, but rather chose with Copernicus a representative of his own age. The portrait on the Cathedral clock attests—just as the reception of the Copernican teachings in Wittenberg—to the degree to which the Frombork canon was respected among astronomers, not for his cosmological views, but rather as a reformer of Ptolemaic planetary theory. The controversies surrounding Copernicus's work that culminated with the trial and condemnation of Galileo Galilei were still in the distant future.

PART 3

The Technical Components of the Clock

CHAPTER 7

Gear Trains and Layout

7.1 Principal Arrangement

In the course of time, wear and tear took its toll on the Cathedral clock, which was initially remedied by shutting down individual components or removing them in order to keep the clock running, until the clock finally became inoperable in 1788. As the fragments exhibited in the Musée des Arts Décoratifs in Strasbourg are incomplete and two verbal descriptions without any diagrams are the only other sources of information, a detailed analysis of the technical components poses considerable difficulties.

The first of the two verbal sources is a report commissioned by the caretakers of the 'Frauenwerk' and submitted by the clockmaker Johann Philipp Maybaum and his son in 1788 on the state of the clock, which had stopped running, together with suggestions for repairing it.[1] Except for information on the general structure of the clock and the damage to individual components, it contains few technical details.

In 1827, Jean-Baptiste Schwilgué published a short description of the functions and features of the Cathedral clock, and included possibilities for repairing it.[2] Nine years later, Schwilgué inspected the clock more thoroughly, but found that the mechanical components were spread throughout all the tiers of the clock.[3] He put everything back together again, but shortly after that the internal works of the clock were ruined once again; numerous parts were either damaged or removed.[4]

In 1836 Schwilgué presented his proposal for rebuilding the Cathedral clock in four volumes, the first of which took stock of the state of the clock dating from 1571 to 1574.[5] Schwilgué described how the individual assemblies interacted, recorded the number of gear teeth as well as some dimensions, and listed the missing parts. The explicit goal, however, was not repairing the old

1 AMS, 3 OND 589; transcription in Oestmann 1993/2000, 229–232.

2 Schwilgué 1827; in Schwilgué 1857, 86–97, this was essentially a reprint of the original manuscript with a few changes. The original texts published by Schwilgué's son Charles could not be found in the municipal archives.

3 Schwilgué 1857, 219–220.

4 *Ibid.*, 221–222.

5 AMS, 3 OND 589; also in Strasbourg, OND, inv. no. 1108, v. II, 198–216. Transcription in Oestmann 1993/2000, 233–244.

 | DOI:10.1163/9789004423473_009

Cathedral clock, but rather building a completely new one, as can be seen from his notes:

> It is difficult, if not impossible, for me to express, when I entered this cabinet, and when I saw for the first time the mechanisms of this clock, how much I was surprised by the inferiority of its execution, and by the all too ordinary means of its dispositions; a glance at the wheelwork was enough to convince me of the little accuracy it must have had, especially in the revolutions of the planets of the Astrolabe. It took me only twenty minutes to grasp the entire composition of this work, and to recognize in all its parts a great imperfection, compared to the work I had proposed to do. Also, it was only by looking back to the early date of its creation that I was able to explain to myself the mediocrity of this work which, however, was once considered to be so wonderful. [...] But since the public still claimed that this work was inimitable, and that out of respect for its authors, it was necessary to keep it scrupulously in its current state, it was very difficult for me to fight against this opinion, and especially to make people believe that I could build a clock superior to that of Dasypodius. It was in this frame of mind that I drew up a quotation, with a view to a complete refurbishment, noting that I could not agree to take on only one part, since the public, seeing this incomplete reconstruction, might believe that I was not able to make the astronomical part function, and at the same time, I pointed out that restoring a work that was no longer up to the standard of our time was to perpetuate errors.[6]

From the point of view of a historian, the construction of the new clock in 1838–1842 was most unfortunate, as an extremely important astronomical clock was destroyed in the name of the dubious belief in progress.[7] The fact that at least the housing withstood the radical modernization and remained intact for the most part must be credited to a condition set by the magistrate, as Schwilgué begrudgingly stated:

> The obligation which was imposed on me to arrange the new clock, so that it could be placed in the old case, caused me many difficulties, and prevented me from giving free rein to my imagination; it is truly regrettable that I could not build a case, which would have been related to the

6 OND, inv. no. 1108, v. I, preface, 9 and 10–11.

7 "*Impitoyablement remplacée, au nom d'un scientisme stupide*" was Emmanuel Poulle's harsh criticism; Poulle 1980, v. I, 694.

> mechanisms I wanted to execute, and would have made it possible to orient the clock, as well as to make visible from the outside the functioning of the various wheel-trains of this work.[8]

He concluded his report with scathing criticism of the design and the state of the clock.[9] As Schwilgué used the state of the art in mechanical engineering and tower-clock construction of his day as his point of reference, the technical shortcomings of the clockwork from the sixteenth century were naturally in stark contrast to his demands. Despite his ahistorical judgments, Schwilgué's report is, together with the extant components themselves, the most important source on the technical design of the second Cathedral clock. As a result of the problematic situation with the sources, the following descriptions of the individual parts of the clock vary in their levels of detail. In particular, the fundamental structure, the arrangement of the individual assemblies within the housing, as well as their interconnection can only be established schematically.

7.2 The Clockwork

The frame for the clockwork (width 2.05 m, depth 1.02 m, height 1.92 m) was made of rectangular wrought iron bars, held together with iron wedges. The corner uprights were attached at a 45° angle to the sides, gradually tapering to form an ever smaller rectangle with a forged coil at the top. The going train was at the center, with the quarter-striking train to its left and the hour-striking train to its right (*Fig. 7.1*).

Some parts of the frame were made of wood painted black, which were added in the nineteenth century.[10]

The gear teeth invariably had rectangular profiles with a slightly tapering curvature and the gear shafts were supported in rectangular bushings. To reduce the load on the gear bearings for the striking trains for the quarters and hours as well as that on the free ends of the fly vane shafts, wooden support rollers were placed under them; they were originally made of metal, but had probably been lost before the additions were made to the clockwork, which is why they were replaced by wooden dummies that were painted black.

---

8 *Ibid.*, 12–13.

9 *Ibid.*, 19–20.

10 In a list of the works carried out from 1843 Schwilgué remarked: "Old Clock. Cleaned the gears of the old Clock. Built of wood the missing parts of the old mechanism; these parts have been painted black"; Paris, Archives Nationales: Fonds de l'administration des Cultes, F/19/7892, no. 72.

FIGURE 7.1 Main movement of the Strasbourg Cathedral Clock

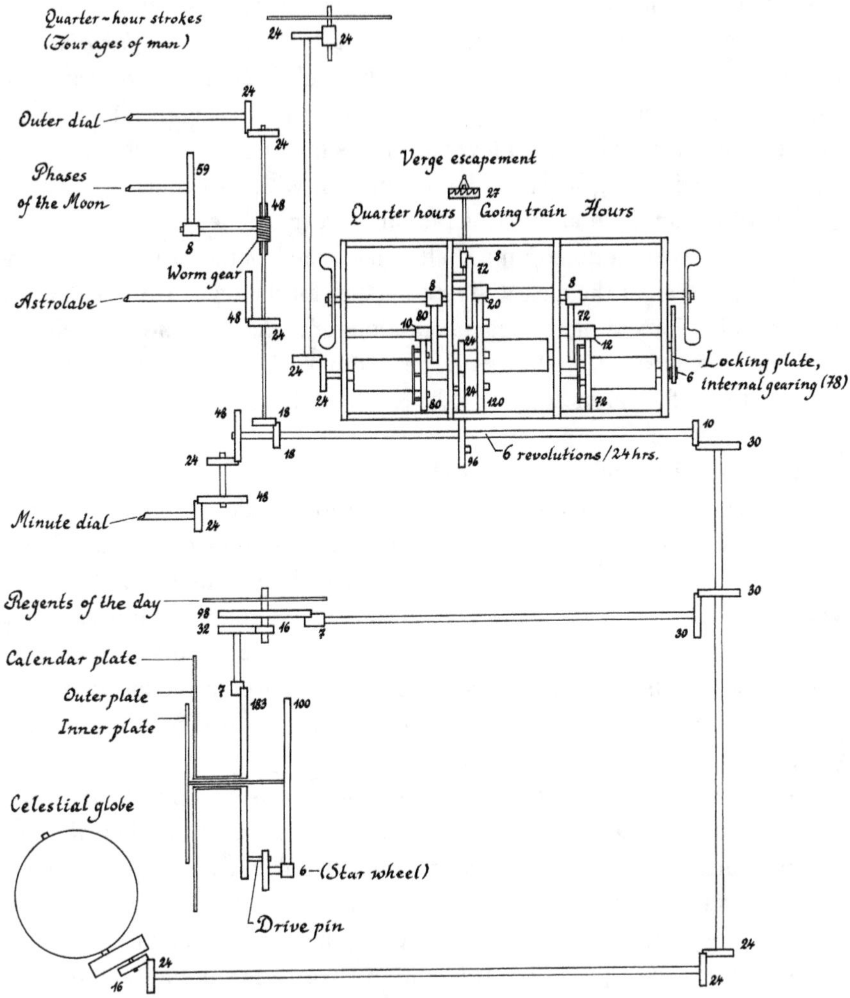

FIGURE 7.2 Gear trains of the Strasbourg Cathedral Clock

The clockwork was mounted on the middle tier of the central clock tower, behind the astrolabe.

7.2.1 *Going Train*

The barrel wheel (N_G = 120) engaged with a pinion (N_P = 20), on the shaft of which was a crown wheel with 72 teeth (*Fig. 7.2*). The crown wheel engaged at a right angle with the 8-trundle lantern pinion of the vertically mounted escape wheel (N = 27). Originally the going train had a verge and foliot escapement mechanism, which the clockmaker Johann Jacob Straubhar replaced in

1732 using a pendulum with cycloid suspension cheeks, as specified in the inscription on the upper support strap, top left: "*Renoviret Durch Johann Jacob Straubhar Anno 1732*" (Refurbished by Johann Jacob Straubhar in the year 1732). Mounted on the barrel wheel were four lifting pins releasing the quarter-hour chimes (the laterally mounted lever for this is missing) as well as a gear with 24 teeth, which engaged with a lantern pinion with 24 trundles for reversing the direction of rotation. This pinion, in turn, also engaged with a gear (N = 96) that was mounted on a shaft which acted as the drive shaft for the celestial globe, all of the dials, the rotating disk for the regents of the day, and the astrolabe. For setting the gears in motion for the cock's crowing there was a pin, no longer extant, on the driving gear with 96 teeth. The going train had a running time of 27 hours and had to be rewound daily.

7.2.2 *Gear Train for Chiming the Quarter-Hours*

The barrel wheel ($N_G = 80$) engaged with a 10-trundle lantern pinion with an 80-tooth gear attached to its shaft. This gear was connected to the pinion ($N_P = 8$) on the fly vane shaft. There was a ring with four notches on the barrel wheel for arresting the train after the chiming of each quarter-hour, as well as two lifting pins for starting the striking train for the full hours and the drive mechanism for the angel who turned its hourglass over at the fourth quarter chime. The lifting piece is missing; the small separate movement for the angel (with three wheels and a fly vane) is extant.

Outside the frame, the barrel wheel supports a wheel with 24 teeth, which engages at right angles with another gear with the same number of teeth, mounted on a vertical shaft. This shaft, of which only the lowest part remains, transmitted the rotary motion of the quarter-hour train to the upper tier of the central clock tower and supported a wheel with 24 teeth that engaged with a lantern pinion with 24 trundles. This lantern pinion was mounted at the center of the wooden disk on which the figures of the four ages of man were mounted such that they could rotate about their axes and be paraded by a series of seven bells, of which the first four were made of metal, the rest of wood. Beneath this disk there were four pins of different lengths, the longest of which was for the first bell, the shortest for the fourth one. By means of these pins and levers of different lengths under the bases of the figures, of which nothing remains, angular motion was produced. The highest base lever of the statue of childhood was caught by the longest pin and resulted in the figure turning to the front (that is, towards the viewer), while at the same time a spring was strained. As the disk continued to turn, the lever was released and the figure was turned back by the force of the spring and struck the first bell. Its further motion was unimpeded because the lever was high enough to pass over the other three pins. Correspondingly, the lower lever of the youth was stopped twice, that of manhood three times, and that of old age, which was the lowest, was stopped by all four pins.[11]

7.2.3 *Gear Train for Striking the Hours*

The barrel wheel (N_G = 72) with 6 lifting pins engaged with a pinion with 12 teeth, on the shaft of which a wheel (N_G = 72) was attached that engaged with the pinion (N_P = 8) on the fly vane shaft.

The end of the barrel wheel shaft carried a 6-trundle lantern pinion, which engaged with the internally toothed locking plate (or count wheel) with 78 teeth outside the frame.[12] As normal, the locking plate had twelve stages on its perimeter at varying intervals which were scanned by a locking piece, thus determining the number of chimes. The locking piece no longer exists.

When the striking of the hour was released by the quarter-hour train, the rotation of the barrel wheel with its lifting pins was transferred to the horizontal wooden disk with the statues of Christ and Death, located above the figures of the ages of man, using a lever, a tension cable, and rockers. Nothing is left of these components for the transmission of torque to the disk, and the disk is also missing.

These two figures could rotate around their axes, just like the statues of the ages of man. The disk itself was returned to its original position by a strong return spring after the rotation was complete.

The sequence of movements was the following: first the disk turned counterclockwise; Christ came out of the left opening in the housing, turned toward the viewer, making a gesture of blessing, and turned toward Death, who was retreating at the same time. As soon as the lever in the chiming train dropped, the return spring sent the disk quickly back to its original position: Christ retreated and Death struck the bell for the hour with his bone.[13] This process was repeated corresponding to the number of strokes. The mechanism for turning these figures was similar to that of the statues of the ages of man: a lever in the base of the figure with a return spring and vertical pins under the disk.

Dasypodius describes the movements of these automata in similar fashion:

> Finally [after the quarter chime], the statue of Death [...] comes forward, striking the hour itself, also with his own bell. At last the statue of the Savior meets the individual statues of the ages, going out to signify

11 AMS, 3 OND 589, 3.

12 The figure 72 indicated by Schwilgué, *ibid.*, 4, is incorrect.

13 *Ibid.*, 4.

> redemption. But Death's statue gives forth the last sound with its own bell (as it were, the last time in life).[14]

Furthermore the angel next to the minute dial was set in motion by the movement for the hour striking, who beat time with his scepter as the hour strokes sounded. Nothing is left of the transmission mechanism for the angel.

7.3 Gear Train for the Mechanical Cock and Carillon

On the top tier of the central clock tower—where the wooden disk with the figures of the four ages of man was located—there were two trains of gears, which:

1. provided for the movements of the iron cock (see *Chapter 1*) using rocker arms and connecting rods,
2. using a lever, operated the bellows that were mounted in the tower crown for the cock's crowing and, finally,
3. started the carillon, which previously had also been here.[15]

---

14 "*Tandem prodit mortis statua,* [...] *horam ipsam suo quoque tintinnabulo pulsans. Salvatoris denique statua, singulis aetatum statuis egredientibus obviat: ad redemptionis significationem. mortis vero (tanquam extrema vitae linea) extremum suo edit tintinnabulo sonitum*"; Dasypodius 1580b, fol. Hᵛ. The description in Fischart's poem for the smaller edition of Stimmer's woodcut, according to which Death turns to the ages of man at every quarter hour, but is driven back by Christ—"*Da kompt der Tod all viertail stund Gegen aim jeden Alter her, Den jagt der Heiland wider fern: Doch laßt den Tod all stund er schlagen, das wir vns sein all stund verwagen* (Death comes each quarter hour then, To face off every Age of men, Then he's chased away by the Savior again: But lets Death the knell every hour sound, that our thoughts of him each hour abound)"—is therefore incorrect. The description in Behr 1747, 76 is also unclear: "*Ueber diesen 4. Glöcklein hängt endlich das Stunden-Glöcklein; auf einer Seiten darneben stehet Christus, auf der andern der Tod: nachdem nun die Viertel-Stund geschlagen worden, so nahet sich allemahl der Tod zum Glöcklein, die Stund zu schla-gen; diesen treibt aber Christus allezeit hinweg, biß zur Zeit da die Stund solle geschlagen werden, alsdann kommt der Tod die Stund zu schlagen, welchem Christus weichet* (Finally above the 4 little bells the little bell for the hour is suspended; Christ stands on one side of it, Death on the other: After the quarter hour has now been chimed, Death approaches the little bell each time, in order to chime the hour; but Christ drives him away each time, up until the time when the hour is to be rung; and then Death comes to strike the hour, to whom Christ yields)". It was probably based on Fischart's poem. And the wheels for chiming the hour could hardly have been positioned in lift (*Hebung*), by which Death would have retreated after the last stroke of the bell and Christ moved toward it, as claimed by Bach 1978, 36.

15 AMS, 3 OND 589, 5–6.

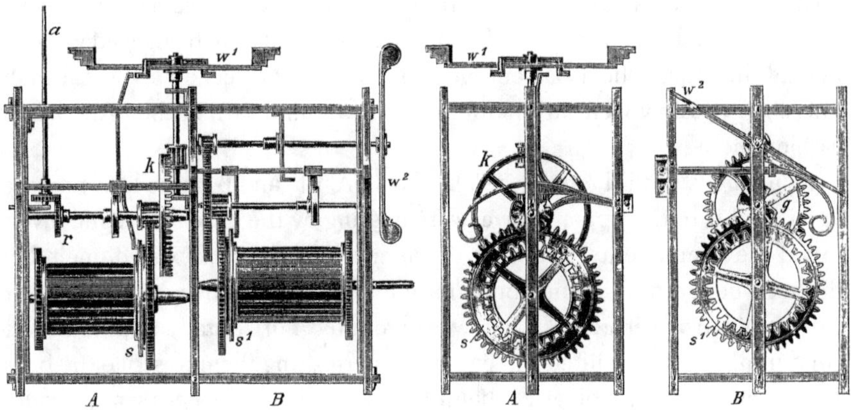

Fig. 2

Fig. 3

FIGURE 7.3 Movements for the cock and carillon

The two gear trains were mounted in one frame and still exist, but numerous components are missing, which makes an exact description of how the mechanisms worked impossible (*Fig. 7.3*).

In one of these gear trains the barrel wheel engaged with a pinion, on the shaft of which a wheel was mounted with a heart-shaped disk, into which a laterally attached lever could drop. This wheel engaged with the lantern pinion on the horizontal shaft of a large-vane fly. This gear train was probably used for the movements of the cock and for operating the bellows.

The other gear train had the same number of gear wheels, but its fly was mounted on a vertical shaft and it had two adjustable, flag-shaped vanes. The years 1572 and 1669 are inscribed on the vanes; the first year indicates when the gear train was installed in the new clock, the second year was that of repairs.[16] The fronts and backs also have a shield each, with the symbols of the masons of the 'Frauenwerk' and of the city of Strasbourg painted on them. The frame and some of the wheels date from the first phase of construction of the new clock in the second third of the sixteenth century.[17] The pinnacle-like corner

16 In 1669, repairs were carried out on the Cathedral clock: "*Im Jahr 1669 ist das Uhrwerck im Münster erneuert, auch mit etlichen Gemählden und Sprüchen ausgezieret worden* (In 1669 the clockwork of the Cathedral were renovated, and were also embellished with numerous paintings and inscriptions)"; Behr 1747, 156; Einundzwanzigerprotokolle, v. 149, 1669, fol. 295^{r}.

17 "*Von welchem werck nichts anders zu disem kommen dann allein das gestell uund etliche wenig meins behaltens redder so das hanen geschrey und die Cymbalen treyben*" (Of which work nothing else comes to this other than the housing alone, and some few wheels retained by me for driving the cock's crowing and the cymbals); Dasypodius 1580a, 6.

posts set at an angle differ greatly from those of the main clockwork. A small wheel with 16 teeth sits on the shaft of the crown gear which engaged with the fly, and this small wheel also engaged with a similar wheel on a vertical shaft. This shaft possibly went to the top of the tower and turned the drum of the carillon there.

Its ten bells, with the pitches C, D, E, F, G, A, B♭, and B in the lower octave, and C and D in the upper octave, were missing by the time Schwilgué inventoried the components, but some of the mechanisms still existed, including the six wooden drums, the configuration and notation of which originated with David Wolckenstein.[18] They were mounted horizontally and had triangular iron pins which lifted the hammers, permitting them to strike the bells. The notary and musicologist Johann M.F. Th. Lobstein (1780–1855) provided some details on the drums in 1839–1840.[19] After that all traces have been lost, and it must be assumed that the rest of the carillon disappeared during the nineteenth century.

Based on Lobstein's specifications, the following can be said about the six drums:

First Drum, with the inscription: "*Aus tieffer Nott schrey ich zu dir Herr Gott erhoer mein rueffen. 130. Psalm*" (KJV: "Out of the depths have I cried unto thee, O Lord"). The drum carried the monogram of David Wolckenstein—W with Δ (delta) in the middle—and that of the Habrecht brothers—I and H, separated by two dashes, intersected by a vertical arrow.

Second Drum, with the inscription: "*Ein Kindelein so liebelich ist uns geboren heute*" (That infant form so sweet, This day to us is born); with the same monograms. It is the second verse of the hymn "*Dies est laetitiae*" by Bishop Benno in Martin Luther's translation "*Der Tag der ist so freudenreich*" (Hail to the day! so rich in joy).[20]

Third Drum, with the inscription: "*Unser Vater im Himmelreich*" (Our Father in Heaven),[21] the Lord's Prayer composed by Luther in 1539.

18 Wolckenstein was a professor for mathematics and music, and was the inspector of the boarding school in the convent for preachers and simultaneously '*musicae director*' of the Strasbourg Academy. He stood out as a composer of psalmody; Vogeleis 1911, 347, 371 and 619.

19 Lobstein 1839a, cols. 417–420; Lobstein 1839b, cols. 893–895; Lobstein 1840, 38–43. There is another brief handwritten description of the carillon that was probably penned by Lobstein, which, however, does not provide any more information than the previously cited texts; OND, box 43, nos. 18–20.

20 The English translations are from Fry 1845, 13–15.

21 The English translation is from Riemenschneider 1933, no. 37.

Fourth Drum, with the inscription: "*O Mensch bewein dein Sünde gross*" (O man bewail thy grievous fall),[22] a passion hymn by Sebald Heyden (1493–1551) or Matthias Greytter (ca. 1550). As Greytter was a canon at the Cathedral, he can be assumed to be the author here.

Fifth Drum, with the inscription: "*Nun bitten wir den heiligen Geist*" (We Now Implore God the Holy Ghost),[23] a Pentecost hymn, with the same two monograms.

Sixth Drum, with the inscription: "*Christ ist erstanden*" (Christ is arisen),[24] a well-known Easter hymn from the thirteenth century.

The first, second and third drums had red and yellow bands; they came from the first Cathedral clock. However, the Reformation hymns were added when the second clock was built.

Traces of an earlier notation, which was later changed, were still visible. The other three drums, which were built in 1571–1574, had black and white bands.

After the carillon played, which occurred every four hours, the cock crowed twice.[25] According to a travelogue from 1606, this occurred at 7 AM, 11 AM, and 3 PM.[26]

7.4 Transmissions

On the aforementioned drive shaft for the going train, two wheels were mounted outside the frame; one of these had 48 teeth and engaged with a 24-tooth wheel at a right angle. On the vertical shaft of this wheel, there was another wheel with 48 teeth, which engaged with a 24-tooth wheel on the horizontal shaft of the quarter-hour dial. Neither of the 24-tooth wheels still exists.

The second wheel mounted on the drive shaft had 18 teeth and engaged at a right angle with another gear with the same number of teeth. Its vertical shaft, of which only the bottom part with a coupling is extant, drove the astrolabe, the displays for the phases and age of the Moon, and the hand of the outer dial in the southern transept. A worm gear was used for transmitting the torque to the Moon dial, which engaged with a helical gear with 48 teeth and an 8-trundle lantern pinion. Engaging with the latter was a 59-tooth wheel for the hand for the Moon. Some of the components of this mechanism for

22 *Ibid.*, no. 24.
23 The English translation is from LH 1941, no. 33.
24 The English translation is from LBW 1978, no. 136.
25 Dasypodius 1580a, 50; also in the travelogue published by Thomas Coryat (1577–1617) in 1611: Coryat 1611, v. II, 188.
26 Walther 1877, 257.

transmitting the torque still existed at the time of Schwilgué's inventory,[27] but have since been lost.

Via a pair of wheels of 10 and 30 teeth respectively, and a vertical shaft extending into the bottom tier of the clock, the wooden disk with the regents of the day, the calendar plate, and the celestial globe were driven (for the number of teeth of each wheel, see *Fig. 7.2*).

The outer ring of the calendar plate (one revolution per year) with a 183-tooth wheel was driven by the gear train for the presentation of the regent of the day. By means of a drive pin, this wheel turned a 6-pointed star wheel together with a 6-trundle lantern pinion, which engaged with a wheel with 100 teeth for the inner part of the calendar plate.[28] The latter therefore rotated once every one hundred years.

None of the wheels and transmission parts just mentioned or the wooden disk for the figures of the regents of the day still exists; the wheels for the calendar plate also seem to have disappeared.[29]

---

27 AMS, Œuvre Notre Dame, XIXe siècle, no. 589, 7.

28 *Ibid.*, 9.

29 In a photograph of the clockwork in EA, V. III, 1296, both the wheels for the movement of the regents of the day as well as of the calendar plate are visible at the extreme left. It was impossible to determine whether these components still exist and, if so, where they are located. The holdings of the Musée des Arts Décoratifs were searched without success.

CHAPTER 8

The Celestial Globe

The celestial globe displayed in front of the calendar plate (Strasbourg, Musée des Arts Décoratifs)[1] has a diameter of 860 mm; its axis is tilted at an angle of 48° relative to the horizontal ring supported on the four wooden columns, which corresponds to the latitude of Strasbourg (*Fig. 8.1*).[2] On the periphery of the ring, the months and saints' days are listed, and on the inside the signs of the zodiac corresponding to the months are shown by symbols with pictorial representations. A ring with markings for degrees, which is perpendicular to the first ring, and on which the climate and latitudes are specified for numerous cities, indicates the local meridian. Moving around the outside are two circular indicators for the Sun and Moon, which previously had the symbols for the two luminaries and had slits so that the exact intersection with the ecliptic could be read.

Next to the south pole is the inscription: "*Cunrad Dasypodius David Wolkenstein describebant: Tobias Stimer pingebat: Isaac Habrecht ferr. fabr:*". A second inscription can be seen below the symbol for Pisces: "*Petrus Reichelt Mathem: Isaac Habrecht Autom. Iohan Mock Pict. Renovarunt Anno* MDCLXX.".[3]

The 48 constellations listed in the star catalog of Ptolemy's *Almagest*, Books VII–VIII, are shown on the celestial globe, with a total of 1,022 stars:

> *Globus Astronomicus* is what the sphere carried by the pelican down below on the floor, is called by the Latinists, in the said sphere all the stars known to astronomers, are described, namely 1,022, which are divided into 48 pictures there, and the great and the small stars are thus set in this sphere according to the longitude and latitude, and also shaped, such as one see such stars in the heavens and firmament, we two and Tobias Stimmer, the painter, have also had very great effort and work with this, before we had brought about this accomplishment, and did not let ourselves desist in the strict meting out of diligence and seriousness, after that through us it was manufactured, designed, and sectioned off,

1 The standard work on the celestial globe is Beyer/Bach/Muller 1960.

2 The exact latitude is 48° 34′ 57″ N.

3 The first of these was Professor Julius Reichelt, already mentioned in connection with the eclipse tables; his given name was incorrectly added during a later renovation; Beyer/Bach /Muller 1960, 104. Isaac Habrecht III lived between 1611–1682 and was a Cathedral clockmaker, just like his grandfather; Ungerer 1925, 132–135.

 | DOI:10.1163/9789004423473_010

FIGURE 8.1 Celestial globe of the Strasbourg Cathedral Clock

and was also painted, we have delivered such to the clockmaker, who had made the drive for it with other components.[4]

---

4 *"Globus Astronomicus wirt die Kugel so unden auf dem bodem der Pellican tregt, von den Latinis genant, in welcher Kugel alle sternen so den Astronomis bekandt, beschriben seind, namlich 1022., die da in 48 bilder abgetheylet werden, und seind also in diser Kugel die grossen und kleine sternen nach der leng und breite gesetzt und auch gestaltet, wie man solche sternen an dem himmel und firmament sihet, haben auch wir zwen und der Tobias Stimmer der Mahler sehr grosse muehe und arbeit mit gehabt, ehe dann wir solches zuwegen bracht haben, und an fleiß und ernst in der scharpffen außtheilung an uns nicht lassen erwinden, nach dem sey auch durch uns verfertigt, gerissen und abgetheilt, auch gemalt wardt, haben wir solche dem Uhrenmacher gelüffert der den tryb daran mit anderen zugehoer gemacht hat"*; Dasypodius 1580a, 18–19.

The model for Stimmer's paintings has not been identified with any certainty.[5]

Twelve semicircles perpendicular to the ecliptic intersect at the two ecliptic poles and form the boundaries between the signs of the zodiac. In addition, the location is given for the supernova that appeared in Cassiopeia at the end of 1572. Dasypodius referred to it as a comet in his description of the celestial globe;[6] however, he set its origin in the fixed star sphere and postulated that this phenomenon was similar to the appearance of the 'comet' for the birth of Christ.[7] Dasypodius shared this opinion with many other authors—such as Calvin's successor Theodore Beza—, who assumed, in analogy to the Star of Bethlehem, that there would be great upheavals, such as the second coming of Christ or the end of the world, as a consequence of this phenomenon.[8] Tycho Brahe, on the other hand, firmly rejected an interpretation of the phenomenon as a comet and, as the parallax could not be measured, felt that it had to be a new star in the fixed star sphere.[9] These conclusions challenged Aristotle's postulate that the heavens were incorruptible and unchangeable.

The gear train was mounted at the bottom of the pole axis (*Fig. 8.2*) and was hidden by a hammered copper pelican, which is no longer extant, but is clearly visible in Stimmer's two woodcuts (*Fig. 8.3*). According to the early Christian *Physiologus*, the pelican tears its breast in order to bring its young, which it had killed, back to life with its blood.[10] Dasypodius took the initial interpretation as a symbol of the resurrection of the dead through Christ's sacrifice and additionally considered it a symbol for eternity:

> [...] thus we have placed the Pelican below, which carries the sphere of the heavens, such a bird signifies the time eternal, which the Latinists call *aeternitatem*, and this Pelican also signifies Christ, inasmuch as his own

5 Brown's assumption that Dasypodius had used Piccolomini's star charts is false, as he erroneously took the nineteenth-century globe built by Schwilgué for an exact copy of the original; Brown 1932, 173–174.

6 Dasypodius 1580b, fol. G^{r}.

7 Dasypodius 1578b, fol. BIII^{v}; Latin version, fol. BIV^{v}. Dasypodius distinguished between two types of comets according to their origin: in the air or in heaven, that is, in the fixed star sphere.

8 Dreyer 1890, 68–69.

9 *De nova et nullius aevi memoria prius visa stella, iam pridem Anno a nato Christo 1572 mense Novembrii primum conspecta, contemplatio mathematica* [...], Copenhagen, 1573, in TBOO, v. 1, 1–72. See Dreyer 1890, 38–69 and 186–197.

10 Seel 1960, 6–7.

FIGURE 8.2 Movement of the celestial globe

FIGURE 8.3 The celestial globe with pelican in front of the calendar disk section of Stimmer's woodcut, 1574

> death, and his blood he did give to the young, to sustain them among the living.[11]
>
> But we have attached this Pelican so that it should be in place of Atlas and represent a symbol of eternity, or even of our Redeemer and Saviour.[12]

In this manner the gears of the globe were out of sight for the viewers and the connection between it and the clockwork was also hidden as the drive shaft was located in a covered channel in the floor. According to Dasypodius, he had been in possession of the globe for more than fourteen years before he voluntarily provided it for use in the new Cathedral clock and had the gears added.[13] The investigations of Beyer, Bach and Muller in 1953, however, brought new findings on the period of origin and the use of the globe. Inside the globe, which was made of papier-mâché on a wicker frame,[14] two documents were found, from which it is clear that the globe had already been assembled in 1546 and the paintings on it were completed in 1549. The first document came from the flour weigher Hans Erstein and contained the following excerpt:

> The sphere was made by the flour weigher Hans Ersten in the year 1546 [...] and it is the sphere in which flour was weighed at the flour scales at the white tower in Strasbourg made by the same weigher. When the sphere was taken apart as far as in [...?] back on the shaft I put the sphere [*in the left margin*: in the year 1549, the 16th of January] and took out that therein [*i.e.* that which was inside], and set in the shaft, there this epistle from Heynrich Zellen was placed in it, who had inspected the work on the 26th day of January of the year 1549 [...].[15]

---

11 "[...] *so haben wir den Pellican unden gesetzt der dise kugel des himmels tregt, welcher vogel ein anzeigung gibt der ewig werender zeit, welche die Latini nennen aeternitatem, auch ist diser Pellican ein anzeigung Christi, dieweil er sich selbers toedt und sein blut den Jungen gibt sie bey dem leben zu erhalten*"; Dasypodius 1580a, 20.

12 "*Pellicanum vero hunc adhibuimus, ut esset loco Atlantis: et ut aeternitatis referret symbolum, aut etiam Redemptoris nostri ac Salvatoris*"; Dasypodius 1580b, fol. G1r.

13 Dasypodius 1580a, 19.

14 The pieces of paper that were partially visible on the inside of the globe could be identified by Ernest Muller as coming from a collection of Reformation hymns published in Strasbourg between 1541 and 1545; Beyer/Bach/Muller 1950, 192ff.

15 "*Dey kugel ist gemach worden von Meyr hanß Ersten Im Ior 1546* [...] *und ist dey kogel In Der Mel wog* [Mehlwaage] *am weyssen thurn zu stroß burg vum selbigen weger gemach worden. Alß dy kugel uß gemach ist worden biß in in* [...?] *der stefft hab ich dy kugel wieder uff gethon: Im Ior 1549 den 16 Ierner und das in geby her uß gethon, und den stefft inngesetz do ist dyse Epistel do rin gelent worden von heynrich Zellen der das werck bescheben hat uff den 26 tag Iernerß des iars 1549*"; *Ibid.*, 107 (facsimile of the document).

According to this, the globe was reopened three years after its completion and a letter written by the cartographer Heinrich Zell (1518–1564)[16] to the Strasbourg bishop Erasmus von Limburg was placed in it. Besides providing interesting contemporary historical and chronological information, this letter contains evidence that Zell had completed the previously geographical globe in March of 1549.[17] The latitudes for various cities listed on the meridian ring also vouches for the original use as a geographical globe.

As Dasypodius claimed to have owned the globe for more than fourteen years and as this claim is included in the first edition of the *Warhafftigen Außlegung*,[18] a large degree of probability can be attributed to the theory that Dasypodius had received the globe from his predecessor Christian Herlin,[19] who died in 1562.

Interestingly, Dasypodius conveyed the materials used in making the globe: "This sphere has three work-shoe lengths in width, and weighs one hundredweight, is artistically prepared from fabric, paper, paste, chalk and of other material".[20] In *Heron mechanicus* he also mentions that, regarding the size, he complied with an ancient proviso according to which a celestial globe should have a diameter of at least three feet.[21] From this we can conclude that he had the globe enlarged to the appropriate diameter before Stimmer painted it, by applying new layers of paste-soaked canvas and chalk to provide a base for the paintings.

To which ancient proviso Dasypodius was alluding could not be determined. For the assembly of a precession globe, no mention is made in Ptolemy's

16 Zell was born in Cologne and studied in Basel and Wittenberg. He supervised the publication of Georg Joachim Rheticus's *Narratio Prima* as his famulus. After 1555, Zell was the librarian of Duke Albrecht of Prussia in Königsberg. For his biography and cartographical works see Burmeister 1978, 427–442 and Burmeister 2015, 661–662. Older literature is provided in Beyer/Bach/Muller 1960, 110, n. 118.

17 "*Supra scripto tempore Corpus Globi huius Geographici perfectum est mense Martio* [...] *Lineamento vero et, quae ad Geographiam pertinent deduxit atque conscripsit Heilrichus Zellius Agrippinas anno aetatis sue 31 currente eodem anno*"; *Ibid.*, 109. Three photographs of the two documents are in the municipal archives of Strasbourg, Sér. II, 128, no. 10, and the originals were probably put back inside the globe. In any case, they could not be found in the municipal archives or in the Musée des Arts Décoratifs.

18 Dasypodius 1578a, fol. DIII^{v}.

19 Beyer/Bach/Muller 1960, 113.

20 "*Dise kugel hat drey werck schuch in der weite und wigt ein Centner, ist künstlich auß tuch, pappeyr, leym, kreyd und anderer matery also zubereit*"; Dasypodius 1580a, 19.

21 "[...] *quam quidam ex antiquis et probatis scriptoribus, volunt tantam esse debere; ut diameter eius ad minus trium sit pedum; quanta nostri Globi* [...] *magnitudo est*"; Dasypodius 1580b, fol. G^{r}.

Almagest of the material, fabrication, or dimensions,[22] and Leontios's *Construction of the Sphere of Aratos* from the seventh century says nothing about its dimensions either.[23] Even if Strabo indicates that a globe with a diameter of at least ten feet would be necessary for a clear depiction of the known world,[24] these instructions can hardly have had any relevance for Dasypodius's proviso, as they refer to a terrestrial globe.

8.1 The Sphere of Archimedes

Dasypodius explicitly mentions the crystal sphere constructed by Archimedes as a model for the celestial globe of the Strasbourg clock:

> Archimedes is said to have enclosed a globe in crystal and on it to have shown all the motions of heaven, and also to have written a book on sphaeropoetica. Although we cannot even in the least part be compared with Archimedes in either erudition or industry, we would not enclose this globe in crystal nor would we show all the motions of all the stars. And, although our work be left far behind that of Archimedes, nevertheless, just as sometimes small things are most rightly compared with great things, we, so far as we could, have performed in proportion to our limitations, and we, so far as we know, were the first to dare to do what no one in our time past before us tried to do in such a work: to make a globular body about ten feet in circumference, and solidly made of various materials; to dare to form it in the shape of heaven, to refine it, with stars distributed and shaped after the arrangement of stars, and to attribute to it a motion similar and proportionate to heavenly motion. That was done not without great effort, an effort several times repeated diligently.[25]

---

22 Ptolemy 1963, v. II, VIII.3, 72–76.

23 Ziegler 1536, 93–104 (first edition: Nürnberg 1531); Houzeau/Lancaster 1882/89, v. I.1, 393, no. 759; in the edition Maass 1898, 561–567. Incidentally Ziegler merely writes the following in the instructions for building a celestial globe, which he wrote himself and placed at the beginning of his text anthology: "*Aptetur sphaera absoluta ad semicirculi circunducti castigationem, super diametro non minori palma et semis, sive ut libitum cuique fuerit, et visum opportunum. Ea Sphaera si ex metallo est, inargentetur. Sic enim profunda circulorum et designationum fient conspectiora*"; Ziegler 1536, 2.

24 Strabo, *Geographia*, II.5, 10; Strabo 1989, 449. In this connection, Strabo mentions the terrestrial globe of Crates of Mallos as an example; however, its dimensions cannot be determined from this context.

25 "*Archimedes dicitur crystallo globum inclusisse et in eo omnes coeli motus monstrasse, librum quoque de sphaeropoetica scripsisse: nos etsi ne minima quidem ex parte, vel*

Archimedes's ancient model of the heavens is only documented in scanty literary references, but enjoyed legendary fame and was discussed in depth by several classical philologists.[26]

During the conquest of Syracuse in 212 BC the Roman general Marcus Claudius Marcellus obtained two spheres produced by Archimedes. One of them was displayed in the Virtus Temple in Rome, the other Marcellus considered to be part of his share of the plunder and was taken to his house. Gaius Sulpicius Gallus saw it there and his fragmentary description was repeated by Cicero:

> Though I had heard this globe mentioned quite frequently on account of the fame of Archimedes, when I actually saw it, I did not particularly admire it; for that other celestial globe, also constructed by Archimedes, which the same Marcellus placed in the temple of Virtus, is more beautiful as well as more widely known among the people. But when Gallus began to give a very learned explanation of the device, I concluded that the famous Sicilian had been endowed with greater genius than one would imagine it possible for a human being to possess. For Gallus told us that the other kind of celestial globe, which was solid and contained no hollow space, was a very early invention, the first one of that kind having been constructed by Thales of Miletus, and later marked by Eudoxus of Cnidus (a disciple of Plato, it was claimed) with the constellations and stars which are fixed in the sky. He also said that many years later Aratus, borrowing this whole arrangement and plan from Eudoxus, had described it in verse, without any knowledge of astronomy, but with considerable poetic talent. But this newer kind of globe, he said, on which were delineated the motions of the Sun and Moon and of those five stars which are called wanderers, or, as we might say, rovers, contained more than could be shown on the solid globe, and the invention of Archimedes deserved special admiration because he had thought out a way to represent

---

eruditione, vel industria, cum Archimede conferri possimus, nec crystallo globum hunc incluserimus aut etiam omnes omnium stellarum motus ostenderimus: ac longo intervallo ab Archimedeo opere, nostrum post relinquatur tamen ut interdum parva magnis rectissime conferuntur: nos pro nostra tenuitate, quod potuimus praestitimus. Et quod antea nos (quod sciamus) in tali opere nemo hoc tempore tentavit primi facere atque globosum corpus, decem fere in ambitu pedum, ex varia materia consolidatum, ad coeli formam formare, et expolire, stellis ad dispositionem stellarum distributis et configuratis, motum coelesti motui simile et proportionalem attribuere ausi fuimus: idque non sine magno, et aliquoties reiterate diligenti labore"; Dasypodius 1580b, fol. G1^{v}; see also Dasypodius 1580a, 21.

26 Schieck 1846; Bunte 1877, 14–18; Heiberg 1879, 33–34 and 41–44; Schlachter 1927, 48–53; Gigon 1973.

> accurately by a single rotation for turning the globe those various and divergent movements with their different rates of speed. And when Gallus moved the globe, it was actually true that the Moon was always as many revolutions behind the Sun on the bronze contrivance as would agree with the number of days it was behind it in the sky. Thus the same eclipse of the Sun happened on the globe as [it] would actually happen, and the Moon came to the point where the shadow of the Earth was at the very time when the Sun […] out of the region […].[27]

According to Cicero, that celestial model—unlike the other looted sphere, which was a celestial globe—indicated the movements of the seven planets as well as the eclipses of the Sun and Moon. It must therefore have been a type of planetarium and Cicero stressed in another passage that its movements were brought about by a single rotation:

> For when Archimedes fastened on a globe the movements of Moon, Sun and five wandering stars, he, just like Plato's God who built the world in the *Timaeus*, made one revolution of the sphere control several movements utterly unlike in slowness and speed. Now if in this world of ours phenomena cannot take place without the act of God, neither could

---

27 *"Nihil novi vobis adferam, neque quod a me sit cogitatum aut inventum; nam memoria teneo C. Sulpicium Gallum, doctissimum, ut scitis, hominem, cum idem hoc visum diceretur et esset casu apud M. Marcellum, qui cum eo consul fuerat, sphaeram, quam M. Marcelli avus captis Syracusis ex urbe locupletissima atque ornatissima sustulisset, cum aliud nihil ex tanta praeda domum suam deportavisset, iussisse proferri; cuius ego sphaerae cum persaepe propter Archimedi gloriam nomen audissem, speciem ipsam non sum tanto opere admiratus; erat enim illa venustior et nobilior in volgus, quam ab eodem Archimede factam posuerat in templo Virtutis Marcellus idem. sed posteaquam coepit rationem huius operis scientissime Gallus exponere, plus in illo Siculo ingenii, quam videretur natura humana ferre potuisse, iudicavi fuisse. dicebat enim Gallus sphaerae illius alterius solidae atque plenae vetus esse inventum, et eam a Thalete Milesio primum esse tornatam, post autem ab Eudoxo Cnidio, discipulo, ut ferebat, Platonis, eandem illam astris stellisque, quae caelo inhaererent, esse descriptam; cuius omnem ornatum et descriptionem sumptam ab Eudoxo multis annis post non astrologiae scientia, sed poetica quadam facultate versibus Aratum extulisse. hoc autem sphaerae genus, in quo solis et lunae motus inessent et earum quinque stellarum, quae errantes et quasi vagae nominarentur, in illa sphaera solide non potuisse finiri, atque in eo admirandum esse inventum Archimedi, quod excogitasset, quem ad modum in dissimillimis motibus inaequabiles et varios cursus servaret una conversio, hanc sphaeram Gallus cum moveret, fiebat, ut soli luna totidem conversionibus in aere illo, quot diebus in ipso caelo, succederet, ex quo et in sphaera solis fieret eadem illa defectio et incideret luna tum in eam metam, quae esset umbra terrae, cum sol […] e regione"*; Cicero, *De republica* 1.14, 21–22; Cicero 1970, 40–43.

> Archimedes have reproduced the same movements upon a globe without divine genius.[28]

Numerous references to the device also appeared at later dates. In the first century AD, Ovid describes an astronomical model in his poem *Fasti*:

> The Earth is like a ball, resting on no prop; so great a weight hangs on the air beneath it. Its own power of rotation keeps its orb balanced; it has no angle which could press on any part; and since it is placed in the middle of the world and touches no side more or less, if it were not convex, it would be nearer to some part than to another, and the universe would not have the Earth as its central weight. There stands a globe hung by Syracusan art in closed air, a small image of the vast vault of heaven, and the Earth is equally distant from the top and bottom. That is brought about by its round shape.[29]

According to the Church father Lactantius (fl. ca. 300) the device also indicated the movements of the fixed stars.[30] According to Cicero, it was made of metal, and Lactantius also spoke of a sphere "devised from hollowed metal".[31] According to Claudian the globe was made of glass.[32] Around 200 AD, Sextus Empiricus claimed that the sphere was made of wood in a passing remark.[33] By relating Claudian's indications that "Some hidden influence within the sphere directs the various courses of the stars and actuates the lifelike mass with

---

28 "*Nam cum Archimedes lunae, solis, quinque errantium motus in sphaeram illigavit, effecit idem quod ille qui in Timaeo mundum aedificavit Platonis deus, ut tarditate et celeritate dissimillimos motus una regeret conversio. Quod si in hoc mundo fieri sine deo non potest, ne in sphaera quidem eosdem motus Archimedes sine divino ingenio potuisset imitari*"; Cicero, *Tusculanae disputationes*, I.63; Cicero 1945, 73–75.

29 "*Terra pilae similis, nullo fulcimine nixa, / aere subiecto tam grave pendet onus: / ipsa volubilitas libratum sustinet orbem, / quique premat partes angulus omnis abest: / cumque sit in media rerum regione locata, / ut tangat nullum plusve minusve latus, / ni convexa foret, parti vicinior esset, / nec medium terram mundus haberet onus. / arte Syracosia suspensus in aere clauso / stat globus, immensi parva figura poli, / et quantum a summis, tantum secessit ab imis / terra; quod ut fiat forma rotunda facit*"; Ovid, *Fasti*, 269–280; Ovid 1989, 338–341.

30 Lactantius, *Divinae Institutiones* II.5; Lactantius 1964, 114. Probably this did not refer to a hollow sphere with stars painted on it, but rather the zodiac mentioned by Claudian in his poem "*In Sphaeram Archimedis*", in *Carmina minora*, 51; Claudian 1922, 278–281.

31 "[...] *concavo aere similitudinem mundi*"; Lactantius 1964, 114.

32 Claudian 1922, 278–281.

33 Sextus Empiricus, *Adversus Physicos* I, 115; Sextus Empiricus 1987, 63.

definite motions",[34] Ovid's description, and the very obscure description from Martianus Capella,[35] with a passage from the *Collectio* by Pappus,[36] the philologist and mathematics historian Friedrich Hultsch (1833–1906) suggested it had a hydraulic drive system.[37]

The literary references mentioned above only provide a vague impression of this important ancient celestial model and hardly permit any conclusions about the technical or structural details. Nonetheless the sphere of Archimedes served as a model and played a significant role as a stimulus for clockmaking in the sixteenth and seventeenth centuries. A famous example is the 'rock crystal clock' made by Jost Bürgi between 1622 and 1628 for Prince Karl I of Liechtenstein (Vienna, Kunsthistorisches Museum).[38] The clock has an octagonal rock crystal housing with a hemispheric lid, under which a rock crystal celestial globe, engraved with signs of the zodiac, revolves, and inside it there is an armillary sphere. This work, the builder of which was praised by William IV, Landgrave of Hesse-Kassel in 1586 as the 'second Archimedes',[39] had to have been based on the celestial model of the Syracuse philosopher. The same applies for the two planetary clocks built by Janello Torriani (*c.* 1500–1585) for Emperor Charles V in 1547–1550 and 1554–1562.[40]

The construction of an enormous globe which was begun in 1650 for Gottorf Castle also followed this tradition.[41] It was not simply by chance that the clock's designer, the mathematician and librarian Adam Olearius (1599–1671) at the court of Duke Friedrich II of Schleswig-Holstein-Gottorf (1597–1659), used a hydraulic drive for the globe based upon a water wheel with reduction gears.[42]

With rhetorical modesty, Dasypodius and Wolckenstein wished to be judged for their work according to the accomplishments and complexity of the model of their great ancient predecessor. Dasypodius valued the celestial globe as the most important part of the astronomical clock:

---

34 "*Inclusus uariis famulatur spiritis astris / et uiuum certis motibus urget opus*"; Claudian 1922, 278–281.

35 Martianus Capella, *De nuptiis Philologiae et Mercurii*, VI, 583–585.

36 "*Mechanicos denique etiam illos vocant, qui globorum fabricationem callent et coeli effigiem per aequabilem et circularem aquae motum construunt*"; Pappus 1875/78, v. III, 1027.

37 Hultsch 1877.

38 Maurice/Mayr 1980, 97–99; Bertele 1961, 54–55; Karlsen 1986.

39 "[…] *diligentiaque et industria Automatarii nostri, Iusti Byrgii, qui quasi alter Archimedes est*"; Letter to Tycho Brahe dated 14 April 1586; TBOO, VI, 52; German translation, *ibid.*, 49.

40 Zanetti 2014; Zanetti 2016, 117–123; Zanetti 2017, 270–296.

41 For very detailed and knowledgeable information on this see Lühning 1997.

42 Oestmann 1999b. On the life and work of Olearius see Baumann/Köster/Kuhl 2017.

> And, to be sure, I value this globe more highly than the remaining parts of the machine, [...] because I know that this globe is the most excellent of the exceptional parts of the whole work.[43]

8.2 Gear Train of the Celestial Globe

The following calculations are based on the technical specifications of Henri Bach.[44]

Movement of the Sun's indicator. The drive shaft *A* in the gearing (*Fig. 8.4*) rotated 3 times in 24 hours. The wheel *B* (N = 20), mounted on *A*, engaged with the Sun wheel *C* (N = 60). This resulted in *C*, and the Sun's alidade, making (3 × 20) / 60 = 1 rotation in 24 hours.

Movement of the Moon's indicator. Another wheel *P* (N = 19) mounted on the drive shaft *A* engaged with the Moon wheel *Q* (N = 59). This resulted in *Q* making (3 × 19/59) = 57/59 rotations in 24 hours. The Moon alidade attached to *Q* therefore lagged behind the Sun alidade with its one rotation in 24 hours by 1 – 57/59 = 2/59 rotations in 24 hours. This delay corresponded to a complete rotation relative to the Sun in 29.5 days.

As the synodic month—the average period between two identical phases of the Moon; for example, two new moons—has 29.53059 days, the indicator for the Moon rotated 29.53059 – 29.5 = 0.03059 day too fast. The year consists of 365.2422/29.53059 = 12.3683 synodic months, so that in one year it rotated 12.3683 × 0.03059 = 0.3783 day too fast, or it was a whole day too fast in 1/0.3783 = 2.64 years.

Movement of the celestial globe. On the wheel *C* for the motion of the Sun there were two planetary wheels *E* (N_G = 30) and *G* (N_G = 32) with their pinions *F* (N_P = 4) and *H* (N_P = 8), which rotated with the Sun wheel *C* around a central pinion *D* with 6 teeth. The pinion *H* of the planetary gear *G* engaged with the wheel *I* (N_G = 73) for the celestial globe.

If the planetary gears had been fixed and the central pinion removed, the wheel *I* for the celestial globe would have turned with the Sun wheel *C* and, just like it, would have rotated once in 24 hours. But the planetary gear *E* revolved around the central pinion *D* once every 24 hours and conferred via *F*, *G*, and *H* an additional rotation to the wheel *I* of (6 × 4 × 8) / (30 × 32 × 73) = 1/365.

---

43 "*Et sane hunc globum maioris facio,* [...] *sed quod sciam ex praecipuis totius operis partibus, hanc esse excellentiorem*"; Dasypodius 1580b, fol. G1v.

44 Beyer/Bach/Muller 1960, 125–127; Bach 1979, 32–34. See also Schwilgué's description; AMS, 3 OND 589, 10–11.

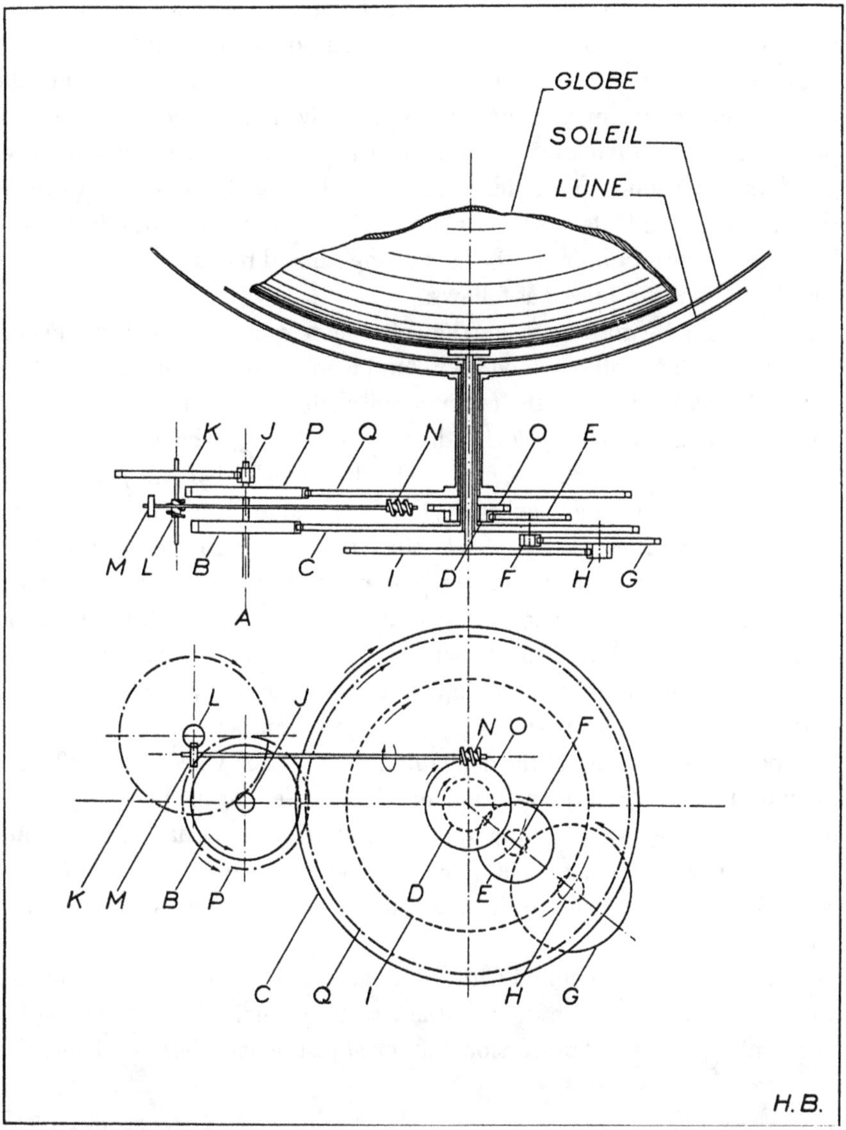

FIGURE 8.4 Gearing of the celestial globe

Therefore after 365 days (or rotations of the Sun wheel) the celestial globe, which displayed the sidereal time, had one additional rotation.[45]

These movements have to be calculated in the direction of rotation of the daily movement; the movement of the Sun relative to the fixed stars in the opposite direction is therefore negative. The period set as a first approximation at 365 days corresponds to the sidereal motion; the actual duration, however, is slightly more than ¼ day (0.256 day) longer. Dasypodius took this into account by having the central pinion *D* rotate in the same direction as the planetary gear *E*, which was achieved as follows.

The pinion *J* with 8 teeth on the drive shaft *A* engaged with a wheel *K* ($N_G = 84$), on the shaft of which a left-hand worm gear *L* was mounted, which drove a wheel *M* with 12 teeth. Its horizontal shaft carried a right-hand worm *N* on the other end, which engaged with a wheel *O* ($N_G = 35$) that was attached to the central pinion *D*. (In *Fig. 8.4* the shaft with the wheel *M* and worm gear *N* is not drawn with the teeth engaging, but displaced to the left for clarity.)

To compute the precise effect, consider that in one day the drive shaft *A* rotated 3 times and the central pinion *D*, rather than being fixed, rotated $3 \times (8 / 84) \times (1 / 12) \times (1 / 35) = 1/1470$ turns. Letting the wheelwork rotate in the same direction resulted in a small lag for *I*, the drive wheel of the celestial globe. If *D* were stationary it would have caused 1/365 extra rotations for *I*, so that in 365 days *I* performed a complete extra rotation, beyond the 365 it would have performed because of the daily rotation of the Sun wheel *C*. The effect of the movement of *D* by the worm *N* therefore caused for the planetary wheel train *DEFGH* $(1/365) \times (1 - 1/1470)$ extra rotations of *I* per day. This accumulates to one whole extra turn in $1/(1/365 \times 1469/1470)$ days $= 365 \times 1470/1469 =$ 365.2485 days, quite close to the sidereal year of 365.256 days—a very ingenious construction.

As the ecliptic was painted with the great circles of the zodiac boundaries and the intersections with the celestial equator were fixed, it was impossible to consider the effect of precession. But the adjustment of the Sun wheel gears

45 A sidereal day is the interval between two successive transits of a fixed star. In that time the Earth rotates once on its axis. The Sun, as a result of the orbital motion of the Earth, changes its position relative to the fixed stars approximately 1° per day and advances from west to east by this magnitude on the ecliptic. For this reason, the mean solar day is not equal to the sidereal day, but has to be longer by the time the Earth with its rotation needs to shift the position of the observer's location by about 1°, an interval of about 4 minutes. A mean solar day is therefore 24h 3m 56s sidereal time; one sidereal day is 23h 56m 4s mean solar time. As a consequence, in the course of a year, the Sun culminates 365 times, but a fixed star 366 times. Therefore the celestial globe had to make one additional rotation per year relative to the Sun's indicator.

by using the sundials mounted outside the Cathedral (see *Chapter 10*) ensured that the Sun wheel's rotation did not differ too much from the tropical course of the Sun. The question of a possible consideration of the equation of time must remain unanswered.

Some points for dividing the rims of the gears are still visible, which were marked with a compass and punched before the individual teeth were filed to shape. The worm gears were also produced using files only. The execution of the gearing, which was confined to an extremely small space, represents extraordinary craftsmanship on the part of the Habrecht brothers, as does the gearing for the astrolabe. Even if the layout of the gearing exceeded the possible accuracy of timekeeping at that time—the inaccuracy per day of a tower clock with a verge and foliot escapement could well exceed a quarter of an hour—an outstanding construction had been achieved.

CHAPTER 9

The Astrolabe Dial

9.1 Layout of the Tympanum

The stone tympanum of the astrolabe was installed in the middle tier of the central clock tower and was, as already mentioned, drawn up during the first phase of the construction of the clock. In 1840, the three-part stone plate with a diameter of 2.2 m was moved to the Musée de l'Œuvre Notre-Dame and served there as a base for the celestial globe until its installation in the "Salle d'Horlogérie" of the Musée des Arts Décoratifs in 1921. As a result, the lines that were originally marked with gold leaf on a blue and green background were completely worn away by visitors stepping on the plate, so that they are only still recognizable near the central hole (*Figs. 9.1* and *9.2*).[1]

For drawing the elevation of the astronomical dial, the stereographic projection of the celestial sphere onto a plane—the plane of the celestial equator—was used. The stereographic projection has two main characteristics:

1. Every circle of the sphere is transformed into a circle on the plane. The great circles, which pass through the center of the projection, appear as straight lines.
2. The amplitude of any angle does not change when transferred from the sphere to the plane. It is therefore a conformal projection.[2]

The design of the dial was based on the system of lines of the planispheric astrolabe (*astrolabium planisphaerium*).[3] This astronomical instrument consists basically of four parts (*Fig. 9.3*):

1. the *mater astrolabii*, the outer rim of which (*limbus*) has degree markings.
2. In the *mater* recess is placed a flat plate (*tympanum*), into which are engraved in stereographic projections the Tropics of Capricorn and Cancer,

1 Ungerer 1925, 110, n. 44.

2 Michel 1947, 16, 27–29. The term 'stereographic projection' was coined by the Jesuit François d'Aguilon (1567–1617) and first appeared in his *Opticorum Libri sex philosophis juxta ac mathematicis utiles*, Antwerpen, 1613. See Delambre 1817, v. II, 457.

3 The literature on this type of instrument is extremely extensive. Besides the two handbook articles Hartner 1968a and Hartner 1968b, a standard source is Michel 1947. Extensive information is also provided by Morrison 2009. For an excellent, easily comprehensible introduction to the history and application possibilities of the instrument see North 1974, and for more detailed information on the geometric design see García-Franco 1945, 70ff, Gunella/Lamprey 2007, and Oestmann 2014b.

 | DOI:10.1163/9789004423473_011

FIGURE 9.1 Astrolabe dial of the Strasbourg Cathedral Clock

FIGURE 9.2 Tympanum

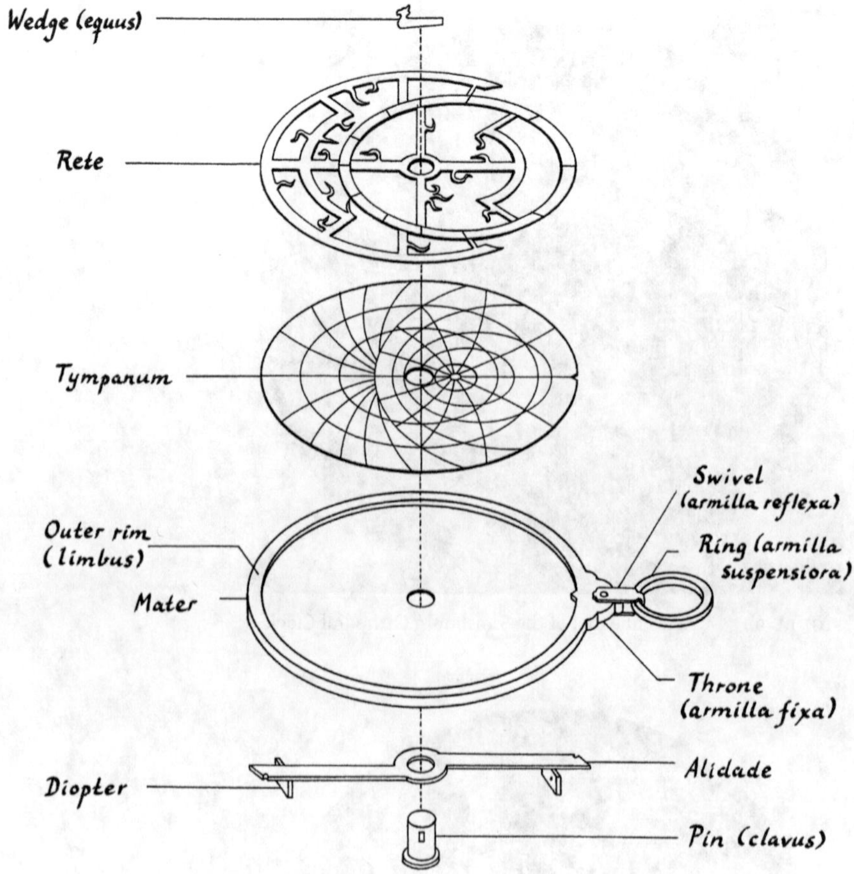

FIGURE 9.3 Planisphaeric astrolabe

the celestial equator, the almucantars (latitude circles), and azimuths (vertical circles, intersecting the almucantars at right angles) for the horizon of the observer. The local meridian (*linea medii coeli*) appears as a straight line (diameter of the plate) and intersects the middle of the mounting hole. The distance between the celestial pole (central hole) and the zenith of the observer depends on the geographical latitude. The azimuths extend radially from the zenith of the observer and end in the line for the horizon. Furthermore the tympanum has lines for the temporal hours below the horizon line and sometimes lines for the equinoctial hours. Usually an astrolabe had a set of *tympana* for various latitudes.

3. A cutout plate, the *rete* ('spider' or 'net'; from the Latin *aranea*), which rotates around the north or south pole (depending on the projection used), that is, around the central hole. In a stereographic projection, the

rete indicates the signs of the zodiac as well as certain important fixed stars. European astrolabes also frequently had a pointer or rule (*ostensor*) which rotated above the rete and facilitated reading the astrolabe.

4. In order to adjust the instrument, the altitude of the Sun or some other star indicated on the rete had to be determined. This was accomplished using the alidade (*regula*), which had two sights (*diopters*) and was attached to the back of the instrument (*dorsum astrolabii*) which had degree markings. The line of sight (*linea fiduciae*) passed through the central hole. On the back of the astrolabe there were engravings for the twelve signs of the zodiac, each in a 30° section, and a calendar scale. This scale was either laid out eccentrically or unevenly divided in order to compensate for the nonuniform movement of the Sun along the ecliptic. With the line of sight of the alidade, the date could be set and the location of the Sun on the ecliptic determined. Frequently other line systems for astrological and geodetic purposes were also engraved in the astrolabe. A shaft with a slot (*clavus*) and a key (*equus*) held the components together.

The system for suspending the astrolabe was designed such that the instrument hung vertically when it was held for sighting. At the top there was a lip or throne (*armilla fixa*) with a hole, into which a pin was inserted (perpendicular to the throne's surface), to permit a loop (*armilla reflexa*) to swivel. A ring (*armilla suspensiora*) passed through the loop.

With this truly universal instrument a multitude of astronomical, astrological, and geodetic problems (such as determining the local time, the culmination of fixed stars, sunrise or sunset, etc.) could be solved mechanically, without calculations or the need for tables. The daily motion of the stars in the heavens could be exemplified by rotating the rete.

Concerning the system of lines on the tympanum of the Cathedral clock in Strasbourg in detail,[4] the central hole represents the center of the projection, in this case, the celestial south pole. The local meridian (*linea medii coeli*) is vertical, with south at the top and north at the bottom. Perpendicular to this is the line connecting east and west, called '*horizon rectus*' (*Fig. 9.4*). As both great circles pass through the center of projection, they are rendered as straight lines on the plane. The celestial equator and the two tropics appear as concentric circles around the center of the tympanum. For a 'northern astrolabe', with the center of the projection at the celestial south pole, the Tropic of Capricorn is on the outside and simultaneously forms the periphery of the tympanum.

---

4 For executing the sketches, the Author used Ritter 1613, 20–22 (Tropics of Cancer and Capricorn, celestial equator), 27–35 (almucantars), 54–61 (azimuth lines); 62–64 (twelve houses), 67–69 (lines for the temporal hours).

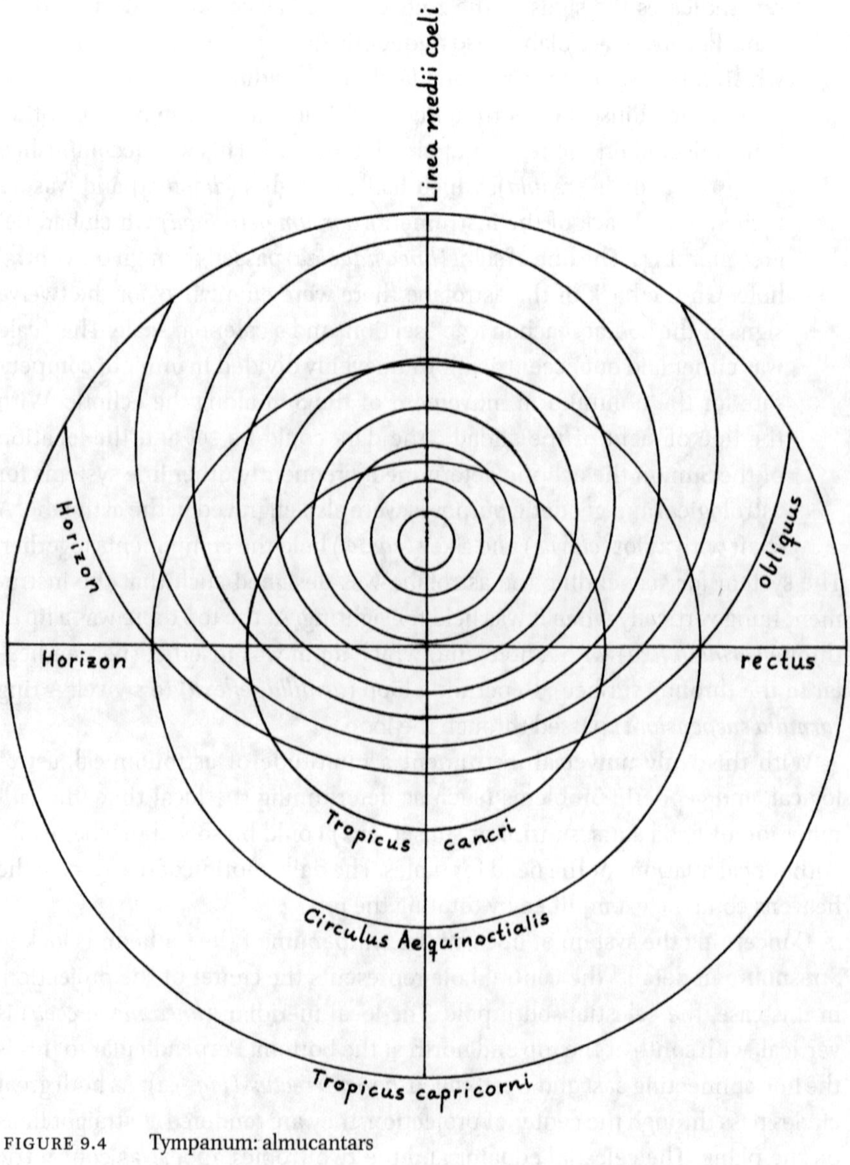

FIGURE 9.4 Tympanum: almucantars

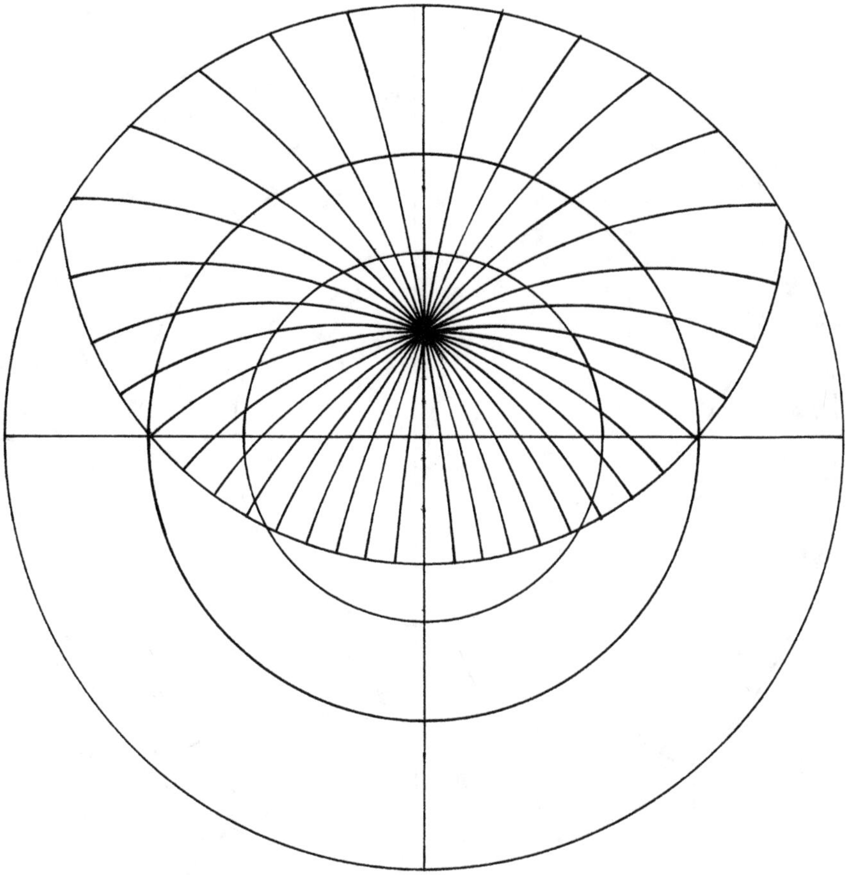

FIGURE 9.5 Tympanum: azimuth lines

Unlike the lines just mentioned, the distance between the celestial pole and the zenith of the observer depends on the geographical latitude; for Strasbourg it is 48.5° N. The curve for the horizon (*horizon obliquus*) passes through the intersections of the *horizon rectus* and the celestial equator and above it are the almucantars in 10° increments, which decrease in size with increasing altitude above the horizon and whose respective centers lie on the meridian. They are intersected at right angles by the azimuth lines,[5] such that a coordinate network is produced (*Fig. 9.5*). On the tympanum of the Cathedral clock, the azimuths are also indicated in 10° increments from the eastern to western point (the two intersections of the *horizon rectus* with the *horizon obliquus* and

5 The azimuth is the angle between the northern or southern cardinal point of the observer and the intersection of the vertical circle of the respective celestial body with the horizon.

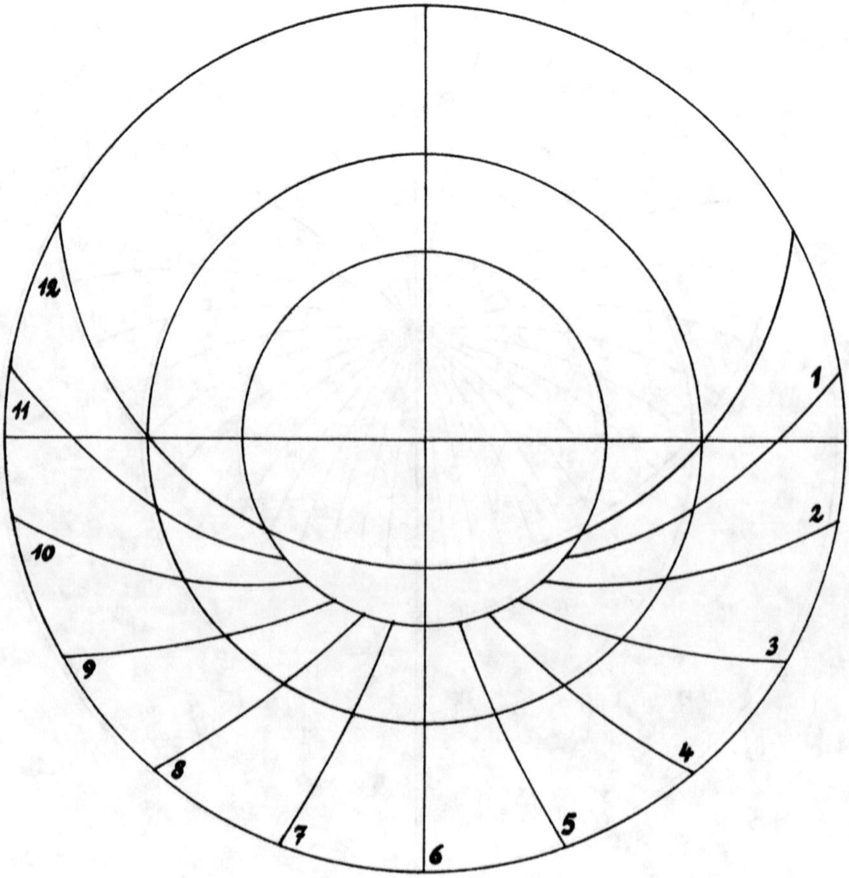

FIGURE 9.6 Tympanum: lines for the temporal hours

the celestial equator) starting in the direction of the meridian and numbered accordingly. Below the *horizon obliquus*, the lines labeled from 1 to 12, starting from the western point, indicated the 'unequal' or 'temporal' hours (*horae temporales*), that is, the hours that differ in length depending on the time of year, which divide the night between sunset and sunrise into 12 equal intervals (*Fig. 9.6*). For constructing these lines, the circles of the celestial equator and the tropics were each divided into twelve equal segments and these points connected with circular arcs. This process only resulted in an approximation, as the curves for the temporal hours are not circular arcs.

These lines were not marked above the horizon on astrolabes, so as not to obfuscate the network of coordinates formed by the almucantars and azimuth lines. This was not necessary, however, as during the day the current temporal hour could be determined by finding the intersection of the extended end of the Sun pointer on the other side with the ecliptic ring. The name "*HORAE*

Taffel der regierung der Planeten nach den vngleichen ſtunden des Tags.

Planetenſtund	1	2	3	4	5	6	7	8	9	10	11	12
Sontag	☉	♀	☿	☽	♄	♃	♂	☉	♀	☿	☽	♄
Montag	☽	♄	♃	♂	☉	♀	☿	☽	♄	♃	♂	☉
Erichtag	♂	☉	♀	☿	☽	♄	♃	♂	☉	♀	☿	☽
Mittwoch	☿	☽	♄	♃	♂	☉	♀	☿	☽	♄	♃	♂
Donnerſtag	♃	♂	☉	♀	☿	☽	♄	♃	♂	☉	♀	☿
Freytag	♀	☿	☽	♄	♃	♂	☉	♀	☿	☽	♄	♃
Sambſtag	♄	♃	♂	☉	♀	☿	☽	♄	♃	♂	☉	♀

E ij

Taffel der regirenden Planeten zu Nacht nach außtaylung der vngleichen ſtund.

Vngleich ſtund	1	2	3	4	5	6	7	8	9	10	11	12
Sontag	♃	♂	☉	♀	☿	☽	♄	♃	♂	☉	♀	☿
Montag	♀	☿	☽	♄	♃	♂	☉	♀	☿	☽	♄	♃
Dinſtag	♄	♃	♂	☉	♀	☿	☽	♄	♃	♂	☉	♀
Mittwoch	☉	♀	☿	☽	♄	♃	♂	☉	♀	☿	☽	♄
Donnerſtag	☽	♄	♃	♂	☉	♀	☿	☽	♄	♃	♂	☉
Freytag	♂	☉	♀	☿	☽	♄	♃	♂	☉	♀	☿	☽
Sambſtag	☿	☽	♄	♃	♂	☉	♀	☿	☽	♄	♃	♂

FIGURE 9.7 Planetary rulers of the hours (Apian, *Instrument Buch*, Ingolstadt, 1533, fols. EII^{r}–EII^{v})

PLANETARIAE" written on the tympanum indicates the importance of the temporal hours for horary astrology, that is, for carrying out or refraining from certain activities depending on the reigning planet at the time. In the bottom tier of the clock, the regent of the day could be seen on his triumphal chariot; this regent reigned for the first hour after sunrise. Corresponding to their orbital times, the other planets reigned as 'chronocrators' for the other hours, starting with the first hour of Saturday, which is ruled by Saturn, the regent with the slowest orbit. Following Saturn comes Jupiter for the second hour, Mars for the third, the Sun for the fourth, Venus for the fifth, Mercury for the sixth, and finally the Moon as the seventh regent of the hour. With the eighth hour the cycle starts over again with Saturn until the first hour ruled by the Sun on Sunday, and so on (*Fig. 9.7*).[6]

Twelve fields can be recognized faintly on the tympanum, the boundaries of which were not marked with lines but set off with alternating blue and green so as not to obscure the arcs of the temporal hours below the line for the

6 Bouché-Leclercq 1899, 476–486; Richards 1998, 268–273.

horizon. These represented the division of the signs of the zodiac into twelve astrological fields of activity, the 'houses' (*loci* or *topoi*). These celestial fields should not be confused with the houses of the planets (*domicilia*), which were already mentioned in connection with the figures of the regents of the day. They are numbered counterclockwise, starting with the degree of the ecliptic rising above the eastern horizon (*ascendant*). They, just as the midheaven (*medium coeli*, the tenth house), are of special importance for the astrological interpretation of a celestial configuration.

Spherical diangles formed by six great circles constitute the houses. The actual house is the arc of the ecliptic that is formed by the diangle, with a beginning and an end. The astrological term for the beginning of a house is 'cusp' (*cuspis*).

There were a considerable number of different methods for the division into houses,[7] and which 'manner' was the right one was a question that has always been contested and has led to controversies among astrologists to this day.

For the Strasbourg Cathedral clock—as for almost all of the portable astrolabes—the procedure attributed to the Franconian astronomer Johannes Regiomontanus (1436–1476) was used. In this case great circles were drawn through the northern and southern cardinal points of the observer's horizon that divide the celestial equator into twelve equal segments of 30° each. The sections of the ecliptic thus produced, however, have different lengths, which is why this is referred to as an 'unequal manner'. The circular boundary lines between the stereographically projected houses, of which the first six are below the horizon and the rest above it, converge at the northern and southern cardinal points; the latter convergence point, however, lies outside the tympanum and is therefore not visible (*Fig. 9.8*).

Regiomontanus's system for defining the houses is referred to as a 'spatial house system',[8] because it departed from the local celestial space, independent of the ecliptic. Dividing the space into equal segments—in this case by dividing the celestial equator—using spherical trigonometry produced the house cusps. As this manner for dividing up the houses, which was also referred to as *methodus rationalis*, was preferred by Regiomontanus in his widespread

7 A comprehensive description of the mathematical aspects and the historical development of the different house systems is given by John North, who outlines seven procedures; North 1986, 46–47. A somewhat chaotic account by Koch and Knappich contains many errors, in particular about the historical interconnections and the attributions of some of the house systems; Koch/Knappich 1959. On the other hand, the two publications by Lorenz 1973 and Holden 1977 are well structured and very useful.

8 Corresponding with the classifications of Holden 1977, 60–61.

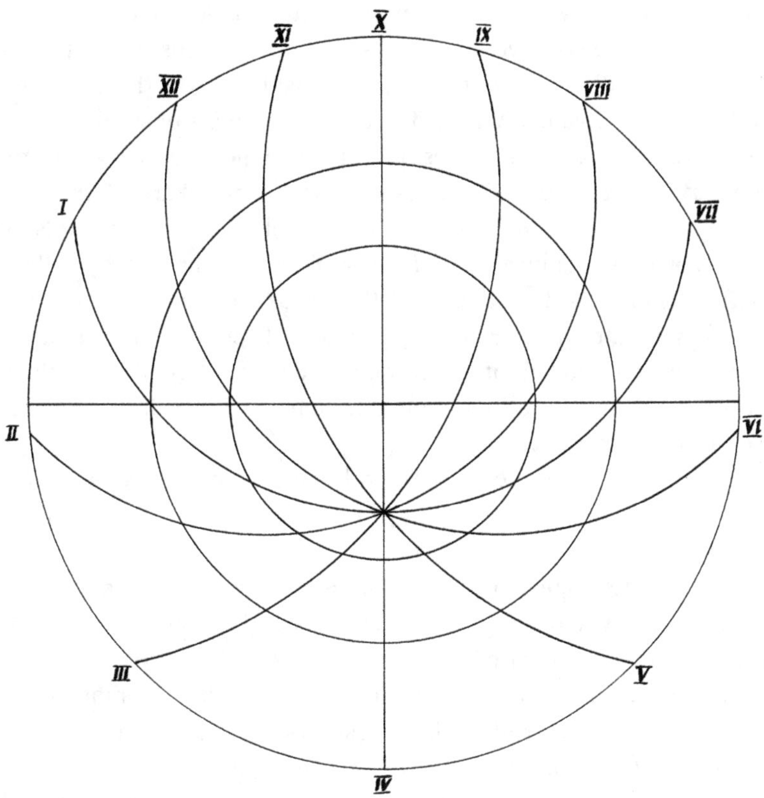

FIGURE 9.8 Tympanum: the twelve astrological houses

Tabulae directionum (1467),[9] it has traditionally been associated with his name. It originated, however, in the Arabic-Islamic culture.[10] Regiomontanus's system was very popular in the sixteenth and the seventeenth centuries, and numerous printed house tables ensured its wide circulation.

The twelve houses were allotted to the different areas of human life and had different significances for astrological forecasts.

First of all there are four 'turning points' or 'corners' (*kentra* or *cardines geniturae*; the Latin *cardo* meaning 'hinge', 'pivot', or 'axis'), namely: the ascendant rising in the east (*horoscope*), that is, the sign of the zodiac observing the hour of birth; the simultaneously descending angle in the west (*descendant*); the

9 Regiomontanus 1490, Problema XIV, fols. B^r–B^v. On the manuscripts and printed editions of this widespread text see Zinner 1968, 339–443.

10 It had already been used by Ibn ʿMuadh al-Jayyani, who lived in Spain around 1070; North 1986, 36–38; Kennedy 1994, 155; Kennedy 1996, 543.

two culmination angles *medium coeli*, or upper culmination, and *imum coeli*, or lower culmination. Decisive for the position of the horoscope is the rising angle of the ecliptic at the birth of a person, which is why the ascendant, but also the *medium coeli* play a more important role in astrological forecasts than the other two houses. The houses II, V, VIII, XI 'following' (*succedens*) the *cardines* are called *epanaphora*. The houses III, VI, IX and XII are referred to as *proanafora* (or *apoclima* or *catafora*), as they are 'falling' (*cadens*) towards the *cardines*. The planets in falling houses are restrained from having good effects or they bring bad ones. On the other hand, the dispositions of the houses rising in the *cardines* are attributed as having favorable effects. Additionally, a distinction is made between four other favorable and four unfavorable positions.

The following compilation lists the twelve houses with their designations and areas of significance:[11]

I *ORTUS* (*ascendant, horoskopus*). The ruler of one's entire life (especially childhood); cornerstone and basis of one's nativity, governs Life, Body, and Soul.

II *PORTA INFERNA*. Gate to the underworld; an unfavorable location, as it is not connected with the ascendant via an aspect.[12] This house determines profits and wealth, but also hope.

III *DEA* is connected with the ascendant via the weak, but favorable sextile aspect and rules over brothers, friends, sisters, and other relatives.

IV *IMUM COELI* (*antimesuranema, hypogeion*). Cardo under the Earth, which signifies concealment and determines the parental home, as well as inherited wealth. It stands in quadrature, that is, in a strong aspect to the ascendant.

V *BONA FORTUNA*. The fifth house is connected with the ascendant via the favorable trine aspect and from it the number of children can be determined.

VI *MALA FORTUNA*. This house falls in front of the western cardo and is unfavorable, as it is not connected with the ascendant via an aspect. Here information is provided on disease, health and servants

VII *OCCASUS* (*descendant, dysis*). This house in opposition to the ascendant signifies women and marriage.

11 Compiled in accordance with Firmicus Maternus 1897/1913, v. I, II.19, 61–65, corresponding to Prugner 1551, 32–33; Uhlemann 1857, 80–87.

12 Ptolemy recognized four geometric relationships between celestial bodies, called aspects, namely opposition, quadrature (or square), trine and sextile. The first two were inharmonious, the last two harmonious; Ptolemy 1980, I.13, 73–75. Conjunction is not mentioned here, although Ptolemy continually treated this configuration as an aspect in *Tetrabiblos*.

VIII *PORTA SUPERNA*. As this house lies above the horizon, it is called the upper gate to the underworld, which stands for life after death, inheritances, as well as death. Because an aspect towards the ascendant is missing, it is unfavorable.

IX *DEUS*. This house, which is opposite to the third house and in a favorable trine to the ascendant, is important for travel, but also for religion and wisdom.

X *MEDIUM COELI* (*mesuranema* or Midheaven) is, according to Firmicus, Manilius[13] and Ptolemy,[14] the most important house, which provides information about places of residence, character, life style, occupation, honors and offices.

XI *BONUS GENIUS*. The personal guardian spirit, which determines benevolence and joy, as well as friends and advisors. The eleventh house is connected with the ascendant via sextile and is therefore a favorable house.

XII *MALUS GENIUS*. This is the locus of the 'evil spirit' with no aspect to the ascendant. It determines enemies, imprisonment, and hardships.

A Medieval mnemonic stanza summarizes which of the twelve houses provided information about which areas of significance: "*Vita lucrum fratres genitor nati valetudo, Uxor mors pietas regnum benefactaque carcer*" (Life, profit, brothers, parent, offspring and health; Wife, death, journeys, power, kindness and prison).

The system of the twelve houses was created around the first century BC and was considered to be a revelation of Hermes Trismegistos. The houses are numbered consecutively counterclockwise in the order in which the signs of the zodiac ascend and, besides mythological arguments, tetrad speculations by Pythagoreans and above all the number twelve for the zodiac played a role.[15] The stars in the lower half of the heavens point to the fate of man determined in future, as all the planets were located there in the hour of the birth of the world.

---

13 Manilius 1977, II, 810ff, 146.

14 Ptolemy places *medium coeli* above the ascendant—which contradicts his assessment of the importance of the ascendant in Book III of *Tetrabiblos*—and considers the celestial regions under the horizon to be unimportant; Ptolemy 1980, III.10, 273. The old teachings of the special importance of the point 90° above the horizon, which originally was not identical with the mathematical middle of the heavens, i.e. the intersection of the ecliptic and the meridian, but simply the highest point on the ecliptic in the heavens, was called '*enenecontameris*'; Firmicus Maternus 1897/1913, v. II, VIII.2, 284 "*De parte nonagesima*". The special importance of the 90° position was also cited by Firmicus as the reason why the signs were counted starting with Aries, as it stood in *medium coeli* when the world was created; *ibid.*, vv. I, III.1, 18, 96.

15 For details see Gundel 1936, 306ff; Gundel 1926, 321–324; Koch/Knappich 1959, 39ff.

With the arrangement of the four ages of man in the spandrels of the astrolabe, a reference is made to teachings of the *kentra* and quadrants,[16] that was developed around the fourth century BC as a predecessor of the system of the twelve houses. The observation that the ascending Sun initially only unfolds weak strength and emits its greatest warmth at the middle of the sky, which then decreases continuously towards sunset, made the transfer to the strength of the impact of the stars on human destiny obvious. In the eastern skies the strength of the stars is initially weak, their maximum strength is attained at the middle of the sky, and it finally expires at the western horizon.[17] The analogous conclusion from equating the microcosm and macrocosm was that the eastern sky with the stars visible there represented childhood and youth, the middle of the sky manhood, and the western sky old age and death. With the recognition of the circular orbits of the stars and the concept of a rotating celestial sphere, this simple analogy led to the concept of a division of the heavens and the ecliptic into four quadrants. In a Hermetic text edited by Wilhelm Gundel,[18] the four equal quadrants[19] of the ecliptic are—taken clockwise, that is, unlike the system of the twelve houses—ascribed to the four ages of man, winds, and points of the compass, and assigned with male and female gender.

The first quadrant extends from the ascendant in the east to the middle of the sky (*medium coeli*); it corresponds to the first age in life, childhood. The second quadrant, from the *medium coeli* to the angle where zodiacal signs set in the west, covers middle age. The third quadrant, down to *imum coeli*, is related to old age. Finally, the last quadrant, between *imum coeli* and the ascendant, represents the last stage of old age up until the hour of death.[20]

In the second book of his didactic poem, Manilius describes the four *cardines*, which he does not conceive as fields of 30° each, but as points: "*tenui partitur limite mundum*".[21] He continues with a description of the four quadrants:

> Nor must you be content with observing the cardinal points; you must note with a retentive mind the spaces between them, which extend over a larger range and possess special powers. The curve which stretches from

16 Gundel 1936, 301ff; Koch/Knappich 1959, 19ff.

17 This view was also shared by Ptolemy 1980, I.24, 117.

18 London, British Library, Cod. Harleianus 3731 (from 1431).

19 The locus of the *medium coeli* and *imum coeli* can simply be derived here by adding 90° to the ascendant. On the astrolabe, however, the two *kentra* are defined by the intersection of the ecliptic with the meridian.

20 Gundel 1936, 50; in the original source fol. 16v.

21 Manilius, *Astronomica*, II, 811; Manilius 1977, 147.

> the orient to the topmost point of the circle claims the earliest age and infant years. The slope which sinks down from the summit of the sky till it reaches the occident succeeds the years of childhood and includes in its province control of tender youth. The portion which appropriates the setting heaven and descends to the bottom of the circle rules the period of adult life, a period tested by incessant change and checkered fortunes. But the part by whose return to the orient heaven's course is done and which with enfeebled strength slowly ascends the backbent arc, this part embraces the final years, life's fading twilight, and palsied age.[22]

Unlike the previously cited text, Manilius, however, attributed the second quadrant to youth and the third to mature adulthood.

Ptolemy also draws parallels between the quadrants and the seasons, regions of the world, winds, and ages of man.[23] Such combinations can also be found with Serapion (a student of Hipparchus), Antiochus and Paulus Alexandrinus.

The doctrine of the *kentra* and quadrants—although in contradiction to the system of twelve houses which evolved from it—continued to be held tenaciously in astrology, as can be seen on the astrolabe on the Cathedral clock. The astrolabe displays the system of twelve houses (*dodekatropos*) on the tympanum and illustrates the old predecessor, the doctrine of the *kentra* and quadrants, in the spandrel paintings. The personification of spring and childhood is consequently shown in the upper left-hand spandrel, followed by middle aged youth, for summer, which passes with the descendant into mature manhood, the autumn of life. With the entrance of winter and old age, *imum coeli* completes the cycle.

Fig. 9.9 summarizes the entire system of lines on the tympanum. With one glance the house boundaries for the latitude of Strasbourg could be read off the astrolabe dial for any point in time.

22 "*Nec contentus eris percepto cardine quoque:/ intervalla etiam memori sunt mente notanda/ per maius dimensa suas reddentia vires./ quidquid ab exortu summum curvatur in orbem,/ aetatem primam nascentisque asserit annos./ quod summo premitur devexum culmine mundi/ donec ad occasus veniat, puerilibus annis/ succedit teneramque regit sub sede iuventam./ quae pars occasus aufert imumque sub orbem/ descendit, regit haec maturae tempora vitae/ perpetua serie varioque exercita cursu./ at, qua perficitur cursus redeunte sub ortum,/ tarda supinatum lassatis viribus arcum/ ascendens, seros demum complectitur annos/ labentemque diem vitae tremulamque senectam*"; Manilius, *Astronomica*, II, 841–855; Manilius 1977, 149.

23 Ptolemy 1980, I.10, 60–65. Ptolemy considered an effect of terrestrial conditions on the influence of the stars, which were thought to be either strengthened or weakened by them.

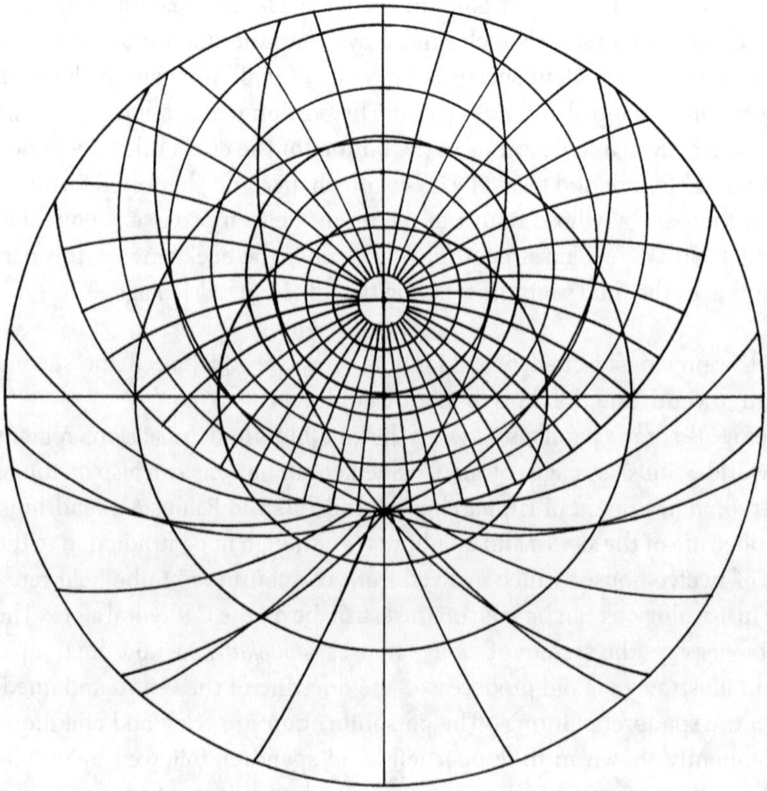

FIGURE 9.9 Complete system of lines and arcs of the tympanum

9.2 Ecliptic Ring and Hands for Planets

The ecliptic ring and the hands for the seven planets and the Moon nodes rotate above the tympanum. Of these, only the hand for the Sun and those for Mercury and Venus, which are attached to it, remain.

All of the hands—as the woodcut by Tobias Stimmer shows—had full figures of the personifications of the planets on one end and the respective symbols on the opposite one (*Fig. 9.10*). A small globe of the Earth sat above the bushings for the hands at the center of the dial.[24]

The ecliptic ring with the multicolored painted zodiacal signs cut out of sheet metal behind it, had a degree scale consisting of fields in alternating colors every 5°, which were repeated on a ring supported by four Y-shaped

24 Dasypodius 1580a, 46.

FIGURE 9.10 The Astrolabe of the Strasbourg Cathedral Clock. Detail of Stimmer's woodcut, 1574

arms. The actual dial, divided into 2 × 12 hours, which no longer exists, was on the outside and could be read by means of a hand that extended from the Sun pointer.

Finally a long dragon can be recognized, which holds the ball of the Moon in its mouth. This hand pointed to the position of the ascending and descending Moon nodes.

The Moon's orbit forms an angle of about 5° with the ecliptic and the intersections are referred to as 'nodes'. The Moon passes through the 'ascending node' (☊) when its orbit crosses the ecliptic moving north, and the 'descending node' (☋) when crossing it moving south. The ascending and descending nodes are joined by a nodal line. They are not fixed, but move from east to west, opposite the direction of the zodiac sequence, in about 18.5 years, which corresponds to a change of about 1.5° every month. In Indian and Arabian astrology, the nodes—because of their varying positions—were treated like planets and included among them.[25]

25 Hartner 1938, 131–134.

Knowing where the node positions are is important for astronomical and astrological purposes because solar and lunar eclipses, which were thought to bring bad luck, occurred during conjunctions or oppositions of the Moon with the Sun—that is, at new moons or full moons—in or near a node. If the hands for the Sun and Moon both stood above the dragon's hand, a solar eclipse was expected; if the Sun lay above the dragon's head and the Moon over the dragon's tail at the same time, or vice versa, a lunar eclipse would occur.

The symbol of a snake- or dragon-like creature that periodically swallowed the Sun or Moon originated in the Middle East and was transmitted to Europe from India by the Arabs.[26] The dragon appears frequently in connection with descriptions of eclipses, both in scholarly works and in calendar illustrations from the fifteenth and sixteenth centuries (*Fig. 9.11*).[27] The terms 'dragon's head' for the ascending and 'dragon's tail' for the descending node have continued to be used up to the present, and the period of time that the Moon needs to return to the original node is referred to as the 'draconitic month'.

The hands for the planets complete the possibilities for indicating astrological information with the dial, with which not only the ascendant, the house cusps, and the distribution of the twelve houses on the ecliptic line, but at the same time the positions of the planets relative to each other (the aspects) and their locations on the ecliptic could be read off.[28]

9.2.1 *Division of the Ecliptic Ring*

In ancient astronomy the ecliptic was used as a reference system and a position was determined using the latitude and longitude on the ecliptic, such that the zero point for longitude was the vernal equinox. The use of this system was suggested by the fact that dealing with the motion of the planets, the paths of which always approached the ecliptic, was seen as the main task. If, however, the time of day was to be determined, a system based on the celestial equator was appropriate—as the daily motion of the celestial sphere rotates around the axis through the celestial poles. Instead of ecliptic longitudes and latitudes the terms 'right ascension', which again used the vernal equinox as the zero

26 The appearance of the Apocalyptic woman clothed with the Sun on the crescent Moon with the dragon (Rev. 12:1–3) can probably be seen in the context of such perceptions, and the arrangement of the two tables on the Moon phase dial is clearly not accidental.

27 Examples in Chojecka 1967, 77–81. On the iconography of the dragon see Kuehn 2011.

28 Describing the details of astrological predictions from celestial configurations is beyond the scope of this book. Concise overviews can be found in Uhlemann 1857, 62ff, and Drechsler 1855. Excellent more recent presentations are given by Eade 1984 and Avelar/Ribeiro 2010.

FIGURE 9.11 The dragon between the ecliptic and the Moon's orbit. Woodcut from *De Magnis Coniunctionibus* by Abu Ma'shar in the Latin translation by Joannes Hispalensis, Venice, 1515

point, and 'declination' were used. Ptolemy showed how to represent this system graphically on a plane in his *Planisphaerium*.[29] Determining day and night times was paramount when using the astrolabe. Sidereal or celestial time (the hour angle of the vernal equinox, which characterizes the position of a celestial body above the horizon) is obtained by adding the right ascension of a star to its hour angle. This also applies for the right ascension of the Sun, but not for its ecliptic longitude. Representing the path of the Sun correctly would imply a

29 Drecker 1927, 262–265. The original Greek text of Ptolemy's *Planisphaerium* has been lost, and only a Latin translation made by Hermannus Dalmata in 1143 from the Arabic version by Maslama ibn Aḥmad al-Majrīṭī is extant. Drecker's German translation was based on the critical edition of the Latin text by Johan Ludvig Heiberg, *Claudii Ptolemaei opera quae exstant omnia*, Leipzig 1907, v. II, 225–259. The *Planisphaerium* was first printed, together with Ptolemy's *Geography*, in 1507–1508 in Rome; Houzeau/Lancaster 1882/89, v. I.1, 630, no. 3070. Further editions were published in Nuremberg (1531), Basel (1536; *ibid.*, v. I.1, 393, no. 759), Venice (1558; *ibid.*, v. I.1, 395, no. 769, with a commentary by Federico Commandino). The extant manuscripts are documented in Thomson 1978, 48, n. 23.

division of the ecliptic ring into sections of 30° right ascension each. This could be executed by using a table that indicated the appropriate angles along the celestial equator starting from the vernal equinox.[30] If such a table was not available, the values could also be determined graphically, as Llobet of Barcelona demonstrated in his treatise on astrolabes in the tenth century.[31] The accuracy of this procedure is more than sufficient for the graphical construction. But in actual practice erroneous methods for dividing the ecliptic were employed very often: in many early treatises on astrolabes—such as that of Hermannus Contractus (1013–1054)—twelve lines were simply drawn from the projection center to divide up the celestial equator, and the resulting ecliptic segments were accepted to be projections of the zodiac signs.[32] This is also the case for the rete of the Strasbourg Cathedral clock. However, the differences between the right ascension and ecliptic latitudes never exceed an order of magnitude of 2° 30′ (*Fig. 9.12*).

The Strasbourg astrolabe was designed as a clock, and the hands for the Sun, Moon and planets were essentially stereographic projections of the circles with the same right ascension on which the respective celestial bodies moved (strictly speaking the pivot point for the hands should have been the ecliptic pole, but accomplishing that mechanically would have been extremely complicated).

Thus the introduction of right ascensions instead of ecliptic longitudes does not simply represent an anomaly in the design of the astrolabe, rather it was a necessity for determining time. The astrolabe as well as the clocks that used it as a model prepared for the use of right ascensions and declinations in astronomy. This was further promoted by Tycho Brahe's attempts to measure meridian transits with extreme precision,[33] and finally generally superseded the old ecliptic coordinates.

---

30 Johannes Stoeffler presents such a table of right ascensions for the division of the ecliptic in his *Elucidatio fabricae ususque astrolabii*, published first in 1513; Gunella/Lamprey 2007, 37–38.

31 Poulle 1955, 122.

32 Drecker 1931, 207–208; Thomson 1978, 64.

33 However, Tycho's attempts were hampered by the insufficient rates of the available clocks in the second half of the sixteenth century; see Ræder /Strömgren/Strömgren 1946, 29–31.

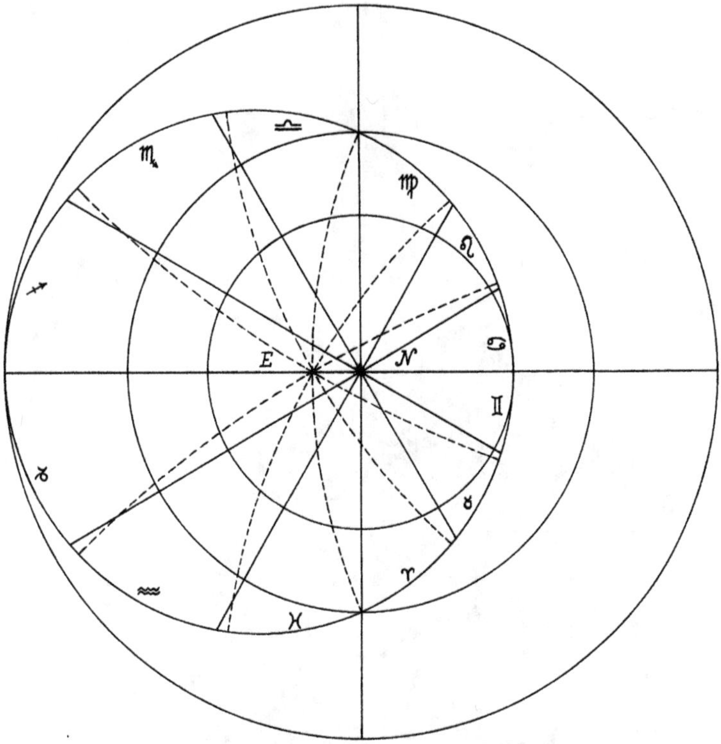

FIGURE 9.12 Divisions of the ecliptic ring. Solid lines: divisions of the celestial equator in 30° segments in right ascension (N = celestial pole); Broken lines: ecliptic boundaries of the signs of the zodiac, which converge at the ecliptic pole E

9.3 Gear Train for the Astrolabe

The gear train of the astrolabe (*Fig. 9.13*) was first analyzed in an article by Henri Bach, which included a very precise schematic diagram.[34] Publications by Lehr and Poulle[35] were based on Bach's specifications. Bach initially used Schwilgué's description, in which the number of teeth of the transmission gear for the hand for Saturn was given as 28, and that for Mars as 128.[36] However, in a later version of his diagram, Bach changed the figures to 29 and 129 teeth

34 Bach 1979, 34–36.
35 Lehr 1981, 348–352; Poulle 1983.
36 AMS, 3 OND 589, 12–13.

FIGURE 9.13 Gearing of the astrolabe

respectively.[37] Poulle on the other hand suggested 130 or 136 teeth for the Mars wheel without taking the gear train apart again.[38]

To resolve these contradictions and determine the structure of the mechanism beyond doubt, I completely dismantled, measured, and photographed the gear train together with Matthias Schramm in October 1989.[39] The dimensions and teeth numbers thus obtained are listed in *Table 9.1*.

The results of this investigation were that:

1. The structure and numbers of teeth given in Bach's later diagram of the gear train were correct;
2. No changes to the gear train—such as might have occurred during later repairs—could be identified;
3. Another arrangement in which gears were exchanged was impossible, due to the design (*Figs. 9.14–9.17*).

In compliance with the graduated lengths and diameters of the tubes for the planets' hands, the sequence in which the wheels can be assembled was fixed and the three transmission wheels for the motion of Jupiter, Saturn, and the

FIGURE 9.14 Set of gears for the planet hands and ecliptic ring

37 EA, v. III, 1301, s.v. "Cathédrale" by Henri Bach.

38 Poulle 1983, 34 and 42.

39 This was only possible thanks to the exceptional kindness of Dr. Jean-Daniel Ludmann, the director of the Musée des Arts Décoratifs at that time. The two of us were also actively supported in this endeavor by the conservator Alain Schmitt, M.A.

TABLE 9.1 Gear trains of the astrolabe of the Strasbourg Cathedral clock

Designation	Number of teeth	Diameter (mm)	Thickness (mm)	Length of the tube (mm)
Moon	118	505	7	353
Sun	114	503	7	300
Driving wheel for the planets' movements	30	186	5	–
Pinion on the driving wheel	4	20.5	–	–
Mars	129	319	5.5	235
Jupiter	72	249	6	206
Saturn	60	249	6	177
Lunar Nodes	74	245	5	148
Rete	73	309	6	120
Gear ring	78	245	7	–
Transmission unit I				
Wheel 1	33	120	8	–
Wheel 2	29	125	8	–
Wheel 3	39	127	12.5	–
Transmission unit II				
Wheel	32	112	6	–
Pinion 1	7	23.4	–	–
Pinion 2	8	42	–	–
Wheel with helical gearing	35	108	–	–
Main driving unit				
Pinion	8	26.5	–	–
Wheel	48	247	9	–
Lantern pinion	38	167	–	–

lunar nodes were fitted exactly to the tapered square shaft. They cannot be exchanged.

With the original data thus determined, the wheel trains will be recalculated in the following. In order to understand the principle of the mechanism designed by Dasypodius, it is expedient to start with the concepts of the mean sidereal and mean synodic movements. Here we can limit our considerations to the three superior planets, as the hands for Mercury and Venus rotated with the hand for the Sun. The clock did not include any mechanism that would

FIGURE 9.15 Set of assembled gears for the planet hands and ecliptic ring

FIGURE 9.16 Gearing of the astrolabe; three driving wheels

FIGURE 9.17 Gearing of the astrolabe; three driving wheels, assembled

have simulated an oscillation of the two inferior planets around the Sun corresponding to their elongations.[40] The clockmaker responsible for servicing the clock, however, was able to adjust their positions manually. In fact, the two hands were not riveted to the Sun hand, but had a loose friction fit allowing for repositioning.

40 The two inferior planets could not distance themselves arbitrarily from the apparent position of the Sun. The angle of separation is called the planet's elongation. The maximum elongation for Venus is 48° and for Mercury 28°. As these two planets are always close to the Sun, they can only be observed above the western horizon shortly after sundown (eastern elongation) or above the eastern horizon shortly before sunrise (western elongation). It would have been mechanically possible to produce oscillation among the inferior planets using a gear with crankpins which engaged with a notched rod mounted on the hand for the Sun. The astronomical clock in the cathedral in Münster has such a mechanism for the Venus hand (*Fig. 9.18*); however, it was not installed until 1929–1932; Werland 1926, 65; Wieschebrink 1983, 78; Peter 2016, 155–157.

FIGURE 9.18 Mechanism for the hand of Venus in the astronomical clock of Münster Cathedral

For representing the celestial movements using gear trains, the simplest possible periodicities are necessary for reproducing the relative motion. The methods of ancient astronomy met this demand in as much as they expressed the mean sidereal and synodic movements as simple whole-number ratios. That even applies for Ptolemy, who took up older periods of this type and only improved them with small correction elements. The designer of an astronomical mechanism would try to select those periodic ratios that satisfied his purposes and his demands for accuracy, and permitted the use of the smallest possible numbers.

There is a simple relationship between the mean sidereal and synodic velocities; the synodic speed corresponds to the velocity of the Sun relative to the planet. The relationship between the two velocities is $1/T_{\text{sid}} + 1/T_{\text{syn}} = 1/a$, where T_{sid} and T_{syn} are the mean sidereal and synodic periods for a planet's orbit, and a is one year, that is, the period for the orbit of the Sun.

This relationship has an important consequence. If in the time interval P a whole number of each of two of these periods is completed, then the third one will also have been completed a whole number of times, as $P/T_{sid} + P/T_{syn} = P/a$.

That means, therefore, that the search for simple ratios for the movements of the planets in which only whole number multiples of the sidereal *Tsid* and synodic *Tsyn* periods occur, as was needed by Dasypodius, could be limited to whole numbers of years. This was a key simplification for the design of such an astronomical mechanism.

When one analyzes the relationships between the movements in a mechanism such as the one designed by Dasypodius, it is vital that the mathematical calculations take these conditions into account. In particular, the reproduction of the orbital times and the velocities must be carried out using such simple numerical ratios. It is of great importance that the calculations are only carried out with simple whole numbers, so that the designer's goal becomes clear. Calculations with multi-digit decimal fractions are liable to lead the analyzer astray.[41]

The mean time actually necessary for passing through the entire zodiac consists of

1. 687 days—a little more than $1^7/_8$ years—for Mars,
2. 4,333 days—a little less than 12 years—for Jupiter, and
3. 10,759 days—a little less than 30 years—for Saturn.

41 Emmanuel Poulle's handling of the mechanisms of the astrolabe is a striking example of the type of false conclusions that can be reached if these demands are not fulfilled sufficiently; Poulle 1980, v. 1, 693–699; Poulle 1983. He took the length of the tropical year from the Alfonsine Tables, which is evidently a methodological mistake, as the rotation of the Sun wheel is to be compared empirically with the actual (or mean) orbital time of the Sun using sundials (see *Chapter 10*), and the Alfonsine figure is of no consequence here. Similar mistakes can be found in the rest of the cases that Poulle dealt with. It is clear that Dasypodius used the simplest ratios and that no conclusions can be drawn as to an interdependence between his figures and astronomical tables.

Those are the mean sidereal periods, that is, those relative to the fixed stars. The corresponding apparent period for the Sun is one year. From such periods, one can calculate the mean velocities of the planets along the zodiac during one year. As the Sun in its apparent motion completes one revolution in one year, its velocity is $V = 1$ per year.

The approximated sidereal velocities V_M, V_J, V_S of Mars, Jupiter and Saturn can then be expressed simply using V:

$$V_M = \frac{8}{15} V; V_J = \frac{1}{12} V; V_S = \frac{1}{30} V.$$

The visibility of a planet depends on its position relative to that of the Sun. Each planet is repeatedly overtaken by the Sun and disappears in its rays; the planet is then in conjunction with the Sun. When and how often this occurs depends on the velocity of the Sun relative to the planet. This relative motion and relative velocity are referred to as 'synodic' (from the Greek σύνοδος, conjunction). Calculations with angular velocities can be carried out in the same manner as with other velocities, and the relative velocities are equal to the difference between the velocities of the bodies involved.

The synodic velocities to be used for Mars, Jupiter and Saturn are therefore:

$$\nu_M = V_M - V = -\frac{7}{15} V; \nu_J = V_J - V = -\frac{11}{12} V; \nu_S = V_S - V = -\frac{29}{20} V.$$

Fig. 9.19 illustrates this relationship. The point E is the Earth, around which the Sun S and a superior planet P orbit. The vernal equinox (♈) determines the direction of the starting point from which the time t is measured. The arcs through which the bodies move are then $V \times t$ and $V_P \times t$; and $\nu_P \times t = (V_P - V)t$, where V_P and ν_P are the sidereal and the synodic velocities of the planet.

Of course all of this can be interpreted in the heliocentric sense of Copernicus also. The geocentric model has the advantage that the lunar nodes (LN), considered to be quasi-planets by astrologers, can also be dealt with in this scheme. They move in about 18.5 years (sidereal) in the opposite direction to the sequence of the signs of the zodiac:

$$V_{LN} = -\frac{2}{37} V; \nu_{LN} = V_{LN} - V = -\frac{39}{37} V,$$

where the negative sign in the first equation indicates the reverse direction of the motion of the nodes.

Finally the synodic velocity ν_F of the fixed stars may be introduced. Because $V_F = 0$, it follows that

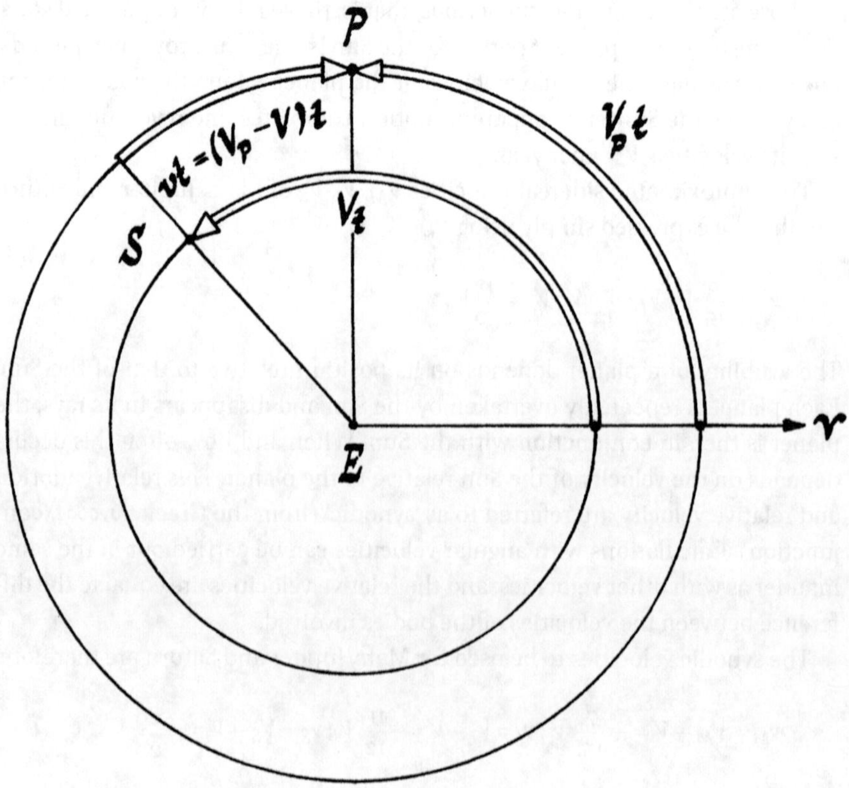

FIGURE 9.19 Sidereal and synodic velocities of a planet

$$v_F = V_F - V = -V.$$

In a system that revolves with the velocity of the Sun, the three superior planets, the lunar nodes, and the fixed stars revolve with the synodic velocities v_M, v_J, v_S, v_{LN}, and v_F. Dasypodius used these synodic velocities as the basis for his gearing design.

The wheels and pinions for the synodic movements were mounted on the Sun wheel S and driven by it. The system of wheels and pinions mounted on S (*Fig. 9.20*) always indicated the mean synodic positions of the planets, lunar nodes, and fixed stars relative to the Sun S to which they were connected. Which rotational velocity the Sun wheel was to receive depends on the pursued goal; Dasypodius designed his mechanism to be a clock and had the hand of the Sun wheel (the mean Sun) rotate once every 24 hours.

For the analysis of Dasypodius's solution, the shafts of the mechanism are listed with the number of teeth on the wheels attached to the shafts, and the Sun wheel S is considered to be a shaft:

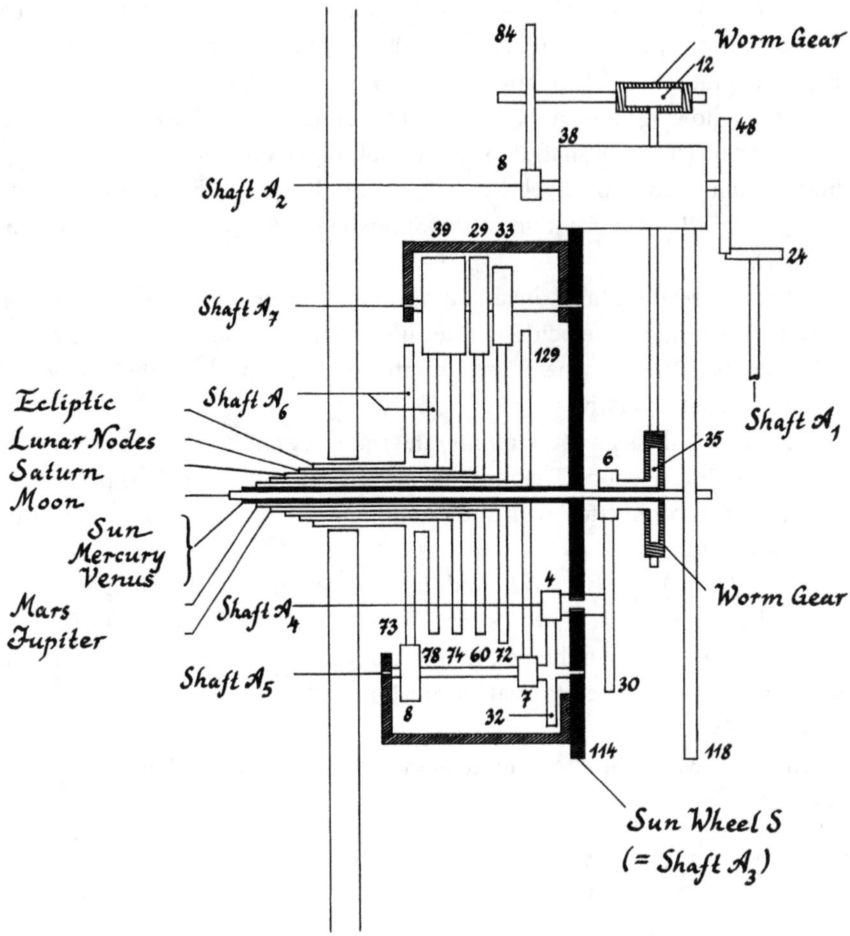

FIGURE 9.20 Gearing of the astrolabe

Shaft	No. of teeth	Rotational velocity
A_1	24	V_1
A_2	48; 38; 8	V_2
$A_3 = S$	114	V_3
A_4	30; 4	V_4
A_5	32; 7; 8	V_5
A_6	73; 78	V_6
A_7	39; 29; 33	V_7

If the rotational velocities ν, ν' of the shafts which carry meshing wheels with numbers of teeth N and N' respectively are compared, and positive and negative rotations are distinguished, then $\nu' = -(N / N')\,\nu$.

In the following, large rotational velocities are expressed using D, one rotation per day. The drive shaft A_1 of the wheel train completes 6 rotations in 24 hours. Then its rotational velocity is $V_1 = -6D$. It follows that $V_2 = -(24 / 48)\,V_1 = 3D$, and that the rotation of the Sun wheel S (or A_3) is $V_3 = -(38 / 114)\,V_2 = -D$.

S thus correctly rotates once in a day; its rotational direction was chosen such that the sidereal velocities of the Sun and planets were positive. As all the other shafts with their wheels are mounted on the Sun wheel, their velocities can be calculated relative to S: $V_4 = -(6 / 30)\,V_3 = -(1 / 5)\,D$.

Concerning the sign (here minus) the transmission by planetary gears has to be taken into account. The minimal movement of the 6-trundle pinion can be ignored in connection with the round figures employed here, but will be dealt with later.

$V_5 = -(4 / 32)\,V_4 = (1 / 40)\,D$, and the synodic velocity ν_F of the wheel that drives the ecliptic ring is $V_6 = -(8 / 73)\,V_5 = -(8 / 73)\,(1 / 40)\,D = -1 / 365\,D = -V$.

The second wheel on the hollow arbor of the ecliptic ring has 78 teeth and drives wheels connected to shaft A_7 for Jupiter, Saturn, and the lunar nodes. $V_7 = -(78 / 39)\,V_6 = 2\,V$.

The Mars wheel, on the other hand, is driven directly by shaft A_5. Its rotation is:

$$\nu_M = -\frac{7}{129}\,V_5 = -\frac{7}{129} \times \frac{1}{40}\,D = -\frac{7}{129} \times \frac{365}{40}\,V = -\frac{511}{1032}\,V,$$

a value decidedly too large, for 511 / 1032 = 0.495. Compare the approximation 7 / 15 = 0.467; while (687 – 365.25) / 687 = 0.468.

For Jupiter we get:

$$\nu_J = -\frac{33}{72}\,V_7 = -\frac{11}{12}\,V,$$

for Saturn:

$$\nu_S = -\frac{29}{60}\,V_7 = -\frac{29}{30}\,V,$$

and for the lunar nodes:

$$\nu_{LN} = -\frac{39}{74}\,V_7 = -\frac{39}{37}\,V,$$

corresponding to the simple values mentioned above for the synodic velocities. In principle, the mechanism reflects the astronomical ratios correctly, except for Mars. For a pinion with 7 teeth as is on shaft A_5, a wheel with 136 teeth would result in the best approximation.[42] Whether Dasypodius miscalculated the rotation of the 7-tooth pinion—as a consequence of the motion of the Sun wheel—, or whether the Habrecht brothers made a mistake in executing the mechanism is open to question.

Dasypodius displayed the motion of the Moon independent of the Sun wheel. The rotation of the shaft A_2 with the velocity $V_2 = 3D$ is transmitted to the wheel for the Moon with the result:

$$V_L = -\frac{38}{118} \times 3D = -\frac{57}{59} D.$$

If this is compared with the velocity $-D$ of the Sun wheel, the synodic velocity is:

$$v_L = -\frac{57}{59} D + D = -\frac{2}{59} D;$$

that is, the Moon wheel has to rotate for 29.5 days in order to catch up with the Sun, thus leading to the common approximation of the synodic month, which elapses between two successive identical phases of the Moon.

The ecliptic ring, which rotates in sidereal time, has to perform an additional rotation compared with the hand for the Sun. This was created—as in the case of the celestial globe—using the two planetary gear pairs 30/4 and 32/8. The fine adjustment was produced using the slowly rotating 6-tooth central pinion, which was driven by two worms acting on helical gears of 12 and 35 teeth. Of these components only the 35-tooth helical gear on the central pinion and a bearing bracket is extant (*Fig. 9.21*).[43]

The movements relative to the Sun wheel depend of course on the velocity of the 6-tooth pinion relative to the 30-tooth wheel, with which it engages. If this relative angular velocity is ω, then the (relative) velocity ω_e of the wheel driving the ecliptic ring is $\omega_e = (6 / 30)\, \omega \times (4 / 32) \times (8 / 73) = 1/365\, \omega$.

If ω corresponds to one rotation a day, then ω_e results in the completion of one rotation in 365 days. If, however, the relative velocity of one rotation a

42 The number 122 given as the number of teeth in the second edition was a regrettable mistake; Oestmann 1993/2000, 119; and unfortunately repeated in the introduction to Dasypodius 2008, 9.

43 See also the descriptions by Schwilgué (1836), 12 and Maybaum (1788), fol. 3^r, both in AMS, 3 OND 589.

FIGURE 9.21 Helical gear with support for the worm gear

day is decreased to (1 – 1 / 1470) rotations a day by means of the additional two worm gears, then ω_e would lead to one rotation in 365 × (1 + 1 / 1469) days (as for the celestial sphere), so that for every 365.2485 days one additional rotation of the ecliptic ring would occur, that is, a total of 366.2485 rotations.[44]

The gear trains displayed the mean synodic motion of the planets and lunar nodes, but do not allow for the effects of eccentricities and retrograde motion, so that the respective ecliptic position of each body was only indicated imprecisely. For the superior planets, knowledgeable users had the possibility of at least going beyond the simple sidereal motion and taking into consideration the oscillation that depended on the synodic period, which resulted in the typical loop in the motion of the planets. This in principle represented nothing more than the displacement corresponding to the motion of the Earth. Ptolemy had taken this into consideration by taking the point rotating around the eccentric deferent with mean sidereal motion as the center of a corrective circle, called the epicycle. He was able to determine that—but not explain why—the end of the epicycle radius of the respective planet always points toward the mean Sun, that is, parallel to the radius from the Earth to the Sun. The epicycle radii for Mars, Jupiter and Saturn, when rounded off, respectively amounted to 2/3, 1/5 and 1/10 of the deferent radius, or, in our case, the length of the hand rotating in front of the tympanum. So it was easy to estimate where the planet was located, and whoever wanted to know the position more exactly could easily use the data from the astronomical dial to execute a construction on paper (*Fig. 9.22*). Certain details in Ptolemy's planetary theory—such as those anomalies explained by introducing the eccentric deferent and the equant—could not, however, be represented in this manner.

The contractual demand that the clock had to agree exactly with the "Mathematical Tables" (*i.e.* tables for the planetary motions),[45] was not met by the relatively simple structure that Dasypodius chose compared with the astronomical clocks of Giovanni de'Dondi and Richard of Wallingford. However, Dasypodius did not want his clock to indicate more than the mean sidereal movements,[46] and he arrived at a felicitous and effective construction. With

44 In order to understand this, one does not need Poulle's cumbersome mathematical formalism, the use of which is inappropriate and obscures more than it clarifies. This is illustrated by the awkward manner in which he calculated the ratio of 0.997270 between the sidereal day and the mean solar day; Poulle 1983, 41–42. His conclusion that there was a connection with the Alfonsine Tables was entirely misguided.

45 "[…] *mit den Tabulis Mathematicis just uber einkhommen*": AMS, 1 OND 15, fol. 5^r.

46 In this respect Poulle's derogatory assessment of Dasypodius's construction seems somewhat inappropriate: "Dondi's tradition has not been lost with him, and the following centuries have kept the taste of such prestigious achievements. However, it must be noted

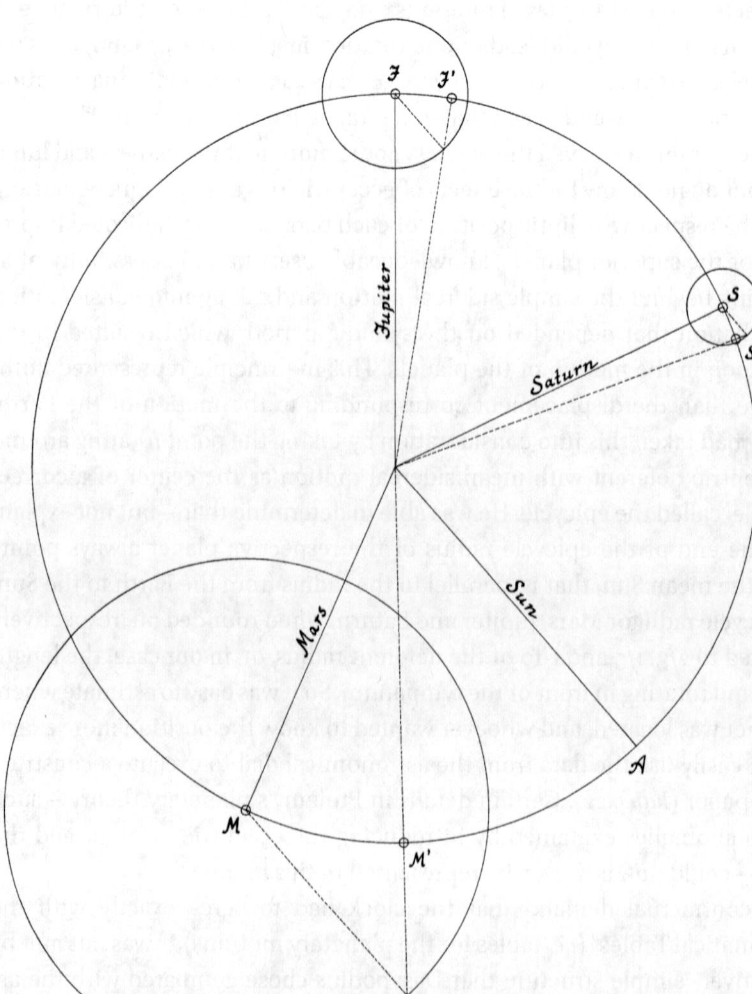

FIGURE 9.22 Readings of the planet positions on the ecliptic ring. *A* indicates the mean position of the Sun; *M* the mean position of Mars, *J* that of Jupiter, and *S* that of Saturn. The end of the epicycle radius always points to the mean Sun; for Mars it is approximately 2/3, for Jupiter 1/5, and for Saturn 1/10 of the deferent radius, or in the case of the clock, of the pointer rotating in front of the tympanum. A rough estimate of the actual position of the respective planet can thus be determined; in the illustration Mars, for example, would be located at *M'*.

TABLE 9.2 Synodic periods and discrepancies in the Strasbourg Cathedral clock

	Synodic period (days)	Synodic period (Strasbourg)	Deviation per year	Error increasing to 360° after
Moon	29.53	29.5	4.68°	77 years
Sun	365.2425	365.2485	0.006°	61181 years
Mars	779.94	737.6447	9.67°	37 years
Jupiter	398.88	398.4529	0.35°	1019 years
Saturn	378.09	377.8432	0.23°	1585 years
Lunar Nodes	346.62	346.518	0.11°	3224 years

the exception of the totally mismatched gearing for the Mars movement the representation of the mean synodic positions of the other planets was satisfying (*Table 9.2*).[47]

There is a constraint however, because the Moon's position advanced about 5° per year and, due to the fast lunar motion of about 12.3 orbits per year, the error quickly added up to a considerable deviation. Thus deducing eclipses from the position of the Moon and the dragon hands was only possible for a short period.

The limitations for indicating the planetary movements seems to have given rise to some criticism. For example, Jacob Cuno (c. 1526–1583/84)[48] planned to

that not all planetary clocks were, ipso facto, epicyclic mechanisms; the authors of most of them, in fact, were satisfied with a rather appalling approximation, with a representation of the movement of the epicycle's centre around the equant's centre, confused for the purposes of the cause both with the centre of the deferent and with the centre of the world. This is certain for the clock in Strasbourg Cathedral; what remains of its mechanism at the City Museum is far too incomplete to be validly questioned, but we have Dasypodius' own admission on this subject: in the short description he published in 1580 of the work that was to establish his fame, he acknowledges that he had represented, in terms of planetary movements, only the mean movements; the Strasbourg clock therefore reduced Ptolemaic astronomy to a system without eccentricities and without epicycles! [...] The memory remains of a certain number of tower clocks whose astronomical dial included the presence of planetary hands: Ungerer's classic book makes it possible to draw up a list of about ten such clocks. They probably belonged to the same type as that of Dasypodius, and therefore offered only a caricature of planetary theory"; Poulle 1974, 60–61.

47 Scheurenbrand/Schwarz/Hahn 2005, 114, here reproduced with rounded figures and a corrected value for Saturn in column IV.

48 Cuno (Küne) was born in Döbeln, Saxony. He studied from 1544–1554 in Leipzig and Wittenberg and was appointed as court mathematician by elector Joachim II (or Hektor

build an artistic clock modeled after, but superior to, the Strasbourg one. In a letter to Albrecht V, Duke of Bavaria dated 1 April 1579, Cuno wrote:

> Now the clockwork in Strasbourg has recently stood among many in great renown, having received great praise both in writing and vocal acclaim, to which I in its place would agree and no one wishes to deprive it of any of its just praise. However, persons knowledgeable about art will upon seeing the Strasbourg clock vouch for the fact and establish that it has more illusion than art, and after a poor fashion. This automaton of mine, however, is based on and built according to the correct astronomical art, and includes much which cannot be found in the Strasbourg work or anywhere else.[49]

Johannes Kepler mentioned the Strasbourg clock and Cuno's project in a letter to Michael Mästlin in connection with his plans for a Copernican

---

of Brandenburg) in 1555. From about 1562 Cuno lived in Frankfurt on the Oder; see the article "Cuno d. Ä., Jacob", in: Klaus-Dieter Herbst, Biobibliographisches Handbuch der Kalendermacher von 1550 bis 1750, www.presseforschung.uni-bremen.de/dokuwiki/doku.php?id=cuno_d._ae._jacob (accessed 12 August 2019).

49 "*Nun ist das Uhrwerck zu Strasburg unlangst wol auch bei vielen in grossem rhum gestanden und beide in schrifften und gemeinem geschrey sehr gelobt worden, welchs ich an seinen ort gestelt und niemand an seinem gebürenden lobe etwas entzogen haben will. Es wird aber bei verstendigen der kunst der augenschein bezeugen und darthuen, daß die Strasburger Uhr an ihr selbst mehr scheinlich denn khünstlich und nur schlecht und recht. Dieses mein Automaton aber auf rechte astronomische kunst gegründet und erbauet, auch viel stück in sich helt, welche weder im Strasburgischen wercke noch sonst irgens gefunden werden*" (Hartig 1925, 247). Around 1580 Cuno published a description of the clock in German and one in Latin: *Kurtze Beschreibung eines künstlichen Newen Astronomischen Uhrwercks*, n.p., n.d.; *Brevis descriptio artificiosi, novi, et Astronomici automati horologici, cuius simile antehac non exstitit*, Frankfurt am Oder, n.d.; see Zinner 1941, 408–409, nos. 5215–5217, and for a brief summary of content Baillie 1931, 244–245. In 1583 Cuno was staying with his son in Augsburg and offered to make several "astronomical automatons" for Emperor Rudolf II, and for an elaborate clockwork displaying the celestial movements (*automatum motuum coelestium*) he was paid 400 guldens in advance by the Bohemian Court Treasury. Shortly after that Cuno died before being able to complete the work, and his son offered to carry out the contract. On 22 May 1584 Rudolf II set half a year as the deadline for the completion of the clock; however, five months later Cuno's heirs offered the emperor two other clocks in return for the advance payment, which were accepted. It is hardly likely that the clock was ever completed. The original records were published in the *Jahrbuch der kunsthistorischen Sammlungen des Allerhöchsten Kaiserhauses* 5, 1887, v. II, CLXXIII–CLXXIV, nos. 4566–4569, 4571; *Ibid.* 12, 1891, v. II, LXIV–LXVI, nos. 8218, 8220–8222, 8224, LXVIII, 8231.

planetarium,[50] in which he makes some critical observations about the clocks with their astronomical indications and automata:

> The to no lesser extent widely famous Strasbourg work in the Cathedral, as was reported to me, is not supposed to indicate more than *motus medios, et quidem, ab invicem separatos*, which otherwise is adorned with all sorts of fantastic *parergis* to fill the eyes of the common man. [...] And I must assume that the majority of such clocks [...] only have *motus medios* alone, as the common *astrolabia*, and those prepared in abundance in Nuremberg and Augsburg indicate *motum Solis, Lunae, et Capitis medium*. [...] I wish to believe that so far the commotion about a clockwork is much greater than the clockwork in itself.[51]

From the perspective of astronomers, automata may have appeared to be only *parerga* for uneducated observers. In the Strasbourg Cathedral clock—as will be shown below—they were incorporated into a complete program.

---

50 On this see Prager 1973, 385–392.

51 "*Wie nicht weniger auch das weitberhuembte strassburgerische Werckh im Münster, als ich berichtet werde, nicht mehr dan motus medios, et quidem, ab invicem separatos, anzeigen solle, das aber mit sonst allerlay wunderlichen parergis dem gmeinen man die augen zu füllen, gezieret ist.* [...] *Und mueß ich vermueten, das der mehrertheil sollicher Uhren* [...] *nur allein motus medios haben, wie dan die gewohnliche, und zu Nürenberg und Augspurg haufenweiß zugerichtete Astrolabia, motum Solis, Lunae, et Capitis medium zeigen.* [...] *Ich will auch glauben, das je bißweilen das gschray von eim Uhrwerck vil grösser sey dan das Uhrwerck an im selber*"; KGW, V. XIII, 221–222.

CHAPTER 10

Calendar Disk and Sundials

10.1 The Calendar Disk

The wooden calendar disk (diameter ca. 2.9 m, Musée des Arts Décoratifs, Strasbourg) consists of three parts, of which the plate at the center with the map has been already discussed in *Chapter 5* (*Fig. 10.1*).

On the outside, the names of the twelve months, 366 days of the months, and letters for the Sundays are listed. Following that are the names of the saints assigned to the days.[1]

On 23 February the outer ring had to be moved forward by hand in non-leap years in order to jump over the intercalary day. The arrow on the statue of Apollo to the left acted as the pointer.

After one revolution of the outer ring, which was continuously moved by the clockwork, the inner plate (diameter ca. 2.34 m), which was divided into 100 radial segments, was moved forward by one segment, such that the information on that plate could be read through a slit in a table attached to the top of the central plate. In its present state, the years from 1670 to 1769 are listed on the plate; it was repainted shortly before the end of the hundred-year period for which Dasypodius made his calculations.[2]

The original data for 1573 to 1673 are contained in tables in the *Warhafftigen Außlegung*.[3] The following calendrical data was included on the inner plate, from the outside to the center:

I. The years from 1573 to 1673;
II. The years calculated from the creation of the world, from 5535 to 5635;
III. The day of the vernal equinox, 9 or 10 March;
IV. The hours of this day;
V. The minutes of the hours of this day;
VI. The date for the Sunday fifty days before Easter (*Estomihi* or *Quinquagesima*);

1 The Name "*Lutherus*" was listed under 18 February, the date of Luther's death in 1546; Grandidier 1782, 291.

2 As the Julian calendar was still used as the basis in both cases, the date of the vernal equinox was 9 or 10 March. The Gregorian calendar was not introduced in Strasbourg until 5/16 February 1682.

3 Dasypodius 1580a, 25–27.

 DOI:10.1163/9789004423473_012

FIGURE 10.1 Calendar disk of the Strasbourg Cathedral Clock

VII. The date of Easter;
VIII. The date of Christmas;
IX. The weeks in the minor interval (*intervalli minoris hebdomadae*), from Christmas to *Estomihi*;
X. The concurrent days (*dies concurrentes*), *i.e.* the rest of the days in the minor interval;
XI. The weeks in the interval between Christmas and Easter (*intervalli ad Pascha hebdomadae*);
XII. The concurrent days;

XIII. The weeks in the interval between Easter and Christmas (*intervalli à Pascha ad Nativitatem hebdomadae*);
XIV. The concurrent days;
XV. Dominical letters in a common year;[4]
XVI. Dominical letters in a leap year.

10.2 The Sundials on the Gable of the Cathedral's Southern Transept

Three vertical sundials are mounted on the gable of the southern transept of the Cathedral (*Fig. 10.2*), for which Dasypodius and Wolckenstein executed the drawings. They were to be used as a *directorium*, that is, as the standard time for those who had to adjust the Cathedral clock.[5]

In 1572 Tobias Stimmer completed a fresco painting there, which can be seen on a copper engraving by Isaak Brunn of 1615 (*Fig. 10.4*) and is no longer extant (the year stood above the roundly tapered arch of the ogival window in the center). Within the band of numbers of the central sundial was a kneeling Atlas holding the globe. In the fields above the two side dials a female figure was depicted on each side, with their legs hanging down and cloth fluttering (personifications of the winds?). On the left was the inscription "*Veritas temporis filii*",[6] on the right "*Tempus edax rerum*".[7] In 1669, renovations were carried out on the dials and the lines of the three sundials—that originally were probably painted on a white background—were chiseled into the stone. The two inscriptions were probably added at that time.[8] Strasbourg Cathedral is not

4 From 1 January on, the days of the week are assigned the letters A, B, C, D, E, F, G. This sequence is repeated until the end of the year. If, for example, the first Sunday in the year falls on 5 January, the letter for that Sunday is E and all of the days of the month labeled E on the outer ring of the calendar disk are Sundays. In the Julian calendar (and also in the Gregorian calendar) the intercalary day was inserted after 23 February and both days were assigned the same Sunday letter. For a leap year, two Sunday letters were necessary, the first for the time before 24 February, the other for the rest of the year.

5 Dasypodius 1580a, 52–53. The sundials are also mentioned in Frischlin's poem in praise of the Cathedral clock; Frischlin 1575, fols. H II^{v}ff.

6 *Noctes Atticae* of Aulus Gellius, XII.1, 7: "*Alius quidam veterum poetarum, cuius nomen mihi nunc memoriae non est, Veritatem Temporis filiam esse dixit*". The quote was also used by Erasmus in his *Adagia*. On its tradition see Saxl 1936, 200, n. 1.

7 Ovid, *Metamorphoses*, XV, 234.

8 According to Rohr 1971, 110–112, based on a view of the southern transept (385 × 265 mm) that was probably engraved around 1660 by Abraham Aubry, who had been living in Frankfurt am Main since about 1653; OND, two copies; see also Schmitt 1930, 244–245, pl. VII. A preliminary sketch of the view by an unknown artist (J.J. Arhardt?) also exists (see *Fig. 10.3*), as well as a mirror-image silverpoint drawing, 390 × 270 mm, Stuttgart, Staatsgalerie (Graphische

FIGURE 10.2 Bernhard Jobin, facade of the southern transept; woodcut, 1566, from a sketch by Daniel Specklin, 521 × 370 mm

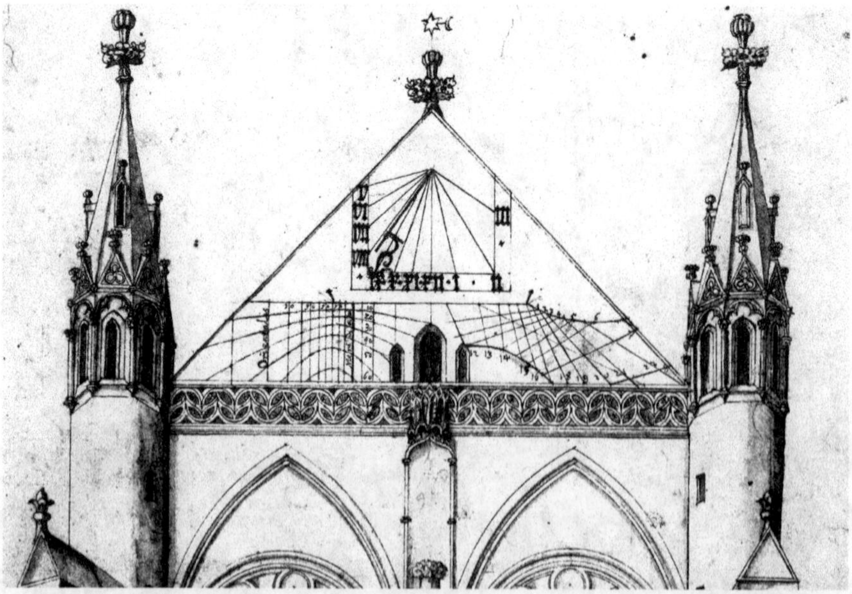

FIGURE 10.3 Unknown artist (J.J. Arhardt?), gable on the facade of the southern transept of the Cathedral; washed pen and ink drawing, 390 × 266 mm, detail

oriented exactly in line with the points of the compass. Its deviation amounts to 29° 40′ (see *Fig. 3.1*), which had to be considered in the design of the three vertical sundials.

1. The top sundial,[9] whose gnomon is parallel to the Earth's axis, indicates true solar time between 5 AM and 2 PM. Originally the numbers were arranged on a U-shaped band and the dots between them indicated the half hours. The time scale also included 3 PM, but that line was unfortunately not chiseled into the stone in 1669 with the others.[10]
2. The vertical and horizontal coordinates of the Sun, altitude and azimuth, can be read off the sundial on the lower left. The Sun's altitude above the horizon is indicated by hyperbolic curves in 10° increments from 0° to 70°, its azimuth by the vertical lines, likewise in 10° increments, counted from the east-west axis, formerly denoted "*orientalis*". The gnomon is

Sammlung), inv. no. C 120. The fresco painting by Stimmer was apparently no longer extant in 1660.

9 Description of the three sundials according to Werkmeister 1912, 70–74 and Rohr 1971, 110–114. See also Staub 1992.

10 On the possible reasons for this lapse see Rohr 1971, 111.

FIGURE 10.4 Isaak Brunn, Facade of the southern transept; copper engraving, 1615, image size 520 × 358 mm, detail

perpendicular to the surface of the wall and ends in a ball (*nodus*), whose shadow indicated the altitude and azimuth.

3. The sundial on the lower right presents two hyperbolic curves which form the boundaries for the hour lines and mark the diurnal arc of the Sun on the winter and summer solstices. A straight line denoting the course of the Sun at the vernal and autumnal equinoxes was originally also painted on the wall. The sundial includes two sets of hour lines. The set from 11 to 20 indicated *horae ab occasu solis* (hours since sunset), also referred to as 'Italian' or 'Bohemian hours', for which there are 24 equal hours per day.[11] The set of hour lines from 0 to 8, indicated *horae ab ortu solis* (hours since sunrise), also referred to as 'Babylonian hours', for which there are also 24 equal hours per day. If one subtracts the Italian hours indicated on the dial from 24, the result is the number of hours until the next sunset. If the resulting difference is added to the number of hours following sunrise, the sum is equal to the length of the respective day between sunrise and sunset. The gnomon on this dial is also perpendicular to the surface of the wall and ends in a ball or *nodus*, whose shadow indicates the point to be read on the hour lines.

---

11 Bilfinger 1892, 185–195.

PART 4

Programmatic Meanings

CHAPTER 11

Clock Construction and Architectural Theory

11.1 Dasypodius's Reception of Vitruvius

The main reference work for a reconstruction of an overall plan of the Cathedral clock is Conrad Dasypodius's *Heron Mechanicus: Seu De Mechanicis artibus, atque disciplinis. Eiusdem Horologij astronomici, Argentorati in summo Templo erecti, descriptio* (Strasbourg, 1580), which provides comprehensive information on the concept behind the clock and its builder's perception of himself. Even if this text includes a description of the clock, the descriptive elements, which dominated Dasypodius's 'expositions' of 1578 and 1580, clearly give way to more programmatic explanations here.

As the title indicates, this text was divided in two parts, namely a programmatic discussion of the mechanical arts,[1] followed by the actual description of the clock, to which were attached the particulars of the two eclipse tables and two dedicatory poems.[2]

From this text it is clear that Dasypodius was familiar with not only the architectural theory of Vitruvius but also the technical writings of the Greek authors, in particular Alexandrine mechanics.

In the following, the argumentation used in Dasypodius's programmatic discussions will first be presented, followed by an explanation of his relationship to the Greek authors on mechanics.[3]

•••

1 Dasypodius 1580b, fols. A^{r}–EIV^{v}; German translation in Leupold 1725, 2–16; reproduction in Oestmann 1993/2000, 212–227.

2 *Ibid.*, fols. F^{r}–HIV^{v}; fols. I^{r}–KI^{r}. Brief summary of the content of the second part in Kästner 1796/1800, v. II, 216–220.

3 Surprisingly, Dengler 2011, 192 remarked: "Even before the first printed editions of Hero's *Pneumatica* appeared, Dasypodius published his *Heron Mechanicus* in 1580, the programmatic meaning and function of which has hardly been recognized so far"; and—also strange to say—additionally commented: "Even Oestmann, who has dealt extensively with the text and discussed it in relationship to other publications in Dasypodius's library, places it in the context of the construction of the monumental work and recognizes that it is an example of the early reception of the ancient scholars in the sixteenth century. However, he does not analyze their importance and their significance for Dasypodius's argumentation."

 | DOI:10.1163/9789004423473_013

The discourse on the mechanical arts starts by ascertaining that there are numerous arts and disciplines; among them are not only those of a theoretical nature, but also those in which something can be planned and built. Among these, the mechanical arts play by no means the least significant role.[4]

'*Mechanica*' is '*scientia machinalis*'. According to Vitruvius,[5] the concept '*machina*' is defined as a stable, self-contained structure that is capable of moving loads. A person who had mastered the science of machine construction was called '*mechanicus*'.[6] A distinction must be made between *scientia machinalis* and the art of producing tools, '*organica et instrumentalis ars*'. Here again, Dasypodius explicitly refers to Vitruvius and elaborates by saying that machines put forth their effects by using multiple workers and greater forces. Tools, on the other hand, fulfill their purpose by being used by a single person.[7]

According to Pappus, Hero, Proclus, and other philosophers and mathematicians not mentioned by Dasypodius, a distinction has to be made between the two branches of *scientia machinalis*: '*scientia machinalis logica*' and '*cheirurgica*'. The former refers to the plans for constructing machines of various types using the mathematical disciplines: geometry, arithmetic, and astrology; whereas the latter comprises the manual skills performed with diligence and talent.[8]

This is a paraphrased citation of a passage from Book VIII of *Collectio* of the Greek mathematician Pappus,[9] who was probably citing from the lost introduction to the *Mechanics* by Hero.[10]

---

4 Dasypodius 1580b, fol. A^{r}.

5 Vitruvius 1983, X.1, 275.

6 Dasypodius 1580b, fols. AIII^{v}, AIV^{r}.

7 *Ibid.*, fols. AIV^{r-v}.

8 *Ibid.*, fol. B^{r}.

9 On Pappus (ca. 300–ca. 350) see Heath 1921, v. II, 356; Cantor 1907, v. I, 441ff; DSB, X, 293–304. Book VIII of *Collectio* deals primarily with the problems of *mechanica*; Heath 1921, v. II, 427–439; Cantor 1907, v. I, 449–450. The work by Pappus was not published until eight years after *Heron Mechanicus*, when it appeared in a Latin translation by Federico Commandino, but Dasypodius had possessed a manuscript of it from 1573 and quoted directly from it: "*Recte ergo Pappus Alexandrinus insignis mathematicus lib. octavo, ita scribit* (Rightly therefore Pappus of Alexandria, a famous mathematician, in his eighth book thus writes)"; Dasypodius 1580b, fol. DIV^{v}. The origin of that manuscript will be dealt with in more detail on pp. 208–209.

10 "*Iam vero ii qui Heronem sectantur mechanicae alteram partem in demonstratione mathematica, alteram in manuum opera versari existimant, et illa quidem parte, quam rationalem dicunt, geometriam, arithmeticam, astronomiam* [demonstrationem] *physicam contineri, ad alteram autem partem, quae manuum opera indiget, referendam esse artem aerariam ferrariamque, aedificatoriam, lignariam, picturam etiam et quaecunque in exercitatione manuum versatur. Et eum quidem qui his disciplinis a prima aetate incubuerit* [et in his

The second author to whom Dasypodius refers is the neo-Platonist Proclus with his commentary on Euclid. In a section on the mathematical disciplines, mechanics is allocated to the branch of mathematics—along with astronomy, optics, geodesy, canonics (the study of harmony), and logistics (computational technology)—which deals with objects perceived by the senses. Proclus in turn referred to Geminos's classification (see *Appendix 5*).[11]

Dasypodius's distinction between *scientia machinalis logica* and *cheirurgica* corresponds to the concepts '*ratiocinacio*' (reasoned, theoretical considerations) and '*fabrica*' (manual accomplishments, providing form) used by Vitruvius.[12]

Without any knowledge of the sciences, a craftsman being led only by his 'natural instincts', can discover and build things. Cleverly conceived activities, such as the crafts of smiths, builders, painters, and sculptors, can be learned without knowledge of writings.[13] The activities of the '*mechanicus logicus*' are quite the opposite. He first consults the writings and opinions that have been passed down by the philosophers and mathematicians before starting to plan a work to be completed by craftsmen.[14] As had already been demanded by Vitruvius, a *mechanicus logicus* must have a good memory, diligence, intelligence, good judgment, and comprehensive education.[15] He was imbued with erudition and should be an intelligent master of all the things that could be brought about by genius, skillfulness, and crafts. The '*mechanici*' Philon, Hero, and 'Hero Mechanicus'[16] were of this mindset.[17]

Different categories of machines are dealt with. Dasypodius first distinguishes between '*scansoria*', machines for climbing or scaling; '*spiritalia*', pneumatic machines, and '*tractoria*', traction machines that pull or lift. Following this he differentiates between: 1. *trochleae*, different categories of pulleys; 2. *Organopoeticae*, war machines; 3. *mechanopoeticae* and *hydraulicae*, hydraulic machines that can produce sound in a whistle using water or air pressure;

---

artibus exercitatus sit] *ac versatile ingenium habeat, optimum mechanicorum operum inventorem* [et architectum] *futurum esse dicunt*"; Pappus 1875/78, v. III, 1023–1025.

11 Proclus 1945, 189–192. The commentary was first published together with the original text of Euclid's *Elements* by Simon Grynaeus in Basel in 1533; a Latin edition edited by Francesco Barozzi appeared in Padua in 1560.

12 Vitruvius 1983, I.1, 7.

13 Dasypodius 1580b, fol. B^{v}.

14 *Ibid.*, fols. B^{v}–BII^{r}.

15 Vitruvius 1983, I.1, 9ff.

16 'Hero Mechanicus' refers to 'Hero of Byzantium' or 'Hero the Younger', an anonymous Byzantine Christian author from the tenth century; see RE, XV, cols. 1074–1077, s.v. "Heron von Alexandreia" by Karl Tittel.

17 Dasypodius 1580b, fols. BII^{r}–BII^{v}.

4. *thaumasopoeticae*, thaumaturgic objects, which enchant; as well as 5. *dioptrica* and *sphaeropoetica*, the art of producing measuring instruments and globes.[18]

Dasypodius derives information from two sources: the machine types *scansoria, spiritalia* and *tractoria* are taken from Vitruvius;[19] the subdivision of mechanics into five fields is based on Pappus.[20]

In his later writings Dasypodius introduced the following diagram for his subdivisions of *mechanica*:[21]

	Logica	
Mechanica	*Cheirurgica varias habet artes*	*Manganaria ars* *Organopoëtica* *Mechanopoetica* *Thaumatopoëtica* *Sphaeropoëtica*

After this excursus Dasypodius proceeds to architecture, which is designated as leader and queen of all the arts. It includes three main areas, namely the construction of buildings, clockmaking, and machine construction.[22]

18 *Ibid.*, fols. BII^{v}–C^{r}.

19 Vitruvius 1983, X.1, 275.

20 "*Omnium autem artium quae ad mechanicam spectant maxime necessariae ad vitae usus sunt hae: ars manganariorum, qui ipsi quoque secundum veteres appellantur mechanici* (*hi enim magna pondera, quae natura immobilia sunt, sursum tollunt minore potentia moventes*), *tum eorum qui tormenta ad bellum necessaria construunt atque ipsi etiam mechanici vocantur* (*tela enim et lapidea et ferrea aliaque id genus instrumentis catapulticis, quae ab his fabricantur, in longum spatium mittuntur*), *deinde ars eorum qui proprie machinarum fabri dicuntur* (*nam ex magna profunditate instrumentis, quae illi ad aquam hauriendam construunt, aqua facilius in altum evehitur*). *Sed mechanici a veteribus etiam mirabilium artifices vocantur, quorum alii spiritalium artem diligenter tractant, ut Hero pneumaticis, alii per nervos et funiculos motiones animatorum imitari videntur, ut Hero automatis et ζυγωοις sive aequilibribus, alii etiam per ea quae aqua vehuntur, ut Archimedes οχουμένις, vel per horologia aquaria, ut Hero υδρείοις, quam quidem disciplinam cognatam esse apparet rationi horologiorum gnomonicorum sive solariorum. Mechanicos denique etiam illos vocant, qui globorum fabricationem callent et coeli effigiem per aequabilem et circularem aquae motum construunt*"; Pappus 1875/78, v. III, 1025–1027.

21 Dasypodius 1596, fol. 46^{r}. On '*elementa mechanicae*' and the subdivisions of *mechanica* see fols. 31^{v}–34^{v}, 41^{v}–43^{r}.

22 Dasypodius 1580b, fols. C^{r}–C^{v}; see also Vitruvius 1983, I.3, 33.

Should an architect understand nothing of the mechanical arts, he would not deserve his title, as he would have no knowledge of this great area of architecture, which is an indispensable prerequisite for constructing buildings and clocks. As proof of this, Dasypodius lists numerous examples to demonstrate that architecture could not yield any results without *mechanica*, citing the walls of Babylon, the Egyptian pyramids, the temple to Diana in Ephesus, and the bridge over the Danube built by Trajan.[23]

In order for such works to be completed, both the generosity of great men with strong personalities, who are not afraid of large expenses, and also the educated planning of architects and *mechanici*, as well as the skill of the craftsmen who are to construct the work are necessary. The craftsmen carry out all the work that can be seen, whereas the architect devises, plans, and sets up everything, which is why most of the praise should be awarded to him.[24] The *idiotae* (uneducated) take in the completed work with their eyes, but do not comprehend how such great splendor, and great magnificence, so great and so suitable a fitting arrangement of all and individual parts, and an elegant appearance are applied to the work.[25]

Without a good theoretical *mechanicus* or architect—Dasypodius considered the two to be the same—the undertaking would be in vain, no matter how large the financial investment or the applied skill of the craftsmen. The *opifices* (workmen) need an experienced architect, a *mechanicus*, who determines the shape, symmetry, elegance, appropriateness, and the structural soundness of the work for them, to the extent that necessity and the application of the work demand.[26]

Dasypodius bases his claims here on the three properties of the work of architects demanded by Vitruvius: *firmitas*, *utilitas*, *venustas*;[27] and added the esthetic category '*symmetria*', with no further comments.

The fact that Vitruvius was explicitly referred to in Strasbourg, in particular while the astronomical clock was being built, was certainly a result of the attempts of the physician and mathematician Walther Ryff to confer a scientific basis on architecture,[28] to free it from being understood as just a craft, and thus upgrade its status. Ryff was involved with the 1543 publication in Strasbourg

23 Dasypodius 1580b, fol. C^{v}–CII^{r}.

24 Dasypodius 1580b, fols. CIV^{r}–CIV^{v}.

25 *Ibid.*, fol. D^{r}.

26 *Ibid.*, fol. D^{v}.

27 Vitruvius 1983, 35.

28 On the life and the work of Walther Ryff (Gualtherus Hermenius Rivius), who was born in Strasbourg in 1500 and died in Würzburg in 1548, see Benzing 1959, Dann 1988, and Jachmann 2006.

of the first edition of Vitruvius outside of Italy, which was based on the two editions of Fra' Giovanni Giocondo (Florence 1513 and 1522),[29] and published the first German translation of Vitruvius in Nuremberg in 1548.[30] In the commentary on Book I, he discussed the position and qualifications of architects in detail, completely in the spirit of Vitruvius.[31]

What Dasypodius says in his discussion of the status of architecture, the various fields of architecture, and the superior position of the *mechanicus logicus* relative to the craftsmen is explained in *Heron Mechanicus* using the beginning of the second phase of construction of the Cathedral clock as an example.

A new clock was to replace the old Cathedral clock to confer honor, fame, and dignity on the Senate and citizenry of Strasbourg. Dasypodius was entrusted with this task and he deliberated on how such a magnificent, splendid and artistic work could be realized. First he presented a *modulus* to the *aediles* (the caretakers of the 'Frauenwerk'), according to which such a machine could be constructed. Following that, Dasypodius drew up the ground plan and the elevation of everything that could be seen.[32]

From the sequence of activities from determining the *modulus* to the plan, one can recognize the extent to which Dasypodius was familiar with the architectural teachings of Vitruvius and his inclusion of them in his design of the Cathedral clock. As we will be returning to this topic later, the terminology will be explained here first.[33]

With the help of a basic unit of measure, the *rata pars* (module), the proportions of an entire structure can be developed: "Proportion consists in taking a fixed module, in each case, both for the parts of a building and for the whole".[34] The determination of the module, which was to be derived from the structure itself, was the *quantitas*. For example, the module of a Doric temple was to be determined from the width of the structure, that is, from the size of the construction site.[35] If—as a result of the module—all the components

29 *Herzog August Bibliothek* 1984, 62–63, no. 34; Ebhardt 1918, 69; Jachmann 2006, 25–27.

30 *Herzog August Bibliothek* 1984, 67–68, no. 39; Benzing 1959, 54, no. 185.

31 Rivius 1548, fols. VI^{r}ff. See also Forssman 1956, 56ff.

32 Dasypodius 1580b, fols. FII^{v}, FIII^{r}.

33 On Vitruvius's fundamental architectural concepts see RE, II.1, cols. 545–547, s.v. "Architectura" by Otto Puchstein; Jolles 1906, 9–39; Knell 1987, 30 and 32–33.

34 "*Proportio est ratae partis membrorum in omni opere totoque commodulatio*"; Vitruvius 1983, III.1, 159.

35 *Ibid.*, IV.3, 182. The details and problems of determining a module as well as the derivation of a system of measure from the proportions of the human body in *ibid.*, III.1, 136–138, do not require further explanation in this connection; see Scholfield 1958, "Vitruvius and the Theory of Proportion", 16–32; Knell 1987, 87ff.

are placed in a harmonious relationship to each other and to the work as a whole (*ordinatio*), *symmetria* is produced: "Symmetry is also the appropriate harmony arising out of the details of the work itself; the correspondence of each given detail among the separate details to the form of the design as a whole".[36] A numerical relationship was intended here, that is, the recurrence of a basic unit in the plan for the entire structure. In addition to this mathematically objectified beauty, the qualitative factor was summed up as *eurythmia* (harmonious appearance). It was only achieved when all of the components existed in a harmonious relationship of height to width as well as width to length and corresponded to their respective symmetry: "Proportion implies a graceful semblance; the suitable display of details in their context. This is attained when the details of the work are of a height suitable to their breadth, of a breadth suitable to their length; in a word, when everything has a symmetrical correspondence".[37] Symmetrical ratios were determined with the help of geometry, which was a prerequisite for the *dispositio* (the appropriate arrangement of all the components), which consisted of the ground plan executed true to scale (*ichnographia*), the elevation of the front side (*orthographia*), and the *scaenographia*, a representation of the facade in perspective.[38] These three forms of disposition (plan, elevation and perspective) arise from imagination and invention.[39]

After the plan had been presented to other scholars, the decision was made that nothing was to be changed in the '*inventio prima*'.[40] With rhetorical modesty, Dasypodius postulated that, as a result of his moderate abilities and his insignificant talent, he was not entitled to claim the title *mechanicus logicus* or architect for himself. Nevertheless he would try to carry out the task assigned to him conscientiously and would do everything possible "to enhance the

36 "*Item symmetria est ex ipsius operis membris conveniens consensus ex partibusque separatis ad universae figurae speciem ratae partis responsus*"; Vitruvius 1983, I.2, 27. The concept had a more inclusive meaning for Vitruvius compared with modern usage, in which only the balanced placement of parts on opposite sides of an axis line or plane (bilateral symmetry) is understood. On the history of this term and the concept behind it, see Schramm 1991 and Hon/Goldstein 2008, 93–106.

37 "*Eurythmia est venusta species commodusque in compositionibus membrorum aspectus. Haec efficitur, cum membra operis convenientia sunt altitudinis ad latitudinem, latitudinis ad longitudinem, et ad summum omnia respondent suae symmetriae*"; Vitruvius 1983, I.2, 27; see also VI.2, 21–23.

38 *Ibid.*, I.1, 24, and I.2, 25–27.

39 "*Hae nascuntur ex cogitatione et inventione*"; *ibid.*, 27.

40 Dasypodius 1580b, fol. FIII^{r}.

splendor, magnificence, elegance, and advantage of the work".[41] Apparently, when using the terms '*splendor*', '*magnificentia*' and '*venustas*', Dasypodius was referring to the category '*decor*'. Vitruvius defined 'decor' as the faultless appearance of a structure composed of approved details.[42] The type and character of the embellishments were to be derived from *consuetudo* (custom, usage), *statio* (placement), and *natura* (nature). In general the functions of a structure and the shapes determined by custom, location, and iconography are preconditions for decor.[43] Discussing the scintillating interpretations of this concept in the history of architectural theory and its roots in ancient rhetoric goes beyond the scope of this work. In the context of the planning process for the Cathedral clock it is sufficient to say that Dasypodius—and also Rivius[44]—used decor in the sense of embellishment (*ornatus*).

David Wolckenstein and the painter Tobias Stimmer, "whose skill and industry in that craft are so great that he can easily be compared with some

41 "*Et pro viribus meis, omnia quae ad splendorem, magnificentiam, venustatem atque commoditatem operis spectare videbantur: priori meo adieci modulo*"; *ibid.*, fol. FIII^v.

42 "*Decor autem est emendatus operis aspectus probatis rebus compositi cum auctoritate*"; *ibid.*, 27–29.

43 Therefore elegant interior rooms, for example, have to be augmented with appropriate *vestibula*; Vitruvius 1983, I.2, 29. Rooms must have a certain orientation with respect to the cardinal points: bedrooms and libraries facing east, bathrooms and winter chambers facing south, picture galleries facing north; *ibid.*, VI.4, 35 and I.2, 31. In a similar manner, the plans of architects must take into consideration the choice of a healthy location for building city walls, *ibid.*, I.4, 35ff; for the orientation of streets with respect to the winds, *ibid.*, I.6, 53ff, and for determining the eastern orientation for temple complexes with their altars, *ibid.*, IV.9, 247. From the connection between cosmology and architecture, Vitruvius derives his demand that architects master astronomy and gnomonics.

44 "*Aber Decor mag die zierlich anschawung des volbrachten wercks genennet werden, wo alle angehörige theil mit höchster geschicklichkeit der massen volbracht werden, das solches schön und zierlich anzuschawen wer* [...]. *Wo nun alle glider sich nach der Eurithmia also wol reimen und nach der Symmetria in iren glidmassen sich zusamen schicken, entspringt darauß Décor, das ist das schön herlich ansehen solches baws* (But decor can be named the graceful appearance of the completed work, in which all the member parts are completed with the highest skillfulness in dimensions, so that such would be beautiful and graceful to look upon [...]. Now when all the elements rhyme well with each other according to eurythmia, and fit together in their appendages in accordance with the symmetria, decor will arise from this, that is the beautifully glorious appearance of such a structure)"; (Rivius 1548, 25, 31). Petrus Dasypodius provides comments in his *Dictionarium latinogermanicum* (Strasbourg 1537) under the keywords "Zierde" (embellishments): "*Ornatus, Decor, Venustas, Elegantia*"; and "decor": "*Zierd, huepsche*" (embellishment, beautiful); cited from Horn-Oncken 1967, 80, n. 157.

ancient painters",[45] are contracted, as are—together with carvers and other craftsmen—the *fabri ferrari* (blacksmiths) Isaak and Josias Habrecht.

Dasypodius explains what the new clock is to exhibit with the following words:

> And so we have rendered this astronomical clock in this form which it has. And on this clock we exhibit eternity, the century, the orbits of the planets, the yearly and monthly revolutions of the Sun and the Moon, the divisions of the week, days, hours, parts of hours, minutes; all these, I say, we exhibit to be seen. We have added also, for the sake of adornment, splendor, and admiration, various contrivances, pneumatic, spheropoetic, and automatic, everything from history and the tales of the poets, and also from sacred and profane writings in which there is or can be some delineation of time. And we show these things by paintings, pictures, statues, and other works similar to these.[46]

The divisions demonstrated in various ways are connected to the terms *symmetria* and *eurythmia* from architecture theory:

> Since there are in all the parts of the whole work eurythmia and symmetry, and the upper parts correspond to the lower and the lower to the upper; also a neat arrangement and agreement is seen in individual items, so that, as it were, one seems not to be able to exist without the other. For through the Pelican eternity is shown, a century through the wheel of 100 years, a year through the calendar, the seasons of the year by paintings, the months by the movement and phases of the Moon, the weeks by statues of the planets, the hours and half-hours by the outer limb of the astrolabe, the quarter-hours and the minutes are shown by a tiny tablet [dial]. Finally, so that these lack nothing to increase elegance, delight,

45 "*Cuius magna est, in ea arte solertia, et industria: ut antiquis nonnullis facile comparari possit pictoribus*"; Dasypodius 1580b, fol. FIII^v^.

46 "*Horologium itaque Astronomicum in hanc quam habet formam redegimus: et in eo aeternitatem, saeculum, periodos planetarum, Solis, et Lunae annuas, et menstruas conversiones, septimanarum, dierum, horarum, partium horariarum, ac minutorum distinctiones: haec inquam omnia, conspicienda exhibemus: adiecimus etiam ornatus, et splendoris, atque admirationis gratia, varia machinamenta, pneumatica, sphaeropoetica, automatopoetica, ex historiis, atque poetarum fabulis, sacris et prophanis scriptis quoque, omnia: in quibus temporis aliqua est, aut esse potest descriptio. Eaque ostendimus tabulis, picturis, statuis, et aliis his similibus operibus*"; Dasypodius 1580b, fols. FIV^r^–FIV^v^. This remark can also be found in the German description of the clock: Dasypodius 1578a, 45.

> and admiration, there were added automata and sphaeropoetica. There were also added index hands to show the mean motions of the planets.[47]

Therefore we have a decidedly original interpretation of Vitruvius, as Dasypodius does not use the concept *modulus* in the sense of a certain unit of measure for determining the proportions of the clock housing—which already existed for the most part—but takes time, or more precisely the smallest unit represented in the clock, the minute, as the basic unit for his plan (this is indicated on the small dial above the calendar disk). The entire repertory of artistic and technical forms used in the Cathedral clock is made up of multiples of this smallest unit all the way to eternity, the largest conceivable unit of time.

11.2 The Emblematic Painting on the Weight Drive Tower

From what has been said so far, the high distinction Dasypodius placed on the activities of the architect and *mechanicus logicus* becomes clear. This attitude was coupled with a pejorative assessment of the abilities of craftsmen and in particular of the clockmakers of his time, which is expressed in especially strong terms in the preface to the *Warhafftigen Außlegung* of 1580:

> But whether our craftsmen are as artistic as Heron, Archimedes and others, that is something that can by no means be admitted to, as they were scholars and experienced personages in all the arts, not alone that such artistic work was invented, it is claimed and was made and set up by themselves, but rather they brought forth the reason and the demonstrations for each and every work obviously from the mathematical disciplines and have explained and can therefore be called Philosophers and Mathematicians, but had they lacked in the same, they would have remained poor craftsmen. Hence I say that our clockmakers and craftsmen are not so artistic as the old *mechanici mathematici* inasmuch as

47 "*Atque haec omnia commemoratione digna sunt: cum in omnibus totius operis partibus, eurythmia, et symmetria sit. ac superiora inferioribus, inferiora superioribus congruant: concinnitas etiam et convenientia in singulis conspiciatur: ut unum absque altero quasi esse non posse videatur. Nam Aeternitas per Pellicanum, Saeculum per rotam 100. annorum, Annus, per Calendarium, anni tempora picturis, Menses motu, et phasibus Lunae: Hebdomadae planetarum statuis, Horae et dimidiae horarum partes, limbo Astrolabii exteriore: partes horarum quartae et minuta: parva tabella monstrantur. His tandem ne quid deesset ad venustatem, ad delectationem, ad admirationem augendam. apposita sunt et automata, et sphaeropoetica: additi etiam sunt, qui medios planetarum motus ostendant, indices*"; Dasypodius 1580b, fol. Hʳ.

> they have not carried out the same in the arts and disciplines, and, as they themselves must admit, they lack for these and can do no more than what their craft allows [...]. As those who in our time are clockmakers and are called such are in the Latin language (not to mention Greek) completely inexperienced, understand nothing of the philosophical arts, [...]. And these same foolishly want to set up such crafts as art and raise them higher to be called liberal arts, but between experience and disciplines there is a great difference, and the empirical [mathematical] disciplines by far surpass the other.[48]

With this background the meaning of the emblematic painting on the right side of the weight drive tower can be understood (*Fig. 11.1*).

The picture (195 × 80 cm), painted on a reddish-brown background, has a lighter border about 15 cm wide around it. It depicts a loose arrangement of objects and figures, positioned one on top of the other and held together with garland-like ribbons. At the bottom there is a blindfolded putto carrying a second putto on his back. Then a number of tools follow: a painter's palette on the left, a brush, a hammer, and a file, a joiner's mallet, and a carver's knife. To the left of the two putti, a painter's stick and a bell are visible. The tools are arranged crossing each other in the form of a bundle which is held together by a ribbon, the ends of which fall down in loops. Above this bundle the bottom of a horizontal gear is visible, on which a pyramid, a sphere, and a book are located. At the top of the composition is an angel's head holding two ribbons ending in volutes in the angel's mouth, on the left of which there is a compass and on the right a pair of compasses.

48 "*Aber ob unsere Handtwercks leut so künstlich seyen als der Heron, Archimedes und andere, das ist gar nicht zu zugeben, dann dise seindt gelehrte und in allen kuensten erfarne leut gewesen, nicht allein in dem das sey solche künstliche werck erfunden, angeben und selbers gemacht und aufgericht, sondern auch den grundt und die demonstrationes aller und jeder werck augenscheynlich auß den Disciplinis Mathematicis herfür gebracht und dargethan haben und derhalben Philosophi und Mathematici genennet, wa sie aber der selben gemanglet, weren sie schlechte handtwercks leut gebliben. Derhalben sag ich das unsere Uhrenmacher und Handtwercks leut nit so künstlich seyen als die alten Mechanici Mathematici, dieweil sie denselbigen noch nicht in artibus et disciplinis nachgethan haben, und wie sie selbst bekennen müssen, so manglet es inen an disem, und koennen nichts weyters, dan was ir Handtwerck vermag* [...]. *Dan die so zu unserer zeit Uhrenmacher seindt und genennet werden seind der Lateinisch sprach (will der Griechischen geschweygen) gantz unerfaren, verstehn sich auch nichts in den Philosophicis artibus,* [...]. *Und thun gar toerig diejenigen so solche handtwerck woellen denen künsten fürsetzen, auch hoeher halten, welche liberales artes, genennet werden. da doch zwischen der erfarnuß und den disciplinis ein grosser underscheidt ist und die disciplinae experientiam weyt übertreffen*"; Dasypodius 1580a, Preface, fols. B^{r}–BII^{v}.

FIGURE 11.1
Emblematic painting on the weight drive tower of the Cathedral Clock in Strasbourg

A possible source of inspiration for this composition was a woodcut in two works by Gualtherus Rivius,[49] showing a putto standing on a block surrounded by tools and books (*Fig. 11.2*). A purse hangs over the wrist of his left hand, his raised right hand holds two wings. This depiction is derived in turn from the emblem "*Paupertatem summis ingeniis obesse ne provehantur*" (Poverty prevents the advancement of the best of abilities; *Fig. 11.3*) in the widespread and influential emblem book *Emblematum Libellus* by Andrea Alciato (1492–1550);[50] the epigram reads: "*Dextra tenet lapidem, manus altera sustinet alas, Ut me pluma levat, sic grave mergit onus. Ingenio poteram superas volitare per arces, Me nisi paupertas invida deprimeret*" (My right hand holds a rock, the other bears wings. As the feathers lift me, so the heavy weight drags me down. By my mental gifts I could have flown through the heights of heaven, if malign poverty did not hold me back).[51]

With the inscriptions "*Aurum probatur igni, ingenium vero Mathematicis*" (Gold is tested by fire, the intellect, however, by mathematics) and "*Vivitur ingenio, caetera mortis erunt*" (Intellect will live, all the rest will die),[52] Rivius emphasizes the primacy of theoretical considerations in architecture. The activities of the architect are therefore carried out between material bondage—symbolized by the purse and the stone on which the putto stands—, practical activity, and soaring intellect.

Furthermore, Alciato's emblem "*Mutuum auxilium*" (mutual help) may have been used as a template for the picture on the weights tower. The blind man carries the lame (*Fig. 11.4*).[53]

Fischart mentions this picture in his descriptive poem for the smaller woodcut of the clock with the following words: "*An der andern seit stahn trei Künst Die zum Werck thaten grose dienst*" (On the other side [of the tower for the weights] stand three arts, which rendered great service for the work; *Fig. 6.6*).

The three arts intended here are clockmaking, fine art (painting and sculpture), as well as architecture—the last being illustrated by geometrical bodies, books, squares, and compasses. More generally, however, the three major areas

49 Rivius 1547, title page verso; Rivius 1548, fol. XI^r with the caption: "*Circkels, Richtscheid und aller gebreuchlichen Geometrischen Instrument kuenstliche fuerbildung*" (Dividers, straightedge, and all means of customary geometrical instruments artistically represented).

50 The illustration on fol. H VII^r of the allegorical novel *Hypnerotomachia Poliphili*, published in 1499 by Francesco Colonna, is given as a second source; Roettinger 1914b, 33–34.

51 Andrea Alciato, *Emblematum libellus*, Paris 1534; Henkel/Schöne 1967, cols. 1022–1023.

52 This is the motto of the portrait of Willibald Pirckheimer by Dürer; Białostocki 1989.

53 "*Loripedem sublatum humeris fert lumine captus, Et socii haec oculis munera retribuit: Quo caret alteruter, concors sic praestat uterque, Mutuat hic oculos, mutuat ille pedes*"; Henkel/Schöne 1967, cols. 990–991; Châtelet-Lange 2010, 27–28.

FIGURE 11.2 Emblematic representation of architecture (Rivius, *Der* […] *gantzen Architectur angehoerigen Mathematischen und Mechanischen kuenst eygentlicher bericht* […], Nuremberg 1547)

46 AND. ALC. EMBLEM. LIB.

Paupertatem ſummis ingenijs obeſſe ne prouehantur. XV.

Dextra tenet lapidem, manus altera ſuſtinet alas.
Vt me pluma leuat, ſic graue mergit onus,
Ingenio poteram ſuperas uolitare per arces,
Me niſi paupertas inuida deprimeret.

FIGURE 11.3 Emblem "Paupertatem summis ingeniis obesse ne provehantur" (Andrea Alciato, *Emblematum libellus*, Paris 1542, p. 46)

of architecture named by Vitruvius—building, gnomonics and mechanics[54]—may also be represented here. The two putti can be interpreted as personifications of craftsmanship and architecture in their mutual interdependence. On the one hand, the putto with his blindfolded eyes must be guided; craftsmanship requires the guidance of a learned person, the *mechanicus logicus*, who knows the writings of philosophers and mathematicians, and is capable

54 "*Partes ipsius architecturae sunt tres: Aedificatio, Gnomonice, Machinatio*"; Vitruvius 1983, I.3, 33.

60 AND. ALC. EMBLEM. LIB.

Mutuum auxilium. XXII.

Loripedem ſublatum humeris fert lumine captus,
Et ſocij hæc oculis munera retribuit:
Quo caret alteruter, concors ſic præſtat uterq;,
Mutuat hic oculos, mutuat ille pedes.

FIGURE 11.4 Emblem "Mutuum auxilium" (Andrea Alciato, *Emblematum libellus*, Paris 1542, p. 60)

of developing a plan for the work with the help of geometry by using the compass and square.[55] On the other hand, architecture relies on craftsmanship, which carries it, for completing its work, but by placing the book, pyramid, and sphere at the top above the gear, the entire arrangement of the picture clearly indicates a hierarchy which places the intellectual activity of the architect on a higher level than all the manual crafts, no matter how cleverly conceived and necessary they are for completing the work.

---

55 *Ibid.*, I.1, 9 and I.2, 25.

11.3 The 'Late Gothic' Architecture of the Clock Housing

As Conrad Dasypodius insisted so vehemently that architects should have a humanist education and should not be considered one of the artistic craftsmen (*künstlichen Handtwerkern*), to whom Rivius had dedicated his translation of Vitruvius—as architects in the Vitruvian sense did not exist in sixteenth century Germany!—, one would expect the enthusiastic advocacy of Renaissance architecture, which is strangely contradicted by the design of the housing of the Cathedral clock.

Dasypodius introduces the "*Querelle des Anciens et des Modernes*" right at the beginning of the *Warhafftige Außlegung*: "It is a question with which scholars concern themselves [...] namely whether the men of ancient times—the Jews, Greeks, Romans and other peoples—were more scholarly, more judicious, and in addition more ingenious than those of our times".[56]

Various examples are listed and assessed: there were magnificent buildings in antiquity, such as the Temple in Jerusalem or "exquisite buildings in Rome",[57] but from more modern times there were Strasbourg Cathedral, St. Stephen's Cathedral in Vienna, and the Cathedral in Florence. There were differences between all of these works, and the question arose: "which of these was the most artistic, those who lived in our time and are still alive, or rather those from ancient times?"[58] What is more, by this time many new arts have been discovered, such as the printing press and various improvements in weapons of war, which the ancients did not know. Dasypodius comes to the conclusions that God distributes his uncountable talents according to unfathomable counsel and "therefore one must definitely conclude, that there is no difference as to which were the more scholarly or artistic, the ancients or those living of our time, with the exception of those were highly honored by the ancients, but now have fallen into disrepute".[59] This qualification applies to clockmaking and related handicrafts.

---

56 "*Es ist ein frag under den gelehrten* [...] *namlich ob gelehrtere und verstendigere, auch künstlichere leut vor alten zeiten bey den Juden, Griechen, Römeren und anderen völckeren gewesen seyen, oder aber jetz zu unseren zeiten*"; Dasypodius 1580a, Preface, fol. AII^{r}.

57 *Ibid.*, fol. AII^{v}.

58 "[...] *welche die künstlichsten seyen, die zu unser zeit gelebt und noch leben, oder aber die gar alten*"; *ibid.*, fols. AII^{v}–AIII^{r}.

59 "[...] *derhalben darauß endtlich zu schliessen ist, das kein underscheid ist welche gelehrter od*[er] *künstlicher gewesen seyen, die alten od*[er] *die zu unser zeit leben, doch etliche außgenommen die bey den alten hoch geehrt gewesen, aber in abgang kommen*"; *ibid.*, Preface, fols. AIV^{r}–AIV^{v}.

A 'Renaissance mind-set' can only be construed from this in part. At most the 'revival' of antiquity applied only to certain fields. Therefore the juxtaposition of different styles in the clock housing was compatible with these views—the designation 'mixed style' would not be completely appropriate. The amalgamation of traditional Gothic architectural style and decoratively conceived Renaissance forms—referred to in art history texts as 'German Renaissance' or 'late Gothic'—was, for German writers on architectural theory at that time, a modern architecture that had developed out of tradition. It had an independent development, which—without giving up established traditions or assessing them negatively—integrated new forms.[60] The view dividing history into three periods—antiquity, decline and obscuration in the barbaric 'Gothic' Middle Ages, Renaissance—as construed by the Italian humanists and architecture theoreticians, did not play a role here. Following Philipp Melanchthon the division of history into four monarchies played a far greater role for the Protestant historians. Within the fourth monarchy, the Holy Roman Empire of the German Nation, a continuous cultural development took place.[61] The example of the Cathedral clock gives practical expression to the attempts at merging the new 'southern European' embellishments with their arrangement of columns and the highly developed Gothic tradition of the Strasbourg builders' lodge. The rebuilding of the *Frauenhaus* by Uhlberger in 1579–1585 represents a rich repertoire of forms comparable to that of the Cathedral clock, which resulted in a virtuosic combination of Gothic elements (like mullions, tracery, and reticulated vaulting) and 'modern' Renaissance forms (orderly arrangements of columns and a stepped gable with entablatures).[62] With the awareness of classical sources, the local traditions were advanced and enriched. A direct reference to this attitude towards history and to the architectural development of that time has to be recognized in the presence of the four ages of the world and the 'Great Image' or colossus from Daniel's prophecy, which here certainly go beyond a 'depiction of time' as intended by Dasypodius.

60 This has been shown by the investigations of Hermann Hipp on '*Nachgotik*'; Hipp 1974; see esp. v. I, 572–583, 660–661 and 700–702; Hipp 2008.

61 Hipp 1974, v. I, 530–531; Menke-Glückert 1912, 44–48. On the text *De quatuor summis imperiis* by Johannes Sleidan (1556) see *ibid.*, 85–86, and on Melanchthon's strong influence over the latter see Neddermeyer 1988, 14 and 68ff.

62 Of particular note is the intricate staircase inside the building with its spiral tracery and Corinthian columns; Hipp 1974, v. I, 303 and v. II, 1624; Winckelmann/Schmitz 1894, 263–266.

CHAPTER 12

Influences of Alexandrine Technology

The fact that Dasypodius had a broad knowledge of the technical literature from antiquity can be seen from the numerous citations in *Heron Mechanicus*,[1] in which, besides Ctesibius, Philon of Byzantium, and others, Hero of Alexandria (fl. ca. 62 AD)[2] is mentioned numerous times and his writings on mechanics, catapults, construction of missiles, hydraulics, clasmata, pneumatics, optics, dioptrics, the making of automata, geodesy, arithmetic of various kinds, and geometry are listed.[3] From the particulars provided on the content of Heron's *Belopoetica* (on the construction of ballistic weapons), *Pneumatica*, and *Metrica* (on measurements), it is clear that Dasypodius had read these works himself,[4] and that some of them were in his private library.[5]

In 1572 Dasypodius remarked in the preface of his *Sphaericae Doctrinae Propositiones Graecae et Latinae* (Strasbourg, 1572)—a collection of texts by Theodosius, Autolycus, and Barlaam—that he wanted to publish the "fragments" of Hero,[6] and he also expressed his desire to own more of his texts. On this matter he probably corresponded with the physician and humanist Federico Commandino (1509–1575) of Urbino, whose first complete Latin translation of *Pneumatica* was published posthumously in 1575.[7] The Strasbourg mathematician also exchanged letters with Petrus Ramus (Pierre de la Ramée, 1515–1572), who besides having a large library of Greek mathematicians owned works by Hero.[8]

---

1 *Heron Mechanicus* (even the choice for a title is indicative) was first analyzed under this aspect by the Danish philologist Johan Ludvig Heiberg (1854–1928) in an essentially unknown paper that was nonetheless path breaking and one of the most important publications on the Strasbourg Cathedral clock; Heiberg 1886.

2 On the Alexandrine *mechanici* in particular and the frequently discussed problem of Hero's life time see Drachmann 1948, 1963, 9–10; Drachmann 1969, 2–5.

3 Dasypodius 1580b, fol. E[r].

4 *Ibid.*, fols. E[v]–EII[r], EII[v]–EIII[v].

5 *Ibid.*, fol. EII[r].

6 Dasypodius 1572, fols. AVII[r]–AVII[v]. The work of Hero mentioned here was *Pneumatica* with several quaternions missing; Schmidt 1898, 181. On *Sphaericae Doctrinae Propositiones* see Delambre 1817, v. I, 119–121.

7 Heiberg 1886, 11.

8 *Ibid.*, 10–11. "*Studiose vel curiose potius Heronis opera nobis exquisita sunt, tandemque e variis bibliothecis collecta graece et manu descripta πνευματικα integra, αυτοματοποιητικα multis locis corrupta, fragmenta quaedam geodetica: alia vero pleraque Heronis praedicantur, ut de machinis bellicis, earumque figuris eleganter descriptis*"; Ramus 1569, 35. "*Dasypodius nobis etiam*

 | DOI:10.1163/9789004423473_014

Hugo Blotius (1533–1608) played a decisive role in the acquisition of the manuscripts of ancient authors. He was born in De Lier, near Delft, enrolled for studies in Leuven in 1558, and continued his education in Toledo in 1560. In 1561 Blotius returned to Leuven, went to Orléans five years later, and from there went to Basel in 1567. In 1569 the school masters of the Strasbourg Academy offered him a teaching position for ethics, but in the following year Blotius left for Italy as the preceptor of the Franconian nobleman Ludwig von Hutten. There he remained in Padua most of the time, but undertook trips to Venice, Pisa, and Rome, among other places. In 1575 he was called to the imperial court library in Vienna and in the following year was appointed to a professorship in rhetoric at the University of Vienna. In 1576 he converted to Catholicism and was raised to the ranks of nobility.[9] Blotius acquired books for Dayspodius,[10] assisted in the purchase of manuscripts, and established contact with Federico Commandino.[11]

The transactions between the two scholars can be traced in the numerous letters written to Hugo Blotius by Dasypodius that have been preserved in the Austrian national library. On 1 June 1571, Dasypodius wrote that a friend of Blotius should buy all of the Greek and Latin manuscripts that he could find in Rome.[12] He was particularly interested in a manuscript of the *Collectio* by Pappus, which Blotius had offered to him and on which lengthy negotiations were carried out. In August 1571 Dasypodius demanded that the works of Pappus and Diophant be sent,[13] and at the end of the year the sale of numerous other manuscripts was mentioned.[14] For the period between April 1572 and July 1574 no letters to Blotius remain; however, it is obvious from a letter

---

familiaribus literis notus, Herlinum Argentinae exuscitat, et quas opes animi Herlinus Platonico illo nescio quo zelo occuluerat, in vulgus expromit"; *ibid.*, 67.

9 On the biography of Blotius and his stay in Strasbourg see Knod 1895; Menhardt 1951; Brummel 1965; Unterkircher 1968; NDB, II, 316–317. On his activities in Vienna see Mosel 1835, 33–54; Aschbach 1865/88, v. III, 360–375; Smital 1920, 50–54, and Smital 1926, 772–788. Smital mentions an edition of the correspondence of Hugo Blotius which was prepared for publication by Th. Gottlieb and was to include an extensive biography; Smital 1920, 97, n. 15. Unfortunately, the manuscript has been lost; Leitner 1968, Preface, IV.

10 Leitner 1968, 184–193 and 206–207.

11 ÖNB, Ser. n. 363, fol. 7^r. Letter of recommendation from Blotius to Commandino dated 7 July 1571; Cod. Vin. 9737 z18/V, 335 (undated). Dasypodius inquired of Blotius about Commandino and asked about a number of published works on architecture, mechanics, and mathematics. The page was probably part of the letter of 23 April 1570 in Cod. Vin. 9737 z14/I, 9. On Commandino and his editorial work see Rose 1975, 185–221.

12 ÖNB, Cod. Vin. 9737 z14/I, 48.

13 *Ibid.*, 56: Letter dated 14 August 1571.

14 *Ibid.*, 73: Letter dated 11 December 1571. Dasypodius instructs Blotius to buy the manuscripts and suggests prices.

written by Dasypodius to Heinrich Rantzau on 5 April 1573 that the Pappus manuscript was on its way.[15] The collection of manuscripts thus gathered was to be edited in a multivolume "*Corpus Mathematicum*", the fourth volume of which was to be a compendium of mechanics and architecture.[16] Dasypodius must have started planning this shortly before 1570, as a charter of privilege for this from Emperor Maximilian II and a missive from the same year exist.[17] As a result of his teaching and other duties at the Academy, the work on this project dragged on for years, even though he had requested—probably without results—relief from school business in 1571 and 1574.[18] The large-scale project did not make it beyond a plan that was not published until 1593,[19] and fragments for the first two volumes.[20]

15 In this letter Dasypodius talks about his planned "*Corpus Mathematicum*" and mentioned: "*Inter caeteros libros antiquos etiam nactus sum ex Pathmo insula et iis locis Henonis* [sic] *Alexandrini summi mathematici opera, ita ex Peloponeso eiusdem alia quaedam, praeterea codicem magnum antiqua graeca manu scriptum, qui fuit Cardi. Bessarionis, quod ipsemet propria sua manu testatur, ad haec Diophanti lib: arithmeticos sex pulcherrime scriptos. Taceo alios, quos prolixum esset enumerare. Unum tantum Pappi Alexandrini codicem nominabo, quem his diebus praeteritis Roma ad me perferri curavi et emi 21 coronatis gallicis* (Among the other old manuscripts I have acquired from the island of Patmos and surrounding regions the works of Hero of Alexandria, the most important mathematician, and likewise from the Peloponnese other works by him; furthermore a large handwritten codex in ancient Greek writing which had belonged to Cardinal Bessarion, to which he himself attests in his own hand, as well as six arithmetic books by Diophantus, which are particularly beautifully written, not to mention others, as it would be too much to list them. I will only mention the codex by Pappus from Alexandria which I had brought to me from Rome a few days ago and which I bought for 21 French crowns)"; Schleswig, Schleswig-Holsteinisches Landesarchiv, Gutsarchiv Breitenburg (Depositum), Handschriften, 293, 63–64. The manuscript was burned in 1870; Treweek 1957, 205. (However, 23 May was given incorrectly as the date of the destruction of the Strasbourg library). Treweek did not know who owned it. He assumed that the manuscript that Johann Wilhelm Camerer saw in 1795 and excerpted had been in Strasbourg since the early seventeenth century. There is a copy in Oxford, which includes Book II, which was missing in the Strasbourg manuscript; Bodleian Library, Savile 9 (Summary Catalogue, 6556); Treweek 1957, 203 and 216.

16 First mentioned in the forward to *Sphaericae Doctrinae Propositiones* (Dasypodius 1572, fol. AII^{v}) and later in *Oratio de Disciplinis Mathematicis* (Dasypodius 1579, fols. BIV^{r}–BIV^{v}). The latter work included a Latin translation of excerpts from the mathematical works of Geminos, which were attributed at that time to Hero or Damianus of Larissa; RE, XIII, col. 1039, s.v. "Geminos" by Karl Tittel.

17 AMS, Sér. V, 14, no. 5, 188.

18 AMS, Sér. V, 14, no. 189.

19 Dasypodius 1593b, 2–22; *Appendix 3*. See also the page in a letter from Paul Virdung to Kepler, which was printed in 1599 and dedicated to Landgrave Moritz of Hesse-Kassel; reproduced in *Appendix 4*.

20 Dasypodius 1593a. A second excerpt appeared in 1596.

The reason for the failure of the editorial project was, in the end, sizable financial problems with which Dasypodius was confronted in 1593. He had married a total of four times. His second wife left him with large debts that he had to meet after her death. In addition, Dasypodius felt obliged to return her dowry to her relatives as there were no children from that marriage. As if that were not enough, he listed himself as guarantor for one of his sons-in-law, who was betrayed by his associate and went bankrupt, losing his entire estate, between 4,000 and 5,000 guldens.[21] This development had been anticipated by Dasypodius, which is why—as Paul Virdung reported in a letter to Johannes Kepler[22]—he bequeathed his books and manuscripts to Landgrave Moritz of Hesse-Kassel as his legatee. They were supposed to be incorporated into the landgrave's library, but Virdung was not able to say what finally happened to the library. He was of the opinion that Dasypodius had bequeathed it to his son-in-law and expected more information from a friend.[23]

---

21 ÖNB, Cod. Vin. 9737 z17/IV, 280; Letter to Hugo Blotius of 31 October 1593, old calendar.

22 Little is known about Virdung's career or his biographical details. He studied Medicine and Mathematics (Witte 1677, 391f.), corresponded with Kepler and had stayed temporarily with Conrad Dasypodius in Strasbourg. Virdung traveled France and Italy as *hofmeister* of Wallenstein on his cavalier tour ("*quam cum Illustri Barone a Waldstein per Gallia et Italiam suscepi*"; KGW, XIV, 439) and is said to have awakened Wallenstein's inclination towards astrology; see Geiger 1983, 78–81, 385–390. In a letter to Kepler on 20 March 1603, Johannes Papius (1558–1622) labeled him as a "*Doctissimus vir dominus* [...], *non vulgaris mathematicus*"; KGW, XIV, 388.

23 Letter to Johannes Kepler from 14 February 1604; ÖNB, Cod. Vin. 10703, fol. 257^r; reprinted in KGW, XV, 34. In the letter, Virdung goes into detail on Dasypodius and describes a rather depressing picture of his circumstances and family situation in the last years of his life: "In Dasypodius's library I saw all the books listed there which he had caused to be copied, while he was in France, at great expense in the royal library established by Francis I. He also had certain other books, already printed, namely, mathematical works of Archimedes, Commandino, and certain [other] Italians. But even though I have searched carefully, I have not yet been able to find out where all these things went. For that fine old man was quite unhappy on account of his wife and children, who brought him terrible trouble and impeded his serious projects. For of the four wives whom he had married, the first and the last, who is still alive, were very well behaved and most attentive to family matters. But of the third one, who suffered from a certain incurable disease, he dissolved the marriage after the third month. The second one, however, who was too extravagant, in a short time contracted a huge debt, which he was compelled to pay after her death and at the same time to repay to her relatives her dowry as a wife who died childless. Hence almost all his property would have been dissipated had not the post of ecclesiastical dean [in the Chapter of St. Thomas in Strasbourg], which he held at that time together with his professorship of mathematics, supported his family structure, which was about to collapse. As far as his children are concerned, of his two sons, who were about to enter the age of manhood, he tearfully attended the funeral processions, which occurred within an interval of a few months. One of them had suffered more than fifteen years and had pityfully

The estate, consisting of books, manuscripts, and scientific instruments, however, did not leave Strasbourg,[24] and was purchased by Matthias Bernegger (1582–1640) in 1613.[25]

The scholar, who was born in Hallstatt in Upper Austria and was awarded a professorship for rhetoric at the Strasbourg Academy in 1626–1629 to accompany the position of professor of history which he had had since 1613, had been living in Strasbourg since 1599. He was a friend of Kepler and translated Galileo Galilei's text on the proportional compass in 1612 as well as his *Dialogo* into Latin in 1635.[26] Bernegger was an ardent bibliophile and invested large sums in his "*bibliotheccula*" (little library), as he modestly liked to call his extensive

---

been released from his body, whereas the other was already married and had recently gladdened him with a grandchild. Two daughters are left. One of them, made pregnant by a certain noble friend of his, afterwards married a certain private scribe and was taken away to Meissen. The other, who still lives in Strasbourg, chose a certain craftsman as her husband. When he unhappily entered into a contract, defrauded by his associate, he was deprived of almost all his goods. As I recently learned, almost all his furniture was auctioned off, and I believe that Dasypodius's books, which he had descended to him by heredity, also were on the block, especially since Dasypodius, while still alive, acting as a guarantor, had pledged his credit for his son-in-law. Dasypodius had foreseen all this, and therefore, as he frequently mentioned to me, he had decided to bequeath his manuscripts to the Illustrious Maurice, Landgrave of Hesse, so that they would be added to the Landgrave's library. It is not yet clear to me what happened after Dasypodius's death. Within a short time, however, I expect an answer from a certain friend who, I trust, will explain the particular circumstances to me. It undoubtedly seems that by a secret power of fate those who carefully pursue these excellent studies, removed from the understanding of ordinary people, find it necessary to contend with many sorrows and misfortunes which either delay or completely block their efforts, worthy of praise and immortal fame"; KGW, v. 15, 33–34 (translation in Rosen 1986, 67, 69, 71).

24 This can be seen from the inventory of his estate: "*Item so Ist deß Hn. seeligen liberey vnd bibliotheca auch allerhandt Instrumenta zur Mathematischen Kunst gehörig seindt sampthafft hiehero zugeschlagen für und* [...?] *25* ≤ ⨍ (Moreover the library and collection of the departed, including all kinds of instruments for the mathematical arts, has been auctioned here in its entirety for and [...?] 25 ≤ ⨍)"; AMS, 58 NOT 21. Unfortunately more specific information on the book titles or the instruments was not given. I thank Dr. Liliane Châtelet-Lange for pointing this out.

25 In a letter to the New Latin author Michael Virdung (1575–1637)—a brother of Paul Virdung—Bernegger writes on 8 January 1619: "*De bibliotheca Dasypodii quod quaeris in alteris tuis, eam annis abhinc quinque vel quod excurrit, emi florenis 150, sed libatam curiosa ac paene dixerim furace scioli cuiusdam manu, quae mihi tamen ex inscitia potius quam incuria intactam reliquit nobilissimam partem, scilicet mss. graeca, ex quibus pleraque ασκδοτα, quorum editionem adornassem, si, quod animo votoque destinaram, studio mathematico, quod inde a puero me cepit, ultra vacare licuisset*"; Reifferscheid 1889, 101.

26 Bünger 1893, 61–91. Further biographical information: ADB, II, 412–413; NDBA, II, Strasbourg 1983, 185–186 (Marcel Thomann); NDB, II, 106–107.

collection.[27] The hardships of the Thirty Years' War, however, forced him to sell some of his treasures afterwards. On 13 April 1625 he wrote to the Dutch poet Daniel Heinsius (1580–1655) in Leiden on this matter, who like his brother-in-law, the lawyer, diplomat, and philologist Janus Rutgers (1589–1625), had expressed an interest in a catalog of his manuscripts. According to Bernegger, there was no one in Strasbourg who was competent enough to appreciate their value, so that he took that task upon himself and sent a detailed list to Heinsius, in which extensive information on the content and in some cases on the provenance of the works was noted.

Fortunately, the draft of Bernegger's letter has been saved.[28] According to it, Bernegger was in possession of the following manuscripts which, with the exception of the first, all came from the estate of Conrad Dasypodius:

1. Three books on harmony by Manuel Bryennius.[29] A book by Aristeides on Music.[30] Johannes Pediasimus:[31] survey on the measurement and dividing up of the Earth.

This collection of manuscripts was owned by the French physician and state councillor Louis Demoulins Rochefort (d. 1609).

2. Three books on harmony by Manuel Bryennius. Hero: definitions of terms in geometry; measuring straight lines, surface areas, and volumes (today collectively termed *metrica*). Hero: Introduction (*Isagogica*). Didymos of Alexandria:[32] the measurement of marble and all sorts of wood. Tractate by an anonymous author on measuring surface areas.

3. Hero: two books of *Pneumatica*; *Belopoetica*; *Dioptra* (text on a leveling instrument). Athenaeus[33] and Biton on 'art works' (text on war machines).

27 Reifferscheid 1889, 100.

28 Draft of a letter in Bernegger's handwriting, dated: 3/13 April 1625: Hamburg, Staats- und Universitätsbibliothek, Supellex epistolica Uffenbachii et Wolfiorum, 4°, 31, 169–173; transcription in Oestmann 1993/2000, 249–251. The manuscript catalog was reprinted in Heumann 1722/32, v. II, Book 2, 226–229. On Daniel Heinsius (1580–1655) and Johannes (Janus) Rutgers (1589–1625), see ADB, XI, 653–656 and XXX, 42–44.

29 Byzantine music theorist, lived around 1300; Sarton 1927/48, v. III.1, 745–746.

30 The Greek musicologist Aristeides Quintilianus (third or fourth century) wrote three books on music, first published in 1652; RE, II.1, cols. 894–896.

31 Keeper of the seal of the patriarch of Constantinople during the reign of Andronicus III Palaeologus (r. 1328–1341), rhetorician, mathematician, and Aristotle commentator, also called Galenus; Sarton 1927/48, v. III.1, 682–683; Cantor 1907, v. I, 510.

32 Greek philologist, grammarian, and lexicographer, born around 65 BC; Sarton 1927/48, v. I, 233–234.

33 Greek mechanicus; *ibid.*, v. I, 196.

Hero: Clasmata;[34] On Making Automata (*Automatopoetica*). Euclid: Catoptrics (theory of mirrors).

This collection of manuscripts was passed on to Matthias Bernegger when Dasypodius died in 1613 and was kept under the catalog number C III 6 in the library of the Protestant Seminary. There it was burned during the siege of Strasbourg by German troops on 24 August 1870. On the cover "*Mathiae Berneggeri ex biblioth. Dasypod. 1613*" was inscribed.[35]

4. Ptolemy, Harmony.

A copy of the manuscript from the collection of Sebastian Müg is in the University Library in Uppsala. Dasypodius gave it to David Wolckenstein as a gift on 24 June 1575, exactly one year after the clock was completed.[36]

5. Euclid: Elements (*Stoicheia*); Data; Optics; Catoptrics. Serenos:[37] on conic sections; on sections of a cylinder. Theodosius:[38] three books on spherics. Autolycus of Pitane:[39] on the moving sphere. Euclid: *Phainomena*. Theodosius:

---

34 This was an abridged *Pneumatica*, with some quaternions missing.

35 Descriptions of the manuscript in Vincent 1858, 170–172 and 431; Wescher 1867, 35–36; summarized in Schmidt 1899, v. I, Suppl., 26–27. Both the municipal library as well as the library of the Protestant Seminary were stationed in the choir of the Temple-Neuf (the Dominican church). On the history of their collections and the catastrophe of 1870 see Reuss 1871; Rathgeber 1876; Rott 1971.

36 Ms. gr. 45; Graux 1889, 353. At the end of Ms. gr. 45, fol. 1, was noted: "*Ex bibl. Sebastian Miegii*". A manuscript of *Geometrie* by Johannes Pediasimus, which had also belonged to Dasypodius but does not appear in Bernegger's catalog, is also in Uppsala; Ms. gr. 46, on fol. 1: "*Sum Conradi Dasypodii et amicorum*". Another manuscript was acquired by the classical philologist Richard François Philippe Brunck (1729–1803) and today is in the Bibliothèque Nationale in Paris, Suppl. grec, 335; Graux 1889, 308. A number of other codices from Sebastian Müg's collection are in the University Library in Uppsala, which were probably copied to produce manuscripts for Dasypodius's library; *ibid.*, 355–356. They are: Ms. gr. 48 (Sebastian Müg's copy?): *Conica* by Apollonius (Books I–IV without demonstrations); a reprint of the Apollonius manuscript from the library of Regiomontanus; Nuremberg, Stadtbibliothek, Cent. V. app. 6. Included in the codex are also the *Opuscula* (?) of Serenos; Ms. gr. 49: Pappus (possibly in Müg's handwriting); Ms. gr. 50: Marinus of Neapolis (student of and successor to Proclus Diadochus), Preface to the *Data* by Euclid; Apollonius, Books I–IV of *Conica*; Serenos, on conic sections; on sections of a cylinder; Ms. gr. 51: Works by Hero; tractate by Didymos on measuring marble; Ms. gr. 53: Works by Barlaam, Diophant, Maximus Planudes, Johannes Pediasimus, Geminos, Ptolemy, Aristarchus, Athenaeus, Biton, Anthemius of Tralles and others; Ms. gr. 54: Works by Hero, Johannes Pediasimus, Proclus, Barlaam and others.

37 Greek mathematician from the fourth century. Of his writings, only the two listed texts are known; Sarton 1927/48, v. I, 353–354; Cantor 1907, v. I, 489–491; DSB, XII, 313–315.

38 Theodosius of Bithynia lived in the second half of the second century BC; Sarton 1927/48, v. I, 211; Cantor 1907, v. I, 411; Wolf 1877, 115–116; DSB, XIII, 319–321.

39 Contemporary of Euclid, lived around 300 BC; Sarton 1927/48, v. I, 141–142; Wolf 1877, 113–115; DSB, I, 338–339.

on inhabited places; two books on the length of day and night. Aristarchus:[40] on the size of and the distances to the Sun and Moon. Autolycus: two books on the rising and the setting of the celestial bodies. Hypsicles:[41] *Anaphorica* (On Ascensions, i.e. rising times of the signs). Theon of Alexandria:[42] Logical methods (user's manual) of the *Handy Tables* of Ptolemy in four sections.

Dasypodius acquired this collective manuscript from the Hungarian polymath Johannes Sambucus (János Zsámboki, 1531–1584), who had purchased it in 1556 from the heirs of Lazzaro Buonamici (1479–1552) for 48 gold crowns and a gold coin. Contact with Sambucus had existed since 1570. In that year Dasypodius had received a Greek Euclid codex from him in order to have it published by Episcopius in Basel. However, neither of them undertook anything on this matter for years, and despite repeated demands, going as far as threatening to go to court, the codex was never returned to Sambucus.[43] This was a canonical collection of short astronomical treatises, which the Arabs collectively referred to as 'intermediate books', as the students were to read them between Euclid's *Elements* and Ptolemy's *Almagest*.[44] Dasypodius combined parts of this manuscript in an anthology that was published in 1572.[45]

6. Diophantus:[46] six books on arithmetic; on polygonal numbers. Scholia from Maximus Planudes[47] for Diophantus.

---

40 Greek astronomer, student of Strato of Lampsakus (c. 310–c. 230 BC). Aristarchus taught about the motion of the Earth around the Sun according to a passage in the "Sand Reckoner" by Archimedes. His *De Magnitudinibus et distantiis Solis et Lunae* dealt with the size of and the distance to the Sun and Moon; Sarton 1927/48, v. I, 156–157; DSB, I, 246–250.

41 Alexandrian mathematician and astronomer (fl. ca. 175 BC). His *Anaphorica* contains the first mention of dividing a circle or an ecliptic into 360 degrees, probably taken from the Babylonians; Manitius 1888; Sarton 1927/48, v. I, 181–182; Cantor 1907, v. I, 358–361; DSB, VI, 616–617.

42 Mathematician and astronomer, Theon lived in the second half of the fourth century and was a teacher at the Museum in Alexandria; Sarton 1927/48, v. I, 367; Cantor 1907, v. I, 491–495; DSB, XIII, 321–325.

43 Details in Gerstinger 1968, 110, 188, 190, 192, 197, 199, 217–218, 222, 229, 233, 245–246, 280, 302–303; Gerstinger 1926, 347; Kaestner 1796/1800, v. I, 341–342. On the biography of Sambucus: ADB, 30, 307–308.

44 Steinschneider 1865, 456–498; see also Cantor 1907, v. I, 447.

45 Dasypodius 1572.

46 Important Greek mathematician (second half of the second century). Father of algebra; Sarton 1927/48, v. I, 336–337; Cantor 1907, v. I, 463–488; DSB, IV, 110–119.

47 Byzantine monk, scholar, and translator (c. 1255–1305); Sarton 1927/48, v. II.2, 973–974; Cantor 1907, v. I, 510–513; DSB, XI, 18.

7. Aristoxenus of Tarentum:[48] *Elements of Harmony* in three books. Alypius:[49] introduction to the study of harmony (music theory). Porphyrius:[50] commentary on Ptolemy's *Harmonics*.

The manuscript was in the library of the Protestant Seminary, cat. no. C III 31, and was burned in 1870.[51]

8. Barlaam monachus:[52] *Logisticae* (computational science); Scholia on Ptolemy's *Harmonics*.

9. Anthemius of Tralles:[53] on paradox mechanical art works. Pappus Alexandrinus: Books III–VIII of *Collectio*.

The first and second books of *Collectio* are missing, as in the posthumously published Latin translation by Federico Commandino: *Pappi Alexandrini Mathematicae Collectiones a Federico Commandino Urbinate in Latinum conversae et commentariis illustratae* (Pesaro, 1588). Commandino indicated that he could not find the two books of *Collectio,* even after an extensive search. Hugo Blotius purchased the manuscript in Rome for 40 gold crowns in 1573.

10. Paulus Alexandrinus: Introduction (*Isagogica*) to Ptolemy's *Apotelesmatica* with several scholia.[54]

Bernegger estimated the value of this collection of manuscripts, which had consistently been written on smoothed parchment-like paper (*charta laevigata*), to be two hundred imperial thalers and was additionally prepared to hand over six other works at no additional cost, including autographs of Dasypodius. The transaction was not completed, however, as Rutgers died in

48 Philosopher and mathematician; student of Aristotle from about 375 to about 360 BC; Sarton 1927/48, v. I, 142; DSB, I, 281–283. The edition princeps of his works was published in Leiden in 1616.

49 The introduction to music written by Alypius (third or fourth century) is the most comprehensive and important source of information about the Greek system of music notation. The editio princeps was dated 1616; RE, I.2, cols. 1710–1711.

50 Neo-Platonic philosopher (233–ca. 304), student of Plotinus; Sarton 1927/48, v. I, 334–335; Cantor 1907, v. I, 457–458.

51 Ruelle 1871, 335–339.

52 A monk who was born in Seminara, Calabria, who became an abbot in Constantinople and later the bishop of Geraci in Italy, where he died in 1348; Cantor 1907, v. I, 509–510; RE, V, cols. 23–24, s.v. "Barlaam" by v. Jan.

53 A Byzantine mathematician, mechanicus, and architect (d. ca. 534). Together with Isidore of Miletus, he built the Hagia Sophia for Emperor Justinian in Constantinople in 532; Sarton 1927/48, vl. I, 427; Cantor 1907, v. I, 501–502; DSB, I, 169–170.

54 Alexandrian astrologer from the second half of the fourth century. On the *Isagogica* see Gundel/Gundel 1966, 236–239; RE, XVIII.2, cols. 2376–2386, s.v. "Paulus" by Wilhelm Gundel. This was probably a copy of the manuscript that Dasypodius gave to Heinrich Rantzau as a gift in 1583. Rantzau had the text printed in Wittenberg in 1586 together with a Latin translation by Andreas Schato; Oestmann 2004, 62–63 and 189–190.

1625 and Heinsius was not willing to accept the collection. Bernegger would have gladly had his valuable collection go to a public library,[55] which he succeeded in doing, at least in part, but not until years later. In September 1635 he offered to sell his "*bibliotheca mathematica*" to the University of Strasbourg, together with several scientific instruments—among them an astrolabe, two proportional compasses, and an armillary sphere. Bernegger considered the old Greek codices "to be the greatest treasure, which could not be obtained for money, even if one were to search throughout all of Germany and many other places".[56] Among the manuscripts there were also papers from Dasypodius's estate, on which he had elaborated for many years.[57] About one year later, the rector and the dean of the University appealed to the council for funds for the acquisition, and the collection was finally purchased for 500 guldens.[58] Somewhat surprising was the manner in which the necessary funds were to be provided. The Styrian student Daniel Tauffkircher had died, and had earned a considerable amount of money giving instruction in mathematics. The decision was made to impose a particularly high tax on his estate because his brothers were papists.[59]

Bernegger's estate was acquired by his son-in-law Johann Caspar Freinsheim (1608–1660), who was offered a professorship in eloquence and politics at

55 On 31 March 1627 Bernegger wrote to the physician Caspar Hofmann in Altdorf: "*Hos vel imprimis aliquo tuto loco et in publica bibliotheca repositos, velim ab interitu servari. Nam si manus imperitorum inciderint* (*me mortuo vix fiet aliter*) *actum est de hoc vere thesauro*"; Hamburg, Staats- und Universitätsbibliothek, Epistolica Uffenbachii et Wolfiorum, 4°, 31, 308.

56 "[...] *den grösten* thesaurum, *welcher mit gelt nit zubekommen, wan man gleich gantz Germaniam vnd mehr Ort durchsuchen solt*"; Letter of 12 September 1635 from Bernegger to Johann Georg Dorsch (1597–1659), professor of theology and librarian of the University of Strasbourg; AMS, Archives St. Thomas, 353/10, 183–184.

57 "*Vnder denen, so* recentiore manu *gschriben, ist sonderlich was Dasypodius in vilen Jaren elaborirt, vnd aussgehen zulassen willens gewesen*"; *Ibid.*

58 AMS, Archives St. Thomas, 353/10, 185–187 (3 August 1636).

59 "[...] *man köndte ordinari und extraordinari Schirmgelt fordern, zumahl weiln seine brüder papistisch seindt und ihme von seinem Vätern und mütterlichen guth nichts wollen volgen lassen* [...]. *Erkannt: man soll die mathematische*[n] *bücher und instrumente von H. Berneggern umb die 500 ß annemen, auf die Bibliothec transportiren, bey dem Schirmgericht ein designation des ordinari und extraordinari Schirmgelts von anno 1626 verfertigen für das viertel jahr 3 £ rechnen* ([...] one could demand ordinary and extraordinary protection fees, especially as his brothers were papists and did not want to have him obliged to his paternal and maternal property [...]. Recognized: one should acquire the mathematical books and instruments of H. Bernegger for the 500 ß, transport them to the library, and have the protection court to claim ordinary and extraordinary protection fees from anno 1626 calculated at 3 pounds for each trimester)"; AMS, Einundzwanzigerprotokolle, v. 117, 1636, fols. 147^r–147^v.

the University of Uppsala in 1642.[60] Part of the estate fell into the hands of Sebastian Müg, a member of an old-established patrician family in Strasbourg, and from him it was finally passed on to Johannes Gerhard Scheffer (1621–1679), who had studied in Strasbourg and became Freinsheim's successor in Uppsala in 1648.[61] Freinsheim later moved to Heidelberg and brought a small portion of Bernegger's library back to Germany. In 1719 the books and manuscripts from Scheffer's estate were turned over to the University Library in Uppsala. As a result of these fateful events, the library of Conrad Dasypodius was scattered and those parts of it which remained in Strasbourg were destroyed in 1870 during the Franco-Prussian War. The manuscript catalog attests to the varied interests of the Strasbourg professor of mathematics and show that he was able to profit from primary sources for developing the basic concepts and program for his plan for the Cathedral clock. Dasypodius himself saw the construction of the Cathedral clock as the "the testing and model of a piece and part of the entire mathematical work".[62]

How his occupation with the writings of Hero affected his concept for the clock will be dealt with in the following. At the end of the preface in *Heron Mechanicus* he wrote:

> Since then, I resolved to explain the building of the Strasbourg astronomical clock in which there are many varied and diverse machines, namely pneumatic, gnomonic, spheropoetic, and automatic, so that anyone might know what they are and how each differs, and so that by my efforts the composition of the machine might be understood. For without this no one can gaze upon this mechanical work with intelligent eyes.[63]

"*Pneumatica*" and "*automatopoetica*" can only refer to Heron's treatise on machines powered by water, wind or steam and *On Making Automata*. Dasypodius discussed the "*gnomonica*" at the beginning of his description of the Cathedral clock in part 2, where he names Proclus as an authority, who had

60 ADB, VII, 348–349; NDBA, XI, Strasbourg 1988, 1021–1022 (Jean Rott).

61 ADB, XXX, 680–681; see also Rathgeber 1888.

62 "*Prob und muster eines stuckhs und theils deß gantzen Mathematischen werckhs*"; AMS, Sér. V, 14, no. 189.

63 "*Deinde ut cum Horologii Astronomici Argentoratensis structuram, explicare statuerim: in quo multa, varia, ac diversa existunt mechanica: ut pote pneumatica, gnomonica, Sphaeropoetica, automatopoetica: quid illa sint, et quae, et qua ratione differant: a quovis intelligatur: et nostri operis, atque machinae compositio facilius perspiciatur*"; Dasypodius 1580b, fol. EIV^{v}.

defined it—together with meteoroscopy and dioptrics—as part of astronomy.[64] Following that is a short excerpt from Book IX of Vitruvius's *De Architectura*, in which various sundials and their inventors are discussed.[65]

"*Sphaeropoetica*" probably refers to the lost work of Archimedes on constructing a model of the celestial sphere, the existence of which was documented by the mathematician Pappus, who mentions it at the end of his categorization of mechanics.[66] Furthermore the term "*sphairopoiïa*" (σφαιροποιεῖν), the 'study of spheres', had numerous definitions in ancient Greece. According to Geminos (first century) it could refer to a subfield of mechanics—the design of astronomical models—, a certain astronomical instrument, a spherical theory of the world, or a spherical system that actually exists in nature (here Geminos may have thought of epicycles and eccentrics embedded in three-dimensional spheres).[67]

In his "*Hypotheses Planetarum*", Ptolemy also referred to the *sphairopoiïa*. In this treatise his aim was to deal briefly with the theories treated in more detail in the *Almagest* for "ourselves", as well as for the makers of instruments:

> We have worked out, Syrus, the models [ὑποθέσεις] of heavenly motions through the books of the *Mathematical Syntaxis* [μαθηματικῆς συντάξεως], demonstrating by arguments, concerning each example, both the logicality and agreement everywhere with the phenomena, with a view to a presentation of uniform and circular motion [τῆς ὁμαλῆς καὶ ἐγκυκλίου κινήσεως], which necessarily was to arise in things taking part in [κεκοινωνηκόσι] eternal and orderly motion and that are not capable to undergo increase or decrease in any way. Here we have taken on the task to set out the thing itself [αὐτὸ μόνον] briefly, so that it can be more readily comprehended by both ourselves and by those choosing to arrange the models in an instrument [ὀργανοποιίαν], either doing this in a more 'naked' way by restoring each of the motions to its respective epoch by hand, or through a mechanical approach, combining the models to one another and to the motion of the whole. Indeed, this is not the accustomed manner of sphairopoiïa [σφαιροποιεῖν]; for this [sort of manner],

64 Dasypodius 1580b, fol. F^r; Proclus 1945, 192.

65 Vitruvius 1983, IX.8, 255; Vitruvius 1969. On the different types of sundials in antiquity see Martini 1825, Ideler 1825/26, v. I, 232–239 and v. II, 7–10 and 13–14; Bilfinger 1886; Drecker 1909, 10–20; Gibbs 1976; Turner 1989, Schaldach 1998.

66 "*Carpus autem Antiochensis nescio quo loco Archimedem Syracusium scribit unum tantum librum mechanicum, qui est de sphaerae constructione, composuisse, reliqua autem eiusdem generis non digna habuisse quae describerentur*"; Pappus 1875/78, v. III, 1027.

67 Aujac 1970; Evans/Berggren 2006, 52–53.

apart from failing to represent the models, presents the phenomenon only, and not the underlying [reality], so that the craftsmanship, and not the models, becomes the exhibit; but rather [the manner] according to which the arrangement together with the different motions under our view, along with the anomalies that are apparent to observers, are subject to uniform and circular courses, even if it is not possible to intertwine them all in a way that is worthy of the aforementioned, but having to exhibit separately each [model] in this way.[68]

The makers of instruments addressed here were not mere craftsmen, they were rather those who designed models of planetary motion. Ptolemy does not provide more details as to what they were to be like; he only speaks about a simple arrangement, possibly with rotatable disks, in which a planet could be moved by hand to the appropriate position for a respective date. This type of instrument probably resembled the *equatorium* used during the Middle Ages for determining the position of a planet in terms of ecliptic longitude and latitude without requiring calculations. The other design mentioned by Ptolemy combined numerous individual models using a "mechanical approach", which were presumably connected to each other with gears, such that the motion of the entire system could be represented mechanically. However, Ptolemy does not provide any technical details either.

Accordingly, there was a tradition in the *sphairopoiïa* of the second century, in which Ptolemy distinguished between the construction of complex astronomical models that illustrated the underlying theory of planetary motion, and a "common" branch in which the underlying reality was disregarded and only the phenomena were displayed (the apparent or mean celestial movements) and the focus was on the skill of the craftsmen.[69]

In his *Heron Mechanicus*, Dasypodius lists each of the mechanical arts that he applied. For shaping and dividing up the celestial globe, the art *sphaeropoetica* was used. For imitating the daily celestial movements he used "that art by which we make things move spontaneously and is called automatopoetica".[70] For the embellishment of the bottom tier with painting, eclipse tables, the calendar disk, and automat a figures, Dasypodius made use of all three arts.[71]

68 Cited from Hamm 2011, 72.

69 *Ibid.*, 72–76 and 94–96; Hamm 2016. On Ptolemy's "Hypotheses Planetarum" see Hartner 1964; Murschel 1995.

70 "*Ea arte, qua res facimus sponte moueri, et automatopoetica appellatur*"; Dasypodius 1580b, fol. GII^{r}.

71 "*Haec itaque omnia ex automatopoeticis, gnomonicis, et sphaeropoeticis desumpta: ad exornandam nostram machinam*"; *ibid.*, fol. GIV^{r}.

For the astrolabe and the Moon dial *sphaeropoetica* is indicated, and the former was augmented by *automatopoetica* for the addition of the hands for the planets.[72] The figures of the four ages of man, Christ, Death, the carillon, and the cock are also inspired by *automatopoetica*; the cock's crowing additionally required *pneumatica*.[73]

Dasypodius does not fail to provide a reason for installing such mechanisms. Just as Hero at the beginning of his treatise On Making *Automata*,[74] he gives as his motive awe-inspiring spectacles: "These pneumatica and automatopoetica have great admiration among the common lot of men, when they hear harmonies or the crowing of the poultry cock without the work of any man".[75] As it is the task of the architect to not only pay attention to proportions and symmetry in every part of the work, but also to add a fitting elegance and agreeable magnificence to the work, the cock and the carillon were installed. As in the first Cathedral clock, its crowing was intended to be a reminder of Peter's denial and Dasypodius placed it on the highest point on top of the weight drive tower.[76]

The treatise clearly invoked the authority of ancient technology and Hero's works in particular. The question arises as to whether Dasypodius borrowed directly from the Alexandrian mechanicus.

The mobile table automaton of Hero had a vertically mounted box as a reservoir into which millet or mustard seed—or dry sand for non-mobile automata—was poured. The content of the box flowed out through a hole, such that a lead weight resting on the content was lowered. The weight was attached to a rope that was wrapped around a shaft, thus turning the shaft and moving the automaton forward. Hero explicitly pointed out that sinewy thread (gut string) should not be used for the mechanism as it expands and contracts too much, depending on the ambient humidity.[77] In this connection the description in Frischlin's poem on the Cathedral clock is enlightening:

> There is a niche on the left side of the structure, facing the north side in the Cathedral, sculptured with wondrous work, rising in the form of a tower at the top, but at the lowest part fastened to a great work where within itself the foundation conceals passageways, which have a use, and hidden wanderings. Within, balanced weights hang down from aloft with hempen ropes. Implanted in these a machine in the rear of the right side

72 *Ibid.*, fols. GIV^{v}–H^{r}.
73 *Ibid.*, fols. H^{v}–HII^{r}.
74 Hero 1899, 339; Murphy 1995, 11.
75 Dasypodius 1580b, fol. HII^{r}.
76 *Ibid.*, fols. HII^{r}–HII^{v}.
77 Hero 1899, 345; Murphy 1995, 12.

> of the structure moves toothed orbs and bears around wheels, and delineates hours and times with silent motion and impelled by the great bulk of the little tower on the left it turns round all the constellations.[78]

For technical reasons alone, there was no need for the ancient requirement, as the heavy weights prohibited the use of gut strings. The installation of a weight drive tower to one side of the clock—as is only found in the Strasbourg Cathedral clock—could well have been inspired by Hero's treatise *On Making Automata*.[79] The mechanisms for moving the ages of man, Christ, and Death on their wooden disks suggest this, as Hero describes a circular ring on which figures of bacchants encircled a temple.[80] As, however, the first Cathedral clock had a disk on which the Three Magi stood, it is unlikely that Dasypodius consulted Heron for his mechanism.[81] It is conceivable that the description of worm gears together with helical gears in Pappus[82] led Dasypodius to the idea of using such gearing for his celestial globe and the astrolabe.

There is no way of proving with absolute certainty that the Cathedral clock borrowed directly from ancient designs, but the programmatic references to Hero of Alexandria and other ancient mechanics in Dasypodius's treatise represent an important example of the early reception of Hero in the sixteenth century, as the first edition of *Pneumatica* did not appear until after completion of the clock (1575), and *On Making Automata* did not appear until 1589.[83]

---

78 "*Est domus a laeva fabricae regione, Boraeum / Versus in aede latus, miro caelata labore, / In formam turris consurgens vertice: at ima / Parte adstructa operi magno: qua pervius usus / Fundorum inter se tegitur, coecique meatus, / Intus cannabeis librata rudentibus alte / Pondera dependent: quibus insita machina dextrae / Atergo fabricae dentatos commovet orbes, / Circumfertque rotas, horasque et tempora motu / Describit tacito, magnisque impulsa sinistrae / Molibus aediculae circum signa omnia voluit*"; Frischlin 1575, fols. GII^r–GII^v, lines 23–26, 1–7.

79 Heiberg 1886, 14, was the first to point this out, followed by Schmidt 1898, 191. In the sphere of Islamic culture celestial models were designed using the type of drive described by Hero. The scholar Al-Khazini (f. ca. 1115–1130), on whom see DSB, VII, 335–351, described a celestial globe, lowered halfway into a box, the upper edge of which represented the horizon. The globe was driven by a weight in a vertically mounted reservoir for sand with an outlet. Text and translation of Al-Khazini's text with commentary in Lorch, 1980; schematic diagram *ibid.*, 302.

80 Hero 1899, 393–395; Murphy 1995, 26.

81 Schmidt 1898, 192.

82 Pappus 1875/78, v. III, 1115 and 1123ff.

83 Bernardino Baldi, *Di Herone Alessandrino degli Automati overo machine se moventi, libri due, tradotti dal Greco*, Venice 1589. Baldi (1553–1617) was motivated to translate this by the editorial activities of his mentor Commandino; Rose 1975, 246. On the edition of *Pneumatica* see Boas 1949, 41–42.

CHAPTER 13

The '*Inventio Prima*' of Dasypodius

Notwithstanding all that has been determined so far, the question to the '*inventio prima*' has not been answered. What motivated Conrad Dasypodius in his decision to display so many things, from numerous automata to the sundials to an astrolabe with hands for the planets to the installation of the celestial globe, which additionally was equipped with an extremely precise mechanical drive system? The history of astronomy by the Wittenberg astronomer and physicist Johann Friedrich Weidler (1691–1755) contains the reference to another printed description of the second Strasbourg Cathedral clock,[1] which has never been examined before.[2] However, it contains the fundamental ideas behind Dasypodius's design written by Dasypodius himself, documented in a collection of scholia on Ptolemy's *Tetrabiblos*.[3] The scholia were added

1 "*A 1570, iussu Senatus Reip. Argentinensis, horologii astronomici, quod hodienum in templo cathedrali visitur, typum primamque formam opificibus et mechanicis in manus tradidit, et in ipsa fabrica perficienda Davidis Wolkensteinii, Augustani, opera usus est. Descripsit hoc horologium in appendice scholiorum et resolutionum, s. tabularum in libros IV apotelesmaticos Ptolemaei, ed. Basil. 1578. f. cum Cardani commentariis in Tetrabiblum Ptolemaei. vid. f. 719. 748. ibidem prouocat ad tr. germanicum, quo uberius de eodem horologio disseruit. edidit etiam tabulas eclipsium solis et lunae ab a. 1573 ad a. 1605*"; Weidler 1741, 380–381.

2 The description possibly remained unnoticed as a result of J.E. Scheibel's comment: "*Die auf dem allgemeinen Titel und auf einem besondern angezeigte Beschreibung der berühmten Uhr zu Strasburg, welche Dasypodius angegeben hat, finde ich meinem Exemplar nicht, sondern nach dessen Aphorismis astrologicis, nach welchen sie stehen sollte, folgt das letzte Blatt mit der Anzeige des Jahres* (I cannot find in my copy the description of the famous clock in Strasbourg indicated with the general title and specified with a special title by Dasypodius, but rather, after his Aphorismis astrologicis, after which it should be located, the last page follows with the indication of the year)"; Scheibel 1769/87, v. III, XV–XVI, Stück (1786), 99–100. C.G. Fix indicated, however, in his history of astronomy: "*Die Beschreibung hiervon steht im Anhange seiner Scholien und Resolutionen, oder den Tafeln zu den 4 astrologischen Büchern des Ptolemäus in der Baseler Edition von 1578 in Fol.; worinne auch ein Commentar von ihm ist zu des Ptolemäi tetrabiblum, auf der 719 bis 748 Seite* (The description thereof [of the astronomical clock in the Strasbourg Cathedral] is in the appendices of his scholia and resolutions, or the tables for the four astrological books by Ptolemy in the Basel edition of 1578 in Fol.; in which there is also a commentary by him on Ptolemy's *Tetrabiblos*, on pages 719 to 748)"; Fix 1792, v. I, 283–284.

3 Dengler 2011, 147–148, explained Dasypodius's inclusion of the description of the clock in his commentary on Ptolemy as follows: "If one takes its [the weekday automat's] performative function seriously, of which Dasypodius himself was so proud, it will become clear that the automaton on its own stage made the course of the planets visible and represented them—the planetary deities were not conceived of as chariot drivers for nothing. In this respect, this

 | DOI:10.1163/9789004423473_015

to a commentary by Girolamo Cardano on this work published in Basel in 1578, entitled *In Cl. Ptolemaei De astrorum iudiciis, aut (ut vulgo appellant) Quadripartitae Constructionis Lib. IIII. Commentaria, ab Autore postremùm castigata, et locupletata.* [...] *Item, Cunradi Dasypodii, Mathematici Argent. Scholia et Resolutiones seu Tabulae in Lib. IIII. Apotelesmaticos Cl. Ptolemaei: Una cum Aphorismis eorundem librorum. Denique brevis explicatio Astronomici Horologii Argentoratensis, ad veri et exacti temporis investigationem extructi.*[4]

13.1 Claudius Ptolemy's *Tetrabiblos*

Like the *Almagest* and the *Hypotheses Planetarum* of Ptolemy, the *Tetrabiblos* (also called *Apotelesmatica* or *Liber Quadripartitus*) was dedicated to a certain Syrus, about whom nothing is known.[5] In the preface Ptolemy explains that astronomy and astrology complement each other. Whereas astronomy, described in the *Almagest*, makes calculating the position of the planets possible, the *Tetrabiblos* discusses the influence of the celestial bodies on the Earth, which Ptolemy considered to be of a physical nature.[6]

Ptolemy's authorship has repeatedly been contested in the past,[7] but since Franz Boll's investigations the attribution of the *Tetrabiblos* is beyond doubt.[8] The philosophical views included in it, as well as the language and the applied astronomical fundamental principles agree completely with the other works of

machine, just as the calendar works, is not only ascribed to a geocentric conception of the universe, but in itself performatively illustrates that view. In accordance with the teachings of Ptolemy, which Dasypodius had dealt with extensively, this machine displayed to the viewer of this time machine the orbits of the planets around the Earth. That is why Dasypodius included a description of his time machine in the commentary on Ptolemy of Hieronymus Cardanus." This is, as will be shown in the following, a completely misguided and absurd interpretation.

4 Cantamessa 2011, v. I, 490–491, no. 1418; Houzeau/Lancaster 1882/89, v. I.1, 778, no. 4856. The scholia by Dasypodius are on pp. 717–758 of Cardano's commentary. This text did not escape C.-E. Wagner's extraordinary diligence, as he commented: "Prof. Scheibel [...] noted that he missed the description of the Strasbourg clock in his copy*, although the title pointed to this. *Which, however, was not true (the relevant description is quite difficult to find in Cardanus); the Paris copy [in the Bibliothèque Nationale] has the note: *deest*, but nonetheless the description is there!"; AMS, 90 Z 22, D 57, reprint of the entire text in 90 Z 23, 7, 1–14. Wagner, however, neither commented on nor evaluated the source in any way.

5 RE, 2nd series, VIII, col. 1795, s.v. "Syros" by W. Kroll; Boll 1894, 67.

6 Ptolemy 1980, I.1, 3.

7 Weidler 1741, 181–182.

8 Boll 1894, 111–180.

this astronomer, the attribution of which has never been called into question.[9] Moreover, astrology in the second century was generally accepted and influenced all the sciences.

Numerous manuscripts of the *Tetrabiblos* have been handed down.[10] Plato Tiburtinus completed the first Latin translation from Arabic in 1138, and in the thirteenth century it was followed by another by Aegidius de Thebaldis. The first printed editions (Venice 1484, 1493; reprinted in 1519),[11] were based on these Latin translations. Joachim Camerarius provided the first Greek edition of the text, which was published together with a Latin translation in Nuremberg in 1535 and once again in Basel in 1553, in this case with a Latin translation by Philipp Melanchthon.[12]

Ptolemy distinguished between universal and special (genethlialogic) astrology;[13] the former dealt with the astrological influences which affect all humanity or cities, countries, and races, the latter was concerned with the horoscopes of individuals.

In the first book Ptolemy first provides evidence that astrology is not only possible, but also useful,[14] followed by a presentation of the influences of the planets, fixed stars, and constellations in the zodiac, and concludes with an explanation of the fundamentals of astrology and its terminology.

Besides an astrological ethnography, the second book contains predictions of wars, epidemics, famines, droughts, etc., for entire countries and cities, based on eclipses and comets, and also includes astrological weather forecasts.

The third book deals with predictions for individuals based on the constellation of the stars at the time of their birth, and finally the fourth book considers external life circumstances, such as wealth, marriage, children, etc. The work concludes with a discussion of the seven ages of man governed by their respective planets.[15]

---

9 *Ibid.*, 111–180.

10 For more details on this see Boll 1899, v. 1, 80–87.

11 Hain 13543, 13544; Houzeau/Lancaster 1882/89, v. I.1, 390, no. 743.

12 C. Ptolemy, *De Praedictionibus Astronomicis, cui titulum fecerunt Quadripartitum, Graece et Latine, Libri IIII. Philippo Melanthone interprete. Eiusdem Fructus librorum suorum sive Centum dicta, ex conversione Iouiani Pontani*, Basel, 1553; Houzeau/Lancaster 1882/89, v. I.1, 685, no. 3636. Melanchthon's text, found on pp. 10–251, is also in CR, XVIII, cols. 11–118.

13 Ptolemy 1980, II.1, 117–119.

14 This defense of astrology was taken directly from the Stoic Poseidonius; Boll 1894, 131ff.

15 Unlike the section already cited about the four winds, four ages of man, etc. (Ptolemy 1980, I.10, 60–65), Ptolemy lists seven ages of man here and, in contrast to his custom, he does not view the Sun and Moon as being in a separate category from the other planets; *ibid.*, IV.10, 441–447. Even though this conclusion was apparently not in the original

13.2 Dasypodius's Commentary to Ptolemy's *Tetrabiblos*

Ptolemy begins Book III of his *Tetrabiblos* with the assertion that there are many starting times for predictions in universal astrology. Predictions for an individual can have either one or numerous starting points. The first one is temperament, in the astrological sense: a combination of qualities which result from a mixture of different celestial influences at the time of conception.[16] Other additional influences affect this initial situation (αρχή) with different degrees of intensity (καταρχή), which Ptolemy discusses in more detail in the last chapter of Book IV, where he considers the seven ages of man. The starting point for the horoscope of a person is therefore conception. Even if the celestial influences change during pregnancy, the constellation at the time of conception remains formative, because, during the growth of the fetus, substance is only mixed, which remains the same and therefore reflects the original influence.[17] As the exact time of conception is frequently not known, the moment of birth is generally taken as the starting point. As far as the level of impact of the stars is concerned, it is as strong as the time of conception, as humans arrive in the world under a constellation that is similar to that at their conception. Ptolemy alludes to a rule referred to as '*Trutina Hermetis*', which is attributed to the astrological compendium of the legendary Egyptian King Nechepso and his high priest Petosiris,[18] and can be found, among other places, under number 51 in *Centiloquium* (also called καρπός, fruit):

> Where the Moon is at the moment of birth, that sign of the zodiac represents the ascendant at conception. And where it was exactly at conception, that [sign of the zodiac] (or the opposite sign) will form the ascendant for the delivery of the child.[19]

---

plan for the *Tetrabiblos*, Boll assumed that this was not a case of interpolation, but that Ptolemy had added this himself at a later date; Boll 1894, 123.

16 Ptolemy 1980, III.1, 221–223.

17 *Ibid.*, 223–225. On this see the section "*Du moment fatidique, conception ou naissance*" in Bouché-Leclercq 1899, 373–393. On the complex problem of the meaning and calculation of the time of conception in ancient astrology see the very thorough and in-depth study by Frommhold 2004.

18 Gundel/Gundel 1966, 27–36.

19 Ptolemy 1961, 48–49. The passage "or the opposite sign" is considered to be an interpolation. Demophilos and Hephaiston only mention Petosiris in connection with this rule. A compilation of the textual evidence is in Riess 1892/93, 357–358, fragment 14. Boll proved that the collection of aphorisms *Centiloquium* could not possibly have been written by Ptolemy; Boll 1894, 180–181; RE, XXIII.2, cols. 1838–1839, s.v. "Ptolemaios, Klaudios IIIA 8b"

Now the astrologer's task is to determine the ascendant at the moment of birth as precisely as possible, by means of which the division of the degrees of the signs of the zodiac into twelve houses is carried out. From the positions of the planets in the houses and from the position of the signs relative to the four *cardines geniturae*, many different kinds of astrological predictions were to be derived, the methods and fundamentals of which were described extensively in Books III and IV of the *Tetrabiblos*.

Determining the time exactly was the basis for prediction and presents a problem, as Ptolemy explains at the beginning of the second chapter of Book III:

> An impasse often arises concerning what is first and most authoritative, that is, the portion of the hour at the time of birth. For in general, only observation at the very time of birth by scientific observers with hour-marking astrolabes is able to submit the minute of the hour; whereas nearly all other hour-marking [devices] to which most of the more careful [astrologers] pay attention are frequently capable of being utterly deceived about the truth, the solar [devices] by occasional distortions of the positions and gnomons, and the water clocks by the cessations and irregularities of the flow of water by different causes and through chance. For this reason it would be necessary for it to be explained beforehand, with a natural and consistent account, how one might find the degree of the zodiac that ought to rise, presupposing the [degree] for the given approximate hour, which is found by the method of ascensions.[20]

Sundials and water clocks therefore do not qualify as time measuring devices for astrologers who work scientifically. With water clocks the starting point for measuring—that is, the time when the operation of the clock is started—is frequently imprecise, and deviations occur, which have various causes, resulting from the design of the instrument. The main problem in operating a water clock—of which there were two basic types in antiquity, namely the inflow and the outflow types[21]—was the fact that water does not flow out of a vessel at a constant rate, but rather slows down as its level decreases. Compensation for this effect was provided in outflow clocks by using a suitable shape for the

by Emilie Boer; see also Zinner 1953. In Richard Lemay's opinion, the text was of Arabian origin and was written by Abū Ja'far ibn Yūsuf in Cairo around 922; Lemay 1977/78.

20 Ptolemy 1996, 8.

21 On the development of the ancient water clocks see Bilfinger 1886, 37ff; Schmidt 1912; RE, XVI, cols. 2428–2433, s.v. "Horologium" by Albert Rehm; Schürmann 1990.

vessel. This was to be achieved by reducing the diameter towards the outlet for the water. The result was a cone-shaped vessel.[22] The inflow types were used either a float valve to control the inflowing water or a reservoir that was constantly being refilled in order to maintain a constant water level.

The diameter of the outlet had to remain constant over the time. Should the outlet get bigger, the clock would speed up; if the outlet were to get blocked by corrosion or dirt, the clock would slow down. Therefore, as Vitruvius reported, Ctesibius made the outlets of his water clocks out of "a hollow tube of gold, or pierced a gem: for these materials are neither worn by the passage of water nor so begrimed that they become clogged".[23]

Another key source of errors was the viscosity of water, which is dependent on the temperature, which invariably changes during extended periods of operation, even in closed rooms.[24]

The "*astrolabium horoscopium*" alone can reliably determine the time of birth to the minute. This instrument could be either the "astrolabon organon" described by Ptolemy in Book V of the *Almagest*, or the planispheric astrolabe.[25]

---

22 Mackensen 1978a.

23 "*Primumque constituit cavum ex auro perfectum aut ex gemma terebrata; ea enim nec teruntur percussu aquae nec sordes recipiunt, ut obturentur*"; Vitruvius 1983, IX.8, 259.

24 Mills 1982/83, 51. Test series with a reconstructed Egyptian outflow clock, based on a clock built around 1400 BC, resulted in deviations from 10 to 20 minutes per day with constant water temperature, which is decidedly too large for astrological purposes; Mackensen 1978a, 17. This is probably what Sextus Empiricus was alluding to when he severely criticized the time measurement methods of the 'Chaldeans', although the real cause of this phenomenon, of course, was unknown in ancient times: "For the scheme of the waterpots mentioned above [§ 24] is of no avail to rescue the Chaldeans, since owing to the flow of water, and owing to the mixture of the air, the flow itself and the times parallel to the flow do not correspond. For as regards the motion of the water, it is probable that it is not the same at the beginning, when the flowing water is clear, and later on, when it is turbid and flows less easily; and as to the mixture of the air, it probably opposes the outflow, acting as a kind of block, when it is misty and rather dense, and gives it more aid when it is pellucid and of fine texture. The jar itself, too, will not leak equally when it is full and when it is half-empty or nearly emptied, but more rapidly at one time and more slowly at another, and at yet another time at a medium pace, whereas the celestial motion continues constantly at an even speed"; Sextus Empiricus 1949, V, §§ 75–78, 355.

25 Ptolemy 1963, v. I, 254–258. The opinions differ on this. Hartner and Segonds assumed that Ptolemy was talking about the planispheric astrolabe; Hartner 1968a, 289, n. 1; Segonds 1981, 21–22. Otto Neugebauer on the other hand held the view that it had to be an armillary device; Neugebauer 1975, v. II, 871. On Ptolemy's *astrolabon organon* see Manitius 1904/05, 402–405; Rome 1927, 77–90; Nolte 1922; Stückelberger 1998.

Whatever the case, both Cardano[26] and Dasypodius[27] clearly understood the "horoscopic astrolabe" to be the second instrument.

Besides using instruments, there was also the possibility of calculating the rising times of the signs of the zodiac (*anaphorai*), as Ptolemy described in chapters VII, VIII and IX of Book II of the *Almagest*.[28] This represented the most reliable procedure, provided the indication of time was exact. Frequently, however, the exact time of birth is not known,[29] so that one is forced to use an alternative method in order to determine the degree of the ascendant more exactly. The Arabs later called this method *namūdār* (Latinized *animodar*).[30]

---

26 Cardano 1578, 327–328.

27 Dasypodius, *Scholia in Cl. Ptolemaei quatuor libros Apotelesmaticos*, in Cardano 1578, 748.

28 Ptolemy 1963, v. 1, 80–100. Ptolemy provided rising times in tables for sections of the ecliptic of 10° each for eleven different climates from 0° to 54°. On this see also the examples for calculations given by Manitius, *ibid.*, 423–426.

29 The astrologists' practice of working with the most precisely determined time of birth possible was criticized even in ancient times, as it is essentially impossible to specify a precise point in time for a continuous process such as birth: "For, firstly, the moment when birth should be said to take place is a matter of doubt,—is it when the child begins to emerge into the cold air, or when it has emerged a little, or when it is deposited on the ground? Secondly, not even in each of these cases is it possible to determine the exact time of the birth; for owing to the present state of the soul and the fitness of the body and the predisposition of the parts and the skill of the midwife and countless other causes, the time at which, after the bursting of the caul, the child is emerging, or has emerged a little, or is deposited on the ground, is not the same but different in different cases"; Sextus Empiricus 1949, V, §§ 65–67, 349–351.

30 It was described by the polymath al-Bīrūnī (973–1048) as follows: "Should no observation have been made at the time of birth, the determination of that time is beyond the reach of science, for there is no way of knowing it, but astrologers by estimation and conjecture arrive at one little different in the sign of the ascendant, when an attentive observer employs cautious questioning. But it is necessary that there should be a certain degree for the ascendant, so they find a way, by using an indicator (namūdār) which furnishes one which they assume to be the degree desired. The indicator most in use is that of Ptolemy, which if it does not disclose the exact degree, is the best substitute. The method in question is to ascertain as accurately as possible the time communicated to you, and determine the ascendant, the cardines and the places of the seven planets. Then find the degree of the conjunction of the moon which occurred before the birth if that was in the first half of the month, or else the degree of opposition, if in the latter half. Then determine which planet has the most dignities and testimonies, then the one that comes next, and so proceed with the others till the last and note the result. The most important testimony is being in aspect to that degree, for when two planets are equal in the number of their dignities, the one in aspect whatever that may be, is preferable. Then examine which of the two most dignified planets is nearest to an angle by counting the number of their degrees. Thereafter transfer the angle to the degree of the nearest planet and derive the ascendant from that. If the degrees of the two planets are very distant from an angle,

First the syzygy has to be determined, that is, the conjunction or opposition of the Moon (new or full Moon) preceding the birth. In the case of a new Moon, one determined the common point of the two luminaries; but if there was a full Moon preceding the birth, one could choose between two positions, of which the one located above the horizon was preferable. This preference coincides with Ptolemy's view that only the celestial regions above the horizon were of importance for astrological prediction.

Now the task was to find the planets that reigned over this position at the time of birth corresponding to the five 'zodiacal essential dignities', which were: 1. Trine, triplicity, or trigon;[31] 2. Domicile or house;[32] 3. Exaltation;[33] 4. Term;[34] 5. Phase, aspect, or face.[35]

If a planet stands in a relationship to one, several, or all of the dignities of a position, then it is the almuten of the figure or ruler of the nativity (*dominus geniturae*), the strongest planet in a horoscope.[36] Then one determined the

---

take the next planet in order of dignity, and examine the others till you find that which is nearest to an angle and proceed as before"; Bīrūnī 1934, §§ 525–526, 328–329.

31 The division of the signs of the zodiac according to the four elements: fire trine: Aries, Leo, Sagittarius; earth trine: Taurus, Virgo, Capricorn; air trine: Gemini, Libra, Aquarius; water trine: Cancer, Scorpio, Pisces. See Ptolemy 1980, I.18, 83–87.

32 Allocation of the twelve signs of the zodiac to the seven planets; see *Tab. 5.1*, *infra* p. 73.

33 The particular points in the zodiac where the planets display their greatest strength are their 'exaltations' (*exaltationes*). Directly opposite an exaltation was the planet's 'fall' or 'dejection' (*deiectio*), in which its effect was at its weakest; Ptolemy 1980, I.19, 89–91; Bouché-Leclercq 1899, 192–199.

34 This was the allocation of the 30° for each sign to the five planets—without the Sun and Moon—using different systems; Ptolemy 1980, I.20–21, 91–107; Bouché-Leclercq 1899, 206–215; Heilen 2010.

35 The fifth dignity was used to determine those things that rise at the horizon besides the ascendant. They are the 'paranatellonta', i.e. constellations, groups of stars, or individual stars which rise or culminate with different fields or degrees of the zodiac. Certain fields or regions of the zodiac which were elevated to divine status, namely the Egyptian decans, which covered the 36 10-degree decans of the ecliptic, were also among the paranatellonta. Arranged according to their orbital times, the seven planets ruled over the 36 decans. In the fields allotted to it a planet shows its 'face', which is why the fifth dignity is sometimes also called '*facies*' (*prosopon*); see RE, XXIII, cols. 869–878, s.v. "Prosopon" by Robert Albrecht Böker; XVIII.2, cols. 1215–1275, s.v. "Paranatellonta" by Wilhelm Gundel; Bouché-Leclercq 1899, 215ff.

36 Bīrūnī 1934, § 495, 307–308. The Arabian astrologers had a weighting system for the dignities: domicile makes 5, exaltation 4, term 3, trigon 2 and facies 1; *Ibid.*, § 494, 306–307. However, in some cases trigon was in third place and facies was not weighted. For the sake of clarity, the entire system could be arranged in a table; e.g. in Schoener 1996, 33; Leovitius 1558, fol. EIII^v^. With such a table one could identify the ruling planet as the "winner" with the most points' so to speak. Heinrich Rantzau owned such an aide-mémoire in the form of a folding table made of gilt brass; Oestmann 2004, 75–85; Oestmann 2018, 92–103.

degree of the sign in which it was located and copied that position into the sign which determined the ascendant, found approximately by calculation. If one had to choose among two or more possible rulers, the planet whose degree was closest to the calculated rising degree of the horoscope was to be taken.

If the positions of two possible planets differed only slightly, the one was to be chosen that stood in a relationship to one of the four *cardines genituræ* and the 'sect' (αἱρέσεις) of the Sun and Moon. Planetary sect is one of the basic Hellenistic astrological concepts, according to which a distinction was to be made between day and night births, and the planets were divided into two 'teams', namely the 'diurnal', led by the Sun, and the 'nocturnal', led by the Moon. This division was also carried over to the signs of the zodiac. The diurnal sector of the signs from Leo to Capricorn corresponded with the sect of the Sun; the nocturnal sector, from Aquarius to Cancer, was ruled by the Moon.[37] The statues of Apollo and Diana lateral to the calendar disk of the Strasbourg Cathedral clock may symbolize this old astrological doctrine as—apart from reasons of symmetry—there was no need for two figures as pointers for the calendrical data.

If, finally, the separation between the degree of the ruler of the nativity and the ascendant calculated approximately was greater than that to the *medium coeli*, the process for determining midheaven was used, so that one proceeded to determine the three other cardines.

Dasypodius in his scholium on the second chapter of Book III of the *Tetrabiblos* first briefly discusses the possibilities just mentioned and the problems of determining the exact time of birth, and lists the four methods:

> Indeed the study of the investigation of the true time is extremely important for the birth horoscopes and for other astronomical calculations; it consists—as Ptolemy teaches us, above all of four things: the knowledge of the [horoscopic] astrolabe, the indicator of the hour, the water clock, which are three instruments. Finally of the study of the risings [of the degrees of the ecliptic]; this refers to the astronomical calculation.[38]

At this point Dasypodius took the opportunity for an excursus on the concept for the Cathedral clock conceived by him:

37 Bouché-Leclercq 1899, 103–104, 155–157 and 188; Brennan 2017, 190ff.
38 Dasypodius, *Scholia in Cl. Ptolemaei quatuor libros Apotelesmaticos*, in Cardano 1578, 748. Both Dasypodius's commentary and the English translation are in *Appendix 1*.

> As, however, of these four we possess a singular example, may I now with benevolent consent be permitted a slight digression to explain, without envy or suspicion of the desire for glory, my thoughts and reflections to which I came when the highly respected Senate of the City of Strasbourg entrusted me with the invention and design of an astronomical clock in the main church of the city: the first model and the first conceptualized form of which I have also placed in the hands of our craftsmen and mechanics, so that they have an impression [of the form]. This form, or this configuration and arrangement of the individual parts of the work as it was invented and designed by me has now been produced artistically or displayed for viewing and observation in a publicly accessible location. If one looks upon it with learned eyes, one will certainly notice that the reflections of my mind on this first invention held these four subjects of which Ptolemy taught us in my considerations.[39]

It is interesting to note that Dasypodius distinguishes between the "*inventio*" (invention) of the astronomical clock and a "*prima inventio*" (first invention). In the language used in the sixteenth century *inventio*, unlike modern usage, did not refer to the original inventing, but rather to a particular technical implementation of it. When chronological priority was of importance, then *prima inventio* was used.[40]

Dasypodius does not claim to have invented, in the modern sense, the astronomical clock as an art form, but only considers himself responsible for his special design. He is however of the opinion that he was the first to have come up with a clock which consistently displayed the indications using the possibilities for measuring time described by Ptolemy. This *inventio prima* is surely identical with the "first concept" mentioned in the *Warhafftigen Außlegung* and is explained by Dasypodius as follows:

> Since the true and well-chosen description of time is to be carried out by instruments or calculations, I saw that it was my job to imitate exactly these methods; therefore if you look for the instruments in our work, you will find the astrolabe, skillfully executed with its circles and lines for the hours, as well as the boundaries of the houses with the mean movements of the planets. You also have three sundials [on the south gable on the outside of the Cathedral], of these one is of the equinoctial or equal hours, the second has unequal hours from the ascent (sunrise) to

39 *Ibid.*

40 Rosen 1954, 308–311.

> the descent (sunset) with the houses of the heavens, the third indicates the hours of the vertical or azimuth arcs, etc. constructed with great zeal, effort, and diligence by me and my colleague [David Wolckenstein]. For the third method, instead of water clocks, which are no longer in use, you have statues and pictures of the seven planets, which depict the seven days of the week named for the planets, the indicator for the Moon displaying the lunar months and their duration; statues of the four ages of man, childhood, youth, middle age, and old age, which indicate the quarter hours with bells; a statue of death, which concludes the hours, besides these statures there are also multiple hands indicating both the mean movements of the planets as well as the hours, fractions thereof, *i.e.* half hours, quarter hours, and fractions of the first order [minutes]. All these I declare as being taken instead of water clocks; thus they are not only pleasant to look upon, they also indicate the times and the fractions of times in select manner, to the extent that this can be carried out by instruments.[41]

The automated figures, the hands for the Moon and the planets are substitutes on the Cathedral clock for the water clocks, which by then were no longer in use—a functional assignment that without this textual evidence could not possibly have been proven, no matter how perceptive the analysis. This is exactly the "particular meaning" given by Dasypodius in *Heron Mechanicus* why the automata of the four ages of man were installed.[42]

Finally Dasypodius had to solve the problem of how to represent the most reliable measurement of time, the method for the ascendant times, and in doing so he remembered the globe in his possession:

> Of the last investigations of time by Ptolemy, the theory of rising times is the most certain. When I deliberated on including them in this astronomical work, the use of a celestial globe came into consideration, because there is no other instrument for investigating time and other astronomical subjects that is more certain or exquisite nor more appropriate for astronomical calculations to such a degree. As I had at hand at that time an exquisite globe of the proper size (which I assert has a diameter of three feet), thanks to the diligence and effort of David Wolckenstein (whom, in

41 Dasypodius, *Scholia in Cl. Ptolemaei quatuor libros Apotelesmaticos*, in Cardano 1578, 748–749.

42 "*Non tamen absque singulari quadam significatione* (not however without a particular meaning)"; Dasypodius 1580b, fol. Hv.

> the second year of the work, I had brought to Strasbourg from Augsburg in order to endure and withstand so many and great hardships, such diverse tribulations, both from the mechanics whom we employed for the work as well as other difficulties) we adorned it with pictures [of the constellations], meridians, parallels, declinations, and magnitudes from the Prutenic Tables; together with the zones, parallel circles, climates, and all the other things which are explained in spherical astronomy. Finally we gave to this most exquisite globe, which we had prepared with singular and exacting care, a soul and daily movement, so that it revolves once in the time of one day, in imitation and reproduction of the *primum mobile*. Therefore you have, my dear reader, the theory of rising times in this globe, and I may say, so exceptionally as you could [only] achieve it through calculation, that is, to the extent to which the imitation of astronomical calculation is permitted with instruments.[43]

The celestial globe in front of the Strasbourg clock with its precise gearing therefore corresponds to the astronomical calculations, it is the real precision clock and carries out its movements in synchrony with the "*primum mobile*".

Once again Dasypodius returns to the esthetic category of 'symmetry'. The "*momentum temporis*" (moment in time) was chosen as the module; therefore all the components of the Cathedral clock are in a harmonious relationship taken individually or as part of the entire work, which results in *symmetria*:

> The regularity of these four [components] can also be perceived easily by everyone. The sundials guide the other instruments, once the true moment of time is deduced from the shadow of a gnomon; everything else is controlled from here, so that you can perceive the true and accurately determined time, or the approximately true time, to the extent permitted by instruments, at any instant with the different instruments for the different intervals of time within the entire work; assuming that reliability, attention, and diligence is added, as can be demanded of such a man who takes care of the work.[44]

In order to depict the "absolute description of time" (*absoluta descriptio temporis*), everything that could not be displayed mechanically was depicted by the painting on the clock:

---

43 Dasypodius, *Scholia in Cl. Ptolemaei quatuor libros Apotelesmaticos*, in Cardano 1578, 749.

44 *Ibid.*

> But it was not sufficient for a carefully calculated representation of time to observe only the four rules set up by Ptolemy; it was necessary to add other things, so that the description of time would be complete and unconfined; and therefore we have accomplished everything which we were not able to achieve by means of figures and hands or by means of gears that were artfully made by those [artisans] who come from Vulcan's workshop, all those things we have expressed with pictures and with the remaining items that were lovingly added to the ornamentation and perfection of the work; such as we have represented Eternity, the Creation and the end of the World, the principal kingdoms, and similar things which are of significance for time with most exquisite pictures; and finally we represented this century with a calendar disk depicting the upcoming one hundred years.[45]

The *inventio prima* handed down by Dasypodius could therefore be compared with a seed of crystallization on which first the technical and then the artistic features of the Cathedral clock were deposited.

45 *Ibid.*

Conclusion

> From this short and limited list everyone will easily be able to recognize that this astronomical work, commissioned by the Senate of Strasbourg for posterity, is unique and worthy of glorification.[1]

These proud words of Conrad Dasypodius are certainly not an exaggeration, as the clock in Strasbourg Cathedral is unique in many ways. It was lavishly decorated with paintings, equipped with complex mechanisms, and displayed a large number of astronomical and temporal indications, which provided all the necessary data for astrological forecasts.

But what was the function of such a mechanism in its church setting? Did the clock only serve to help observe the Divine Office punctually in the early morning hours or assist in the punctual observation of the offices of devotion at the canonical hours later in the day?[2] The wealth of indications given by the astronomical clock suggests not, as they went way beyond mere indications of the time of day. In the Strasbourg case, the thoughts of the clock's designer could be traced by investigating manifold historical sources. However, the questions remain unanswered as to why an astronomical clock was installed in the Cathedral and whether there are possible connections between the clock and the church's architecture or embellishments. Immediately striking is the fact that the clock represents a separate architecture within the Cathedral, which combines both the 'modern' Italian styles—such as classical column arrangements, cornices, etc.—and intricate late Gothic elements—crowned towers, arches adorned with tracery, twined rib vaulting—but does not attempt to integrate itself in the southern transept. Thus one can essentially 'detach' the clock from the wall to which it is attached and view it alone, as was done in an eighteenth-century engraving. As the second main attraction, the greatly enlarged astronomical clock stands next to the western facade of the Cathedral with its steeple (*Fig. 14.1*). The emphasis on a separate architecture for the clock housing was surely meant to underline the understanding of the Cathedral clock as a new, exceptional display piece. It was by far the most ambitious construction project for the Cathedral in the sixteenth century as—apart from the construction of St. Martin's Chapel (today St. Lawrence's Chapel) in 1515–1521

1 Dasypodius, *Scholia in Cl. Ptolemaei quatuor libros Apotelesmaticos*, in Cardano 1578, 750.

2 Sauer 1924, 145–146.

 | DOI:10.1163/9789004423473_016

FIGURE 14.1 Johann Peter Tscherning, The Strasbourg Cathedral and the Astronomical Clock; copper engraving from a 1630 engraving by Wenzel Hollar, 137 × 180 mm

and the new vaulting for St. Catharine's Chapel—the work of the Strasbourg builders' lodge was limited to minor repairs and renovations.

In cathedrals (e.g. Chartres, Laon, Reims, or St. Denis) curiosities were kept and displayed such as ostrich eggs, whale ribs, stuffed crocodiles, narwhal tusks ('unicorn horns'), griffin claws, snake teeth, meteorites, ancient vases, cameos, etc.[3] The astronomical clocks with their automated figures can also be considered such *mirabilia*. They served principally to present God's order in its most perfect form to the believer, an order to be imitated by the human life in its social dimensions. It was a sign of "cosmic, informed piety".[4] Astronomical clocks thus served as a religious teaching tool similar to other artistic adornments

3 Lecocq 1867, 288, n. 1; Sauer 1924, 211–214 and 216–217. "*In the church, the real museum of the Middle Ages, the exact image of the great world outside, all of the components of nature were to have their depository, to inspire and teach the church goers, and to add meaning and embellishment for the church*"; *ibid.*, 217. On churches as cabinets of curiosities see also Laube 2011, 62–74, and Cordez 2015, 125–200.

4 Schmitz-Moormann 1989, 101.

of the church. Such devices also became symbols of the power, status, prestige, and authority of the church, which increased the importance and public perception of an ecclesiastic edifice and were meant to attract people. Just as the cathedrals provided a visualization of the history of salvation and of the liturgical year with their architectural structures, their orientation facing east, and their stained-glass windows and sculptures, the clocks placed inside them also fulfilled these functions. The display of automated figures and the very slowly moving hands on the clocks were surely observed by the public with amazement. In a travel report by the law student Arnold van Buchel (1565–1641) from Utrecht, written in 1587, Cologne Cathedral was described as follows:

> In the Cathedral there is a clock, on which small figures of the three Magi appear, driven by means of a certain weight, an event which is watched by the children and by the people with wonder.[5]

In 1407 the Cathedral Chapter of Chartres installed a clock with the declared intention of increasing the authority of the church and attracting people using the spectacular work.[6] Around 1500 Cardinal Philippe de Luxembourg, the bishop of Le Mans, donated an artistic clock of wondrous invention for the Chapel of St. John the Baptist in the Cathedral of St. Julien du Mans.[7] Similar motivations—artistry never seen before, which eclipsed all previous accomplishments, and spectacular works for the glorification of the church—can be assumed for the construction of the first and second clocks in Strasbourg Cathedral, but in both cases the initiative came from the city government, which provided the financial means for embellishing 'their' Cathedral with a mechanized model of the cosmos. With the construction of the new astronomical clock, visible evidence was clearly provided for the accreditability of the recently founded Strasbourg Academy, whose professor of mathematics provided the design for the clock.

5 "*Im Dom ist eine Uhr, an der mittels eines bestimmten Gewichts kleine Figuren der hl. Dreikönige hervortreten, ein Vorgang, der von den Kindern und vom Volk mit Staunen betrachtet wird*"; Keussen 1907, 70.

6 "*Attenta alta fama et auctoritate hujus ecclesie Carnotensis que hactenus pluribus et dulsis extitit jocalibus et notabilibus operibus decorata, capitulum voluit et ordinavit quod de novo zodiacus et quadrens dicte ecclesie refficiatur pulcriori forma, secundum quod magister Henricus Chicoti retulit in capitulo*"; cited from Lecocq 1867, 329.

7 "*Horologium miri artificii et inventionis admirandae* [...] *in quo annos, solis cursus et ejus per virgulam ascensus et descensus, lunae menstrua revolvuntur*"; cited from Legros 1914, 82, n. 3.

The ambitious project was clearly an external indication of the claim of sole administrative control over the Cathedral by the Protestant city council. The fact that the bishop and the Cathedral Chapter demanded their rights when the Interim was lifted was demonstrated by the simultaneous, enormous renovation of the 'Bruderhof' building, which provided storage and living accommodations for them.[8]

Possibly a symbolic meaning comparable to that of the Cathedral clock can be found in the representation of labyrinths in the cathedrals, which—with the repetition of geometric patterns and intricate passageways—were a metaphor for eternity.[9] Plutarch in his *Life of Theseus* described the Minoan Labyrinth at Knossos, whose architect was supposedly Daedalus. In the Middle Ages, the labyrinth was interpreted as man's erroneous passage through the world and Theseus as the prototype for Christ.[10] However, the labyrinths in Reims and Amiens also had a profane meaning: they added to the fame of the master builders of the cathedrals as a "*Daedalus novus*", and the labyrinth became the archetype of artistic structures.[11] Some of that master builder pride appears in Dasypodius's understanding of himself when he wrote: "Namely that this astronomical horological work was set up first to embellish the temple for the fame of the praiseworthy City of Strasbourg, also for the amazement of many of those who see such and to honor our posterity or progeny, and last of all for the enduring memory of ourselves".[12] In the sixteenth century the Strasbourg mathematician did not need to rely exclusively on the direct effect of his creation. His fame was spread throughout the world by the printing press—of which he made extensive use.

As a demonstration model for astronomical events—thus justifying the name '*Schauuhren*' (display clocks)—the Cathedral clock displayed not only the logically ordered cosmos, but also, on a metaphorical level, the social constitution of the free imperial city. In his poem *Das Glückhafft Schiff von Zürich* Johann Fischart describes the journey of some citizens of Zurich down the

8 According to Büheler, *Chronique strasbourgeoise*, in Dacheux 1887, 127, construction started on 17 December 1571. See also Grandidier 1782, 391–393.

9 Holländer 2000.

10 Ladendorf 1963, cols. 772 and 783.

11 Hahnloser 1972, 38–40; Kern 1982, 220 (Amiens), 225–226 (Chartres), 236–238 (Reims).

12 "*Namlich das dises Astronomisch Uhrwerck aufgerichtet würde, dem Tempel zu einer zierdt, einer loeblichen Statt Straßburg zu rhum, auch maeniglichen so solches sihet zu verwunderung und unserer Posteritet oder nachkomenen zu ehren, zu letzst unserer selbst lang werender gedechtnuß*"; Dasypodius 1580a, Preface, fol. c^{r}.

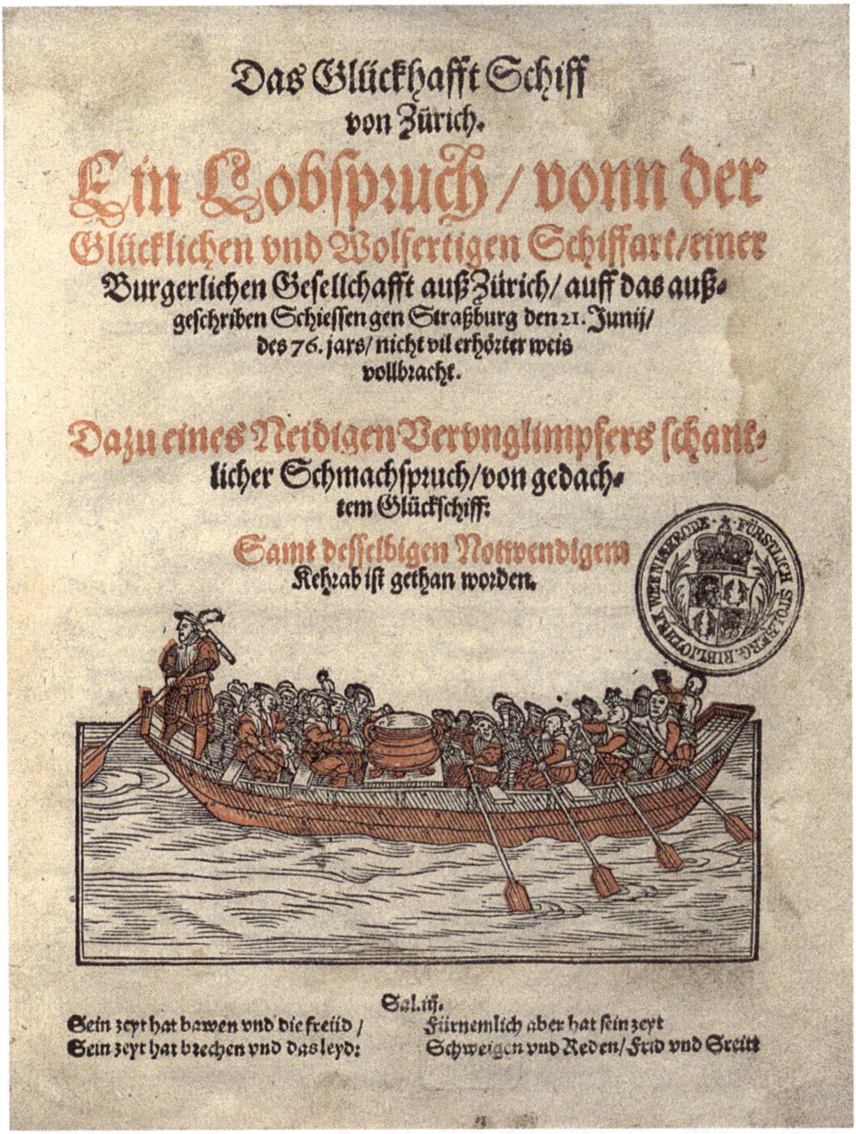

Das Glückhafft Schiff
von Zürich.
Ein Lobspruch / vonn der
Glücklichen vnd Wolfertigen Schiffart / einer
Burgerlichen Gesellschafft auß Zürich / auff das auß-
geschriben Schiessen gen Straßburg den 21. Junij /
des 76. jars / nicht vil erhörter weis
vollbracht.
Dazu eines Neidigen Vervnglimpfers schand-
licher Schmachspruch / von gedach-
tem Glückschiff:
Samt desselbigen Nottwendigem
Kehrab ist gethan worden.

Sal. iij.
Sein zeyt hat bawen vnd die freüd / Fürnemlich aber hat sein zeyt
Sein zeyt hat brechen vnd das leyd: Schweigen vnd Reden / Frid vnd Streitt

FIGURE 14.2 Johann Fischart, *Das Glückhafft Schiff von Zürich*, Strasbourg, 1576, title page

Rhine, who in 1576 rowed to Strasbourg in one day and could hand over a pot of millet that was still warm (*Fig. 14.2*).

In this way the Swiss city demonstrated its willingness to provide help and support at any time in the case of crises. The party of tourists is led into the Cathedral to view the astronomical clock, and Fischart explicitly states that its

'*Künstlichkait*' (artistry or craftsmanship) represented an analogy to the well-ordered government of Strasbourg:

> On Friday one led them later, as is told
> Into the Cathedral where one could behold
> That artistic clock, complete perfection
> The likes of which ne'er finds inspection
> Of it one senses here the artistry
> And with it the city authority
> As naught honors a city as far reaching
> As honest art and proper teaching
> Shall lead them wisely, and guide
> From every class the youth so fine
> Therefore young people so well taught
> Of them vibrant city walls are wrought.[13]

Fischart in his poem praising the civic virtues uses the image of the clock, frequently used in the sixteenth and seventeenth centuries for illustrating hierarchically structured systems.

According to Dasypodius, the "*Beschreibung der Zeit*" (description of time) is what he intended in the iconographic program for the paintings and indications of the Cathedral clock. This statement makes it clear that time meant more for the clock's designer than just a linear continuum that can be described by a digital display. According to this universal claim, an encyclopedic richness of representations of time is assembled, comprising theologically motivated linear timescales,[14] as well as the movements in the heavens of the fixed stars, including those of the Sun, Moon, and planets, displayed by the astrolabe and celestial globe; the "tools of time" (ὄρανα του χρόνου), as Plato called them in his *Timaeus*. Legends and mythologies about stars and the ages of the

13 "*Am Freitag fuhrt man sie darnach / Inn das Münster da man besah / Das künstlich Urwerk ganz vollkommen / Desgleich man nicht vil hat vernommen / Darab man spürt wie Künstlichkait / Auch werd halt dise Oberkait / Dan nichts zirt aine Stat so sehr / Als ehrlich Künst und gute Lehr / Diweil sie weislich führen lenden / Die Jugend fein inn allen Ständen / Daher jung Leut wol angewisen / Das Lebendig Gmäur der Stat hisen*"; Fischart 1977, 33, vv. 905–916.

14 "*Praecipua religionis Christianae capita, depicta sunt: ut Mundi creatio, peccatum originale, redemptio, resurrectio et extremum iudicium. In his enim consumitur temporis uniuersalis descriptio* (Among these, the principal chapters of the Christian religion were depicted, from the creation of the universe, original sin, redemption, resurrection, and the Last Judgment. For in these the entire delineation of time is consumed)"; Dasypodius 1580b, fols. GIII^{v}–GIV^{r}.

world from antiquity and Christianity were merged into a peculiar amalgam, which was by no means considered contradictory. The automata of the four ages of man, of Christ, and the figure of death over the astrolabe challenged the apparently eternal cycles of astronomical and earthly events (the course of the planets through the ecliptic, the alternation of day and night, and the return of the seasons), and every hour the spectator was reproached by the warning '*memento mori*'. In addition to death striking the hour with a bone, the carillon played spiritual songs and the cock crowed, reminding the people of the denial of Saint Peter. By means of automata, a moralizing sermon was addressed to the individual observer.

This part of the program was generally known and had parallels in the sculptures adorning the Cathedral, such as the Last Judgment, which is depicted on the western façade, and the Pillar of Angels near the clock. The situation was different for the equipment for the astronomical-astrological indications, for which Dasypodius started with the four possibilities for determining the time of birth, as described in Ptolemy's *Tetrabiblos*. The rationale behind this could only be understood by competent astronomers and astrologers, and even that group of experts needed a written explanation, which Dasypodius provided in his Latin scholia on the *Tetrabiblos* and had published, but which was not included in either edition of the German descriptions of the clock from 1578 or 1580. That this level of meaning was not understood by his contemporaries is clearly recognizable from the comment by the Cathedral caretakers in their assessment of Dasypodius's complaint about his pay, which he considered insufficient: "But even if we freely admit that we do not understand the astronomical arts, upon which he draws in his work, as well as he possibly does […]".[15] Secreting the central concept behind the clock in an astrological commentary was probably done with the intent of avoiding possible conflicts, as it is unlikely that this 'altar to astrology', which Dasypodius had installed in the Cathedral, would have been fully endorsed by the city authorities and the Protestant clergy. The commentary on the *Tetrabiblos* would surely only have been read by persons who at least sympathized with astrology, and they were the appropriate addressees for the explanations of the clock. The same applies for the inclusion of the architectural theories of Vitruvius and the Alexandrine mechanics in the concepts for the clock, which could only be concluded from the Latin tractate *Heron Mechanicus* and presupposed profound knowledge of the teachings of Vitruvius. The educated observer could not only study the astronomical indications and images in detail, but at the sight of such a

15 "*Dann ob wir gleichwol gern bekhennen, das wir der astronomischen Kunst, die er inn seiner Arbeit anzihet, nicht so hoch als villeicht er verstehen* […]"; AMS, Sér. II, 46, no. 35, fol. 5^{v}.

wondrous work of art also could meditate on the likeness to God of mankind, who as a '*secundus deus*' could approach by means of the mathematical and mechanical arts the perfection lost with the Fall of Man—a leitmotif of the 'machine books' of the sixteenth and seventeenth centuries.[16] Apparently this level of interpretation was intended exclusively for an erudite audience, while the unknowledgeable public could only gaze in wonder at the display side of the work with its automata. This interpretation corresponds to Dasypodius's understanding of himself, as someone who did not think of clockmaking as a mere craft, but as a discipline within architectural theory in the Vitruvian sense and considered his own role to be that of a *mechanicus logicus*. The criticism that was mentioned by Dasypodius in his German description of the clock most probably resulted from a not immediately intuitive understanding of the concept behind the entire clock and the relationship between its individual components.

Dasypodius conceived the iconographic and indicatory program of the clock assuming a multifaceted approach to viewing it, which could extend from amazement without further thought about the movement of the automata, the carillon, and the mechanically produced crowing of the cock to complex interpretations from architectural theory, theology, and astrology. Remarkable is the fact that the automata were included in the concept, without degrading them to being a necessary accessory for the uneducated viewers. Dasypodius did not reduce his plan to mere astronomical and astrological indications.

After a phase of rapid spread of monumental clocks in the fourteenth and fifteenth centuries, which was basically an urban phenomenon and was carried out in part by the urban bourgeoisie, but also in part by the country or municipal nobility or the cathedral chapters,[17] only a few large clocks were built in the sixteenth century, and frequently they were only renovations of older clocks: e.g. Münster 1540–1545, Lübeck 1561–1566, Olmütz 1572–1575, Heilbronn 1580, Ulm 1580–1581, Lyon 1598. Above all this was a result of the Reformation and its purist-iconoclastic tendencies. The clock in Münster, for example, was severely damaged by the Anabaptists in 1534.[18] Among the late monumental clocks, the Strasbourg clock takes a prominent position and the project for constructing a new clock—started immediately after the introduction of the Reformation in

16 Stöcklein 1969, passim and esp. 42–43.

17 Dohrn-van Rossum 1996, 125–172. The list of public clocks preceding 1600 in Baillie 1948 is based in part on Ungerer 1931. However, for some of the dates given, skepticism is called for.

18 Geisberg 1937, 121.

the city of Strasbourg under the supervision of Protestant scholars—indicates that moderate humanistic views prevailed over iconoclastic attitudes.

Monumental clocks represented a model of the macrocosm, which was enlivened by automata in the tradition of Alexandrine mechanics. This relationship was to be reversed with the development of modern science during the seventeenth century.

Philosophical determination based on mechanistic explanations, the development of a mathematical physics, and the view that machines were models of animate systems distinguish the 'modern' philosophers from their scholastic predecessors. While in scholastic natural philosophy bodies were imbued with properties, inner drives, virtues, impulses, and desires, in more modern times quantities, shape, and motion were declared the universally valid 'primary' qualities.

Starting in the middle of the seventeenth century, comparisons of clockwork with the cosmos were common, and Robert Boyle (1627–1691), the most influential representative of mechanistic philosophy in England, saw something in the Strasbourg Cathedral clock completely different from its designer's conception of it. He compared nature to an enormous machine, which in turn was made up of smaller machines. The Strasbourg clock provided him with the appropriate analogue for this. The movements of the clock were so regular that it appeared to be an intelligent creature. Each part of the Cathedral clock fulfilled its function, as if it understood the reason for its motion, but the clock was made of aseptic, lifeless matter, which did not know what it was doing. In his *Essays of the Usefulness of Natural Philosophy*, Boyle wrote:

> That all his works are known to him from the beginning, having resolved, before the creation, to make such a world as this of ours, did divide [...] that matter, which he had provided, into an innumerable number of very variously figured corpuscles [...] that by the assistance of his ordinary preserving concourse, the phænomena, which he intended should appear in the universe, must as orderly follow, [...] as if there were diffused through the universe an intelligent being, watchful over the publick good of it, and careful to administer all things wisely for the good of the particular parts of it, but so far forth as is consistent with the good of the whole, and the preservation of the primitive and catholick laws established by the supreme cause; as in the formerly mentioned clock of Strasburg, the several pieces making up that curious engine are so framed and adapted, and are put into such a motion, that though the numerous wheels, and other parts of it, move several ways, and that without any thing either of

> knowledge or design; yet each performs its part in order to the various end, for which it was contrived, as regularly and uniformly as if it knew and were concerned to its duty. And the various motions of the wheels and other parts concur to exhibit the phænomena designed by the artificer in the engine, as exactly as if they were animated by a common principle, which makes them knowingly conspire to do so, and might, to a rude Indian, seem to be more intelligent than Conradus Diasypodius [sic!] himself, that published a description of it; wherein he tells the world, that he contrived it, who could not tell the hours, and measure time so accurately as his clock.[19]

The ideas of Boyle were closely connected with atomism or rather corpuscularianism. He imagined corpuscles in the form of tiny gears and levers as the fundamental components of the cosmic clock. With its complex indications and automated movements the Strasbourg clock coincided exactly with his concept of nature. Just like the individual components of a clock, bodies also moved along their appointed paths, interacted with their neighboring bodies, and unconsciously carried out the work assigned to them. It was a graphic and haptic world of atoms or corpuscles, which worked in support of and opposing each other, attracted each other, and rotated in order to set the universe in motion. During Creation intelligence was at work, which formed their components with great skill and assigned each a specific site. Since then nature has blindly taken its mechanical course.[20]

Boyle's mechanistic view of the world was therefore not atheistic, but rather he used the metaphor of the clock to explain theism. A complicated clock designed according to a set plan was the work of an ingenious clockmaker and could not possibly have been the result of an arbitrary assemblage of metal parts. The Creator's plan revealed itself in nature.

The emergence of mechanistic explanations of nature therefore benefitted from the existence of complex automata clocks. René Descartes was familiar with the Strasbourg clock and possibly used the mechanical cock as a source of inspiration for his concept of animals as automata without a soul.[21]

The cosmos itself became clockwork and the abstract celestial model became the basis for explaining all natural events. With the introduction of the pendulum for controlling a clock's movements by Christiaan Huyghens, the function of clocks was reduced to merely keeping time. Thus the second

19 Boyle 1772, v. V, 163.

20 *Ibid.*, v. II, 39–40, 45–46 and 48; Mayr 1986, 82 and 94–95; Westfall 1958, 73–75.

21 Mayr 1974, 31.

astronomical clock in the Strasbourg Cathedral, the mechanized model of the cosmos in the church of the citizens of the city, stands with its multifaceted program rooted in the traditional concept of the universe at the end of the period in which monumental clocks were being built, and at the threshold of a century in which time became a mere accidental quality of the motion of matter and was then conceived of as a linear, neutral continuum.

Appendices

Appendix 1

Dasypodius's Commentary on Chapter 2 of Book III of Ptolemy's *Tetrabiblos*: "De parte Horoscopante". *Scholia in Claudii Ptolemaei Quatuor Libros Apotelesmaticos resolutiones* […], in Girolamo Cardano, *De astrorum iudiciis* […], Basel 1578.
Wolfenbüttel, Herzog August Bibliothek: 50.2° Helmst.

English Translation

Chap. 2. On the ascending portion [of the heavens]. [p. 747]

About the horoscope or the ascending portion of the heavens he [Ptolemy] presents two aspects for us. First of all it is important that we must maintain the type [of process] when determining the ascending degree, a procedure that is true and certain, and which he lists in this first part because there are various and different types: the different processes for the investigation of the [ascending] degree.

2. He presents a more certain and more scholarly, a positively select [type of process] [p. 748] that is never to mislead when it is applied using a certain method and in the appropriate manner. This is the case because it is of primary importance that the true and exact time of the nativity (which—the time—is equivalent to the origin of man) is determined.

Therefore it is the following things that are taken into consideration for the investigation of the ascending portion or level: 1. As it often occurs that one cannot determine the true and carefully calculated time of the nativity and the ascending degree, the time is investigated using varying and different methods: either using [horoscopic] astrolabes, which do not provide a small level of certitude, or using sundials and indicators of the hour, which—even if they are accurate—cannot be attributed with the same quality as the first, as they [the determined times] frequently deviate from the true time because of the position of the gnomon or other distorting features of the same, or using sandglasses, water clocks, or other mechanical clocks, because they are produced with the art of stonemasons [in the case of sundials], and are produced by human hands, and are affected by arbitrary pressure or constrictions [of the outflow opening of sandglasses and water clocks]: also those things are subjected to manual, artisanal, and arbitrary restraints, which cannot reproduce natural movement in a selected way and cannot represent the natural movements in appropriate manner or

 | DOI:10.1163/9789004423473_017

the moments of time so exactly, but rather as an approximation; or in the end we use astronomical calculation, as for the investigation of the true and certain time of birth that is the most certain method of all, and no better one than this can be found.

Indeed the study of the investigation of the true time is extremely important for genitures and for other astronomical calculations; it consists—as Ptolemy teaches us—, above all of four things: the knowledge of the [horoscopic] astrolabe, the indicator of the hour [sundial], the water clock, which are three instruments. Finally of the study of ascensions [of the degrees of the ecliptic]; this refers to the astronomical calculation.

As, however, of these four we possess a singular example, may I now with benevolent consent be permitted a slight digression to explain, without envy or suspicion of the desire for glory, my thoughts and reflections to which I came when the highly respected Council of the City of Strasbourg entrusted me with the invention and design of an astronomical clock in the main church of the city: The first model and the first conceptualized form of which I have also placed in the hands of our craftsmen and mechanics, so that they have an impression [of the form]. Admittedly this form, or this configuration and arrangement of the individual parts of the work, as it was invented and designed by me, has now been produced artistically or displayed for viewing and observation in a publicly accessible location. If one looks upon it with learned eyes, one will certainly notice that the reflections of my mind on this first invention held these four subjects of which Ptolemy taught us in my considerations.

Since the true and well-chosen description of time is to be carried out by instruments or calculations, I saw that it was my job to imitate exactly these methods; therefore if you look for the instruments in our work, you will find the astrolabe, skillfully executed with its circles and lines for the hours, as well as the boundaries of the houses with the mean movements of the planets. You also have three sundials [on the south gable on the outside of the Cathedral], of these one is of the equinoctial or equal hours, the second has unequal hours from the ascent (sunrise) to the descent (sunset) with the houses of the heavens, the third indicates the hours of the vertical or azimuth arcs, etc. constructed with great zeal, effort, and diligence by me and my colleague [David Wolckenstein]. For the third method, instead of water clocks, which are no longer in use, you have statues and pictures of the seven planets, which depict the seven days of the week named for the planets, the indicator for the Moon displaying the lunar months and their duration; statues of the four ages of man, childhood, youth, middle age, and old age, which indicate the quarter hours with bells; a statue of death, which concludes the hours, besides these statures there are also multiple hands indicating both the mean movements of the planets as well as the hours, fractions thereof, e.g. half hours, quarter hours, and fractions of the first order [minutes]. All these I declare as being taken instead of water clocks; thus they are not only pleasant to look upon,
[p. 749] they also indicate the times and the fractions of times in select manner, to the extent that this can be carried out by instruments.

Of the last investigations of time by Ptolemy, the theory of rising times is the most certain. When I deliberated on including them in this astronomical work, the use of a celestial globe came into consideration, because there is no other instrument for investigating time and other astronomical subjects that is more certain or exquisite nor more appropriate for astronomical calculations to such a degree. As I had at hand at that time an exquisite globe of the proper size (which I assert has a diameter of three feet), thanks to the diligence and effort of David Wolckenstein (whom, in the second year of the work, I had brought to Strasbourg from Augsburg in order to endure and withstand so many and great hardships, such diverse tribulations, both from the mechanics whom we employed for the work as well as other difficulties) we adorned it with pictures [of the constellations], meridians, parallels, declinations, and magnitudes from the Prutenic Tables; together with the zones, parallel circles, climates, and all the other things which are explained in spherical astronomy. Finally we gave to this most exquisite globe, which we had prepared with singular and exacting care, a soul and daily movement, so that it revolves once in the time of one day, in imitation and reproduction of the *primum mobile*. Therefore you have, my dear reader, the theory of rising times in this globe, and I may say, so exceptionally as you could [only] achieve it through calculation, that is, to the extent to which the imitation of astronomical calculation is permitted with instruments.

The regularity of these four [components] can also be perceived easily by everyone. The sundials guide the other instruments, once the true moment of time is deduced from the shadow of a gnomon; everything else is controlled from here, so that you can perceive the true and accurately determined time, or the approximately true time, to the extent permitted by instruments, at any instant with the different instruments for the different intervals of time within the entire work; assuming that reliability, attention, and diligence is added, as can be demanded of such a man who takes care of the work.

But it was not sufficient for a carefully calculated representation of time to observe only the four rules set up by Ptolemy; it was necessary to add other things, so that the description of time would be complete and unconfined; and therefore we have accomplished everything which we were not able to achieve by means of figures and hands or by means of gears that were artfully made by those [artisans] who come from Vulcan's workshop, all those things we have expressed with pictures and with the remaining items that were lovingly added to the ornamentation and perfection of the work; such as we have represented Eternity, the Creation and the end of the World, the principal kingdoms, and similar things which are of significance for time with most exquisite pictures; and finally we represented this century with a calendar disk depicting the upcoming one hundred years.

I. The years of our Lord from AD 1573 to AD 1673.

II. The years from the Creation of the World, starting from 5535 to 5635.

III. The day of the vernal equinox, when day and night are of equal length.

IV. The hours of [this] day [mentioned in III].

v. The minutes of the hours.
vi. The Sunday *Esto Mihi* [*Quinquagesima*].
vii. Easter Day.
viii. The first Sunday in Advent.
ix. The weeks of the smaller interval.
x. The intervening days.
xi. The weeks of the interval up until Easter
xii. The intervening days.
xiii. The weeks of the interval from Easter until the Nativity [of our Lord].
xiv. The intervening days.
xv. The Dominical letters.
xvi. The letters of the leap years.

[p. 750] This central disk is completed with an outer border in which an eternal calendar is embedded, and to its right, a statue of Apollo points with an arrow to the individual days, to its left Diana or Urania [points] to the opposite day on the inner border, which completes the central disk; it [the central disk] contains geographical information on Germany and a topographical description of the City of Strasbourg.

On both sides, tables are presented with the eclipses up to the year 1605, in which graphical depictions of the individual eclipses are presented together with the calculations. All of this is adorned with the most exquisite pictures executed by the most eminent painter of Germany, Tobias Stimmer of Schaffhausen: they were added to this magnificent work not alone because of their splendor.

From this short and limited enumeration everyone will easily be able to recognize that this astronomical work, commissioned by the Council of Strasbourg for posterity, is unique and worthy of glorification. To that Council and my highly loved native country I, Conrad Dasypodius, Peter's son, dedicate the invention, design, and completion of the entire work in eternal commemoration of this endeavor in lieu of a monument. And as I have profited from the toils and diligence of David Wolckenstein, my most faithful colleague, may posterity accept the integrity, the care, and the diligence of my friend with gratitude and commendation, and also that of Tobias Stimmer, the excellent painter's art and diligence, as well as uprightness, who is to be compared with the ancient painters of Greece and is to be given preference over all our others.

I ask for forgiveness that I have deviated extensively from my intent, which I would not have done if this passage of Ptolemy had not given me the justification for the reflection and consideration of the excellent work and its first invention, as well as the examination of this passage that would have been mandatory to remember and [if not] the matter would have been so extensive that brevity would not have been tolerable. Should, however, someone wish for a more extensive description of our work, then let him read what has been published in domestic [German] language [i.e., the *Warhafftige Außlegung* published by Dasypodius in 1578 and 1589] and the poem by the erudite scholar Nicodemus Frischlin, the laurel-crowned poet.

HIERONYMI CARDANI,

IN CL. PTOLEMAEI

DE ASTRORVM IVDICIIS, AVT (VT VVLGO APPELLANT) QVADRIPARTITAE Constructionis Lib. IIII. COMMENTARIA, ab AVTORE postremùm castigata, & locupletata.

HIS ACCESSERVNT,

EIVSDEM CARDANI, DE SEPTEM ERRATICARVM STELLARVM qualitatibus atq; uiribus liber posthumus, antè non uisus.

GENITVRARVM item XII. ad hanc scientiam rectè exercendam obseruatu utilium, exempla.

ITEM,

CVNRADI DASYPODII, MATHEMATICI ARGENT. SCHOLIA ET RESOLVTIONES SEV TABVLAE IN LIB. IIII. Apotelesmaticos CL. PTOLEMAEI: Vnà cum APHORISMIS eorundem librorum. Deniq; breuis explicatio ASTRONOMICI HOROLOGII Argentoratensis, ad uerì & exacti temporis inuestigationem extructi.

BASILEÆ EX OFFICINA HENRICPETRINA.

CVNRADI DASYPO-
DII, MATHEMATICI ORDINARII ACA-
DEMIÆ ARGENTORATENSIS: SCHOLIA IN CLAV-
DII PTOLEMAEI QVATVOR LIBROS APOTELESMATICOS.
RESOLVTIONES etiam apotelesmatum, & Commentariorum Hie-
ronymi Cardani. APHORISMI quoque aliquot eorundem
librorum. Deniq; BREVIS EXPLICATIO Astro-
nomici Horologij Argentoratensis, ad ueri
& exacti temporis inuestigatio-
nem extructi.

AA

In hoc capite hæc potiſsimũm tradit. 1. Oſtendit diſcrimen eſſe inter uniuerſales prædictiones, & genethlialogicas. 2. Docet quod'nam ſit temporale natiuitatis principium, & quale illud ſit, quæue conſideranda ueniant circa hoc temporale initium. 3. Duos modos proponit huius doctrinæ genethlialogicę. Vnũ antiquorum Aegyptiorum, & Arabum, eumq; ſimplicem: alterum uerò ex ſimplicibus compoſitum, antiquum tanquam infinitum reijcit. Hũc uerò, ut uerum & expeditum recipit, quaue methodo uſurus ſit in hac doctrina indicat.

Vniuerſales prædictiones Aſtrologicas ait natura eſſe priores prædictionib. genethlialogicis. Deinde eas plurimum poſſe in particularibus prędictionibus: ideoq; ordine doctrinæ etiam priore loco explicatam fuiſſe, nihilominus tamen una & eadem eſt uis utriuſq; partis, ſiue effectus conſideres, ſiue contemplationem & doctrinam, quia ex natura, ui & efficatia ☉, ☽ & Planetarum de his inferioribus euentibus naturalibus prædictiones tam generales, quàm particulares inſtituimus. Plura tamẽ ſunt generaliter effectuum initia, quam particularium: nam hominis unum tãtum principale eſt initium, nempe geneſis ipſa, quod quidem initium plura ſubſequuntur alia, eaq; uaria & multiplicia, pro humanæ noſtræ naturæ conſtitutione. Sũt autem directiones, progreſsiones, reuolutiones, & alia quæ in Aſtrologia genethlialogica conſiderãtur, poſt natiuitatis tempus, eiuſq; cognitionem, de quibus in fine horum præceptorum agit.

Temporale natiuitatis uerò initium duplex eſt: unum quidem conceptionis, alterum uerò exitus fœtus ex utero matris. Natura generatio, poſt conceptionẽ prior eſt, ſed natiuitas potentia eſt perfectior generatione. Id quod demonſtrans hoc in loco exemplo ſationis, & alijs quoq; argumentis, ut in textu habetur.

Non abſq; magna, & ſingulari ratione Ptolemæus toties repetit in ſuis hiſce libris, modum doctrinæ, quem ſeruare uult in his præceptis. Vt ἁρμόζοντα τῇ φιλοσοφίᾳ τρόπον, & hoc in loco κατὰ τὸν ἐν ἀρχῇ, τῆς δὲ συντάξεως ὑφηγούμενον ἐπιλογισμὸν περὶ τοῦ δυνατοῦ, τῆς τοιαύτης προγνώσεως. Præterea κατὰ τὸν φυσικὸν στοχασμόν. Hæc ſæpius repeto, propter calumnias eorum, qui cùm artem hanc nunquã deguſtârint, malè non tantum ſentiunt de principaliſsima totius Phyſicæ doctrinę parte, ſed & ſine iuditio, aut peruerſitate quadam animi, & temeritate quadam inepta, euertere conantur dona pręſtantiſsima à Deo Opt. Max. humano generi data, atq; conceſſa. Nam ſi quid in his eſt erroris, ſi quid mali, ſi quid uitij, non rebus, nec rerum cognitioni, ſed imbecillitati noſtræ tribuendum eſt.

Tertius huius capitis locus diligenter notandus eſt ijs, qui in lectionẽ Aſtrologicorum ſcriptorum incidunt: ut Antiquiorum, Aegyptiorum, Chaldæorum, Arabum, Græcorum, Ptolemæi, & recentiorum, qui uel hos, uel illos ſunt imitati. Nam Ptolemæus diligenter proponit duos modos, quibus de hominum natiuitatibus iudicamus. Primus fit κατὰ σύνθεσιν quo antiquiores ſunt uſi, dum ex collectione ſimplicium teſtimoniorum hos & illos prædicebant effectus ſyderum, quem reijcit Ptolemæus, neglecta facilitate, quæ in hoc exiſtit modo. Alterum modum ex ſimplicibus ita compoſitum, ut ſingula conſideremus rerum genera, quę ad quæſtionem ſoluendam faciunt, approbat, nec ſolum approbat, uerum etiam oſtendit methodum huius doctrinę, qua rectiore uia peruenimus ad ueritatis inueſtigationem. Tandem cõcludit, & quę explicaturus ſit de conceptione, ſemine atq; natali tempore proponit.

CAPVT SECVNDVM.

De parte Horoſcopante.

PRoponit nobis duo in Horoſcopo, aut aſcendente cœli parte obſeruanda. Primò eſt, ut modum teneamus inquirendi gradum aſcendentem, eum'que uerum & certum, qui cùm uarius & multiplex ſit, enumerat in hac priore parte: Varias inquiſitionis gradus aſcendentis rationes.

2. Cer-

748 CVNRADI DASYPODII SCHOLIA,

2.Certiorem & magis eruditum,ac exquiſitum modum oſtendit: qui nunquam fallet,ſi certa ratione ac modo bono adhibeatur. Plurimum namq; in eo ſitum eſt,ut uerum & exactum tempus natiuitatis (quod inſtar originis hominis habeatur.

Sunt itaq; hæc quæ conſiderantur in inueſtigatione partis,ſeu gradus aſcendentis. 1.Quia ſæpenumero contingit,ut uerum & exquiſitum natiuitatis & gradus aſcendentis tempus haberi non poſsit,uarijs & diuerſis modis inquiritur: aut per Aſtrolabia, qui modus certitudinem non exiguam habet,aut per horologia ſolaria,atq; horoſcopia,qui etſi certus ſit:tamen primo æqui parari nequit. Nam ſæpius à uero aberrant tempore propter poſitionem gnomonis aut eiuſdē uaria impedimenta,aut per arenaria,hydraulica , aliaq; mechanica horologia, quæ quia arte ſunt facta marmaria,manibusq; ſunt tractanda atq; fortuita intenſione,aut reſtrictione: manuali etiam & artificiali fortuitaq; directione regūtur: naturalem imitari motum exquiſitè nequeunt: neq; ita exactè ſicuti conuenit, temporis momenta repræſentant,ſed proximè,aut deniq; calculo Aſtronomico utimur,in uera & certa temporis natalitij ratione inueſtiganda,quæ omnium eſt certiſsima,nec melior hac inueniri poteſt.

Atqui doctrina hæc de ueri temporis inueſtigatione, ualdè eſt neceſſaria in genethlialogicis,alijsq; ſupputationibus aſtronomicis, quæ ut Ptolemæus decet:in his quatuor potiſsimum conſiſtit,cognitione ἀστρολάβου, ὡροσκοπείου, ὑδρολογίου,quæ quidē tria,ſunt inſtrumentorum. Deniq; πραγματείας τῶν ἀναφορῶν,hæc ad calculum refertur aſtronomicum.

Quoniam uerò horum quatuor,exemplar ſingulare habemus,liceat mihi bona uenia,& abſq; inuidia,atq; captandæ gloriæ ſuſpitione, paululum digredi ad explicationem mearum cogitationum,atq; meditationum,quas habui,cum ampliſſ.Senatus Reip.Argentoratenſis mihi inuentionem & delineationem Horologij Aſtronomici,in templo urbis ſummo extructi demandaſſet: cuius etiam primum typum,primamq; formam delineatam,opificibus & mechanicis noſtris in manus tradidi,ut illam exprimerent.Quæ ſane forma,quæue diſpoſitio & ſingularum partium totius operis diſtributio,ut à me inuenta & delineata fuit,nūc in loco publico affabrefacta,omnibus hominibus uiſenda ſpectandaue eſt expoſita.Quam ſi quis eruditis oculis intueatur,is ſane animaduertet,animi mei cogitationes in prima inuentione reſpexiſſe,ad hæc quatuor de quibus hoc in loco præcipit Ptolemæus.

Cùm enim temporis uera & exquiſita deſcriptio,uel per inſtrumenta,uel per calculum fiat,ideoq; eadem mihi uidebam ad imitandum eſſe propoſita.Propterea ſi inſtrumenta quæras in noſtro Aſtronomico opere,habes Aſtrolabium,affabrè factum, cum ſuis circulis,& delineationibus temporum,atq; domiciliorū diſtinctionibus, medijs etiam motibus Planetarum. Habes etiam ὡρολόγια tria ſolaria,quorum unum eſt horarum æquinoctialium ſeu æqualium, alterum inæqualium,ſeu ab ortu & occaſu cum cœli domicilijs,tertium circulorum uerticalium &c.quæ ſingulari ſtudio labore & induſtria,à me & collega meo ſunt elaborata.Tertio,quia ὑδρολόγια in uſu amplius non ſunt,idcirco habes loco ὑδρολογιῶν,ſtatuas & imagines ſeptem Planetarum monſtrantes ſingulos dies ſeptimanæ à Planetis denominatos,Lunarem indicem, lunationes, & menſtruū ſpatium monſtrantem.Quatuor ætatū pueritiæ, adoleſcentiæ, uirilis ætatis,atq; ſenectutis ſtatuas tintinnabulis ſignificantes horarum quartas partes.Mortis ſtatuam horas determinantem,præter has ſtatuas, iudices complures tam mediorum motuum Planetarum,quam horarum,& partium earundem,ut ſemihorarum,quartarum & ſcrupulorum primorum. Hęc inquam omnia,ut loco ὑδρολογιῶν ſunt habenda:ita etiam uiſa non ſolum ſunt iucunda, ſed & tempora temporumq;

porumq; partes exquisitè, quatenus per instrumenta fieri potest demonstrant.

Postrema temporis inuestigatio, quam Ptolemæus proponit, est πραγματείαν τῶν ἀναφορῶν, quæ omnium est certissima. Eam itaq; cùm in hoc opere Astronomico proponere cogitarem, Globi astronomici incidit mentio, quo non habetur instrumentum ad inuestigationem temporis, aliarumq; rerum Astronomicarũ certius & exquisitius, magisq; calculo Astronomico conuenientius. Cùm itaque tum temporis ad manus haberem elegantem, & iustæ magnitudinis (trium inquam pedum diametrum habentem) globum, opera & auxilio Dauidis Vuol ckenstein, (quem Augusta secundo ab incepto opere anno ad me Argentinam accersiui, ad tot, tantosq; labores, ad multiplices molestias, tam hominũ mechanicorum, quos operi adhibuimus, quàm aliarum difficultatũ sustinendas & perferendas) suis imaginibus exornaui, & singularum stellarum longitudines, latitudines, declinationes, magnitudines ex tabulis Prutenicis adhibui. Vnà cũm Zonis, parallelis, climatibus, cęterisq; rebus, quæ in sphærica explicantur doctri na. Tãdem animam & motum globo huic elegantissimo, & singulari atq; accurata diligentia à nobis elaborato dedimus diurnum: ita ut diurno spatio periodũ faciat suam, ad similitudinẽ & imitationem primi motus. Habes itaq; Lector optime πραγματεῖαν τῶν ἀναφορῶν in hoc globo, adeò exquisitam ferè dixerim, atque per calculum habere posis, hoc est, quatenus instrumentis licet imitari calculum Astronomicum.

Symmetriam etiam horum quatuor facilè quiuis animaduertit. Horologia solaria dirigunt cætera instrumenta, postquam uerum temporis momentum ex umbra gnomonis percipitur: cætera omnia ad illud diriguntur. Vt uerum & exquisitum tempus, uel proximè uerum, quantum per organa licet, quouis momento, per diuersas partes totius operis: diuersarum temporis partium haberę posis. Si ea fides, diligentia, & industria accesserit eius, qui curam operis gerit, quæ in tali requiritur homine.

Verum non satis erat ad exquisitam temporis descriptionem repræsentandã, hæc quatuor Ptolemæi præcepta obseruare: Sed & alia erant adijcienda, ut plena esset, & absoluta descriptio temporis. Itaq; quæcunq; statuis, & indicibus, rotis dentatis, affabrè ab ijs, qui ex Vulcani prodeũt officina factis: efficere non potuimus: illa picturis præstitimus, cæterisq; ornatus, & perfectionis gratia adiunctis. Vt æternitatem, & creationem, finemq; mũdi, monarchias præcipuas, & his similia, quæ temporis significationem habent, picturis elegantissimis expressimus. Postea sęculum ipsum per rotam centum annorum, in qua descripsimus ad annos insequentes centum, hæc.

I. Anni Christi à 1573. usq; ad annum Christi 1673.
II. Anni à creatione mundi 5535. usq; ad 5635.
III. Dies æquinoctij uerni.
IIII. Horæ diei.
V. Minuta horarum.
VI. Dominica Esto mihi.
VII. Dies Paschatis.
VIII. Dies aduentus.
IX. Interualli minoris hebdomadæ.
X. Dies Concurrentes.
XI. Interualli ad pascha Hebdomadæ.
XII. Dies Concurrentes.
XIII. Interualli à pascha ad natiuitatem hebdomadæ.
XIIII. Dies Concurrentes.
XV. Litera Dominicalis.
XVI. Litera Bisexti.

CC Hæc

Hæc media rota, clauditur exteriore limbo, in quo Calēdarium habetur perpetuum, & ad dexterā statua Apollinis, telo dies singulos monstrat, ad sinistram Diana, uel Vrania diem oppositum interiore limbo, quo media clauditur rota, Geographica Germaniæ, & urbis Argentoratensis topographica descriptio continetur.

Vtrinq; ad latera habentur Tabulæ Eclipsium, ad annum usq; 1605. in quibus diagrammata singulorum eclipsium unam cum calculo habentur. Hæc omnia picturis elegantissimis, ab excellētissimo totius Germaniæ pictore Tobia Stimmer, Scaphusiano sunt exornata: præter alia quæ splendoris causa huic magnifico operi sunt adiuncta.

Ex hac igitur enumeratione breui & succincta, facilè quiuis uidere potest, singulare & celebratione dignum esse Amplíss. Senatus Argentoratensis opus Astronomicum, quod POSTERITATI fieri fecit. Cui etiam ego Cūradus Dasypodius Petri filius, & patriæ meæ charissimæ primam inuentionem, delineationem, & absoluti operis perfectionem in perpetuam rei memoriam monumenti loco dicatam uolo. Et quia usus sum, opera & industria Dauidis Vuolckenstein, fidelissimi mei collegæ, posteritas fidem, diligentiam, industriam amici mei, sibi gratam commendatamq; habeat, ut & Tobiæ Stimmeri præstantissimi pictoris artem & industriam, atq; dexteritatem, qui antiquis Græciæ pictoribus conferendus, & nostris omnibus anteferendus est.

Veniam peto, quod longius à proposito sim digressus, quod non fecissem, nisi Ptolemæi hic locus ansam mihi, ut cogitandi & meditandi de tale magnifico opere in prima inuentione, ita nunc in huius loci recognitione memorandi præbuisset, & res tam ampla sit, ut breuitatem ferre nequeat. Sed si quis pleniorē habere uelit nostri operis descriptionem, is legat quæ uernacula lingua sunt edita, & carmen Doctiss. uiri Nicodemi Frischlini Poëtæ Laureati.

PAVLVS MELISSVS FRANCVS, POETA LAVreatus, anno quo absolutum est opus Astronomicum, 1574. hoc scripsit Carmen.

APOLLO TELO ARMATVS AD SPECTATORES MAchinæ Astronomicæ, in templo Argentinensi.

Prodere si uerum sum creditus augur Apollo,
Oraculisq; & sortibus,
Reddere nunc sortes oracula Dodonæis
Hæc ueriora credite.
Qui regit hanc molem, non laus tribuenda magistro est,
Tanquam repertori uafro,
Posteritas norit, dicatq; uoluminis huius
Inuentor est DASYPODIVS.
Quanto maiores hic exantlasse labores,
Compertus est pro patria.
Argentina ipsi socioq; rependere tantò
Maiora debet præmia.

VRANIÆ MVSA.
Ad Senatum, populumq; Argentinensem.

Quod uiuunt hæc cuncta, quod hæc & cuncta mouentur,
Non opus hoc teretum præstitit artificum,
Inuentor CVNRADVS, *& hos animauit & illa,*
Vsus ad hoc socia Dauidis artè sui.

In quos

In quos grata caue minus Argentina feraris:
Nam meruit dignus præmia digna labor.
En rigidis telis alioqui armatus & arcu,
Sæuior ingratis iactat Apollo minas.
Iactat & ipsa minas Diana, quid obsecro, præstat
Iratos meritò pertimuisse deos.
Paulus Melissus Francus, Poet. Laureatus.

Amplißimus itaque Senatus Reipub. nostræ, sempiternam sibi gloriam & nomen perpetuum, hoc Astronomico opore extructo parauit, neque minorem laudem sibi comparauit, dum præmijs eos adfecit, qui suam operam, fidem, & diligentiam, in hoc magnifico opere perficiundo præstiterunt. Vt autem integram descriptionem huius Horologij Astronomici cum picturis, tabulis, calculationibus ipsis, studiosi habere poßint, in lucem propediē ædam, & ita ędam, ut singula non solum ob oculos spectanda proponantur, uerùm etiam ad usum accommodentur Astronomicum, hoc est temporis & motuum inuestigationem.

Sed unde digreßi sumus redeamus. Quæ itaq; sint in natali tempore consideranda, & qua ratione inuestigāda, à Ptolemæo eiusq; interpretibus proponūtur.

CAPVT TERTIVM.

Diuisio eorum quæ in genethlialogicis præcipiuntur.

SVprà huius diuisionis facta est mentio, ideoq; prætereunda hoc loco erit, tantum hoc addam. Ptolemæum quinq; proponere, quæ in unaquaq; genethlialogica quæstione sunt consideranda. 1. Locus de quo quæstio fit, ut de uita, de moribus, aut honoribus &c. 2. Dominus loci, iuxta quinq; dignitates essentiales. 3. Qualitas euentus, ut faustus an infaustus, &c. 4. Vehementia & magnitudo effectus. 5. Tempus quo hæc euenient, quæ per astra significantur.

CAPVT QVARTVM.

De Parentibus.

PRimam partem post genesin & conceptionem dicit esse de parentibus, & statu parentum, ac docet qui Planetæ parentibus familiares sint, patri inquam ♄ & ☉, matri ☽ & ♀, deinde considerandus est horum Planetarum status, & configuratio eorundem inter se. Tertio stellæ quæ circundant hos Planetas. hi enim parentum formam & opes significant, maximè uerò quando stellæ matutinæ ☉ circundant, & Lunam uespertinæ, atq; bono fuerint loco. Hic stellarum positus maximè illustris est.

Est & alius positus Planetarum quem exponit & is medius, quando solus ♄ antecellit ☉, in suo statu, & sic de alijs statibus Planetarum, quos ordine enumerat: & textus cùm per se perspicuus sit non indiget prolixa interpretatione.

CAPVT QVINTVM.

De fratribus & sororibus.

Τὸ καθόλȣ μόνον.) Ex hoc loco intelligimus, non omnia inquiri exactē, & ἐν μέρει, quæ de parentibus, fratribus & alijs traduntur, sicuti Arabes plurimos aphorismos conscripserunt, de ijs, quæ antecedunt natiuitatem: sed tantum generali consideratione facta, tanquam ab hominis natiuitate remotiora inquiri uniuersali quodam modo. Quæ uerò uniuersalia illa sint, breuiter explicat, atq; sic perficit ea, quæ ante partum existunt: nam parentes nati, ut antè fuerint necesse est, fratres uerò & sorores interdum ante fuerunt, interdum ut in primogenitis subsequuntur, aut non subsequuntur, ideoq; primogenitorum natorum de parentum statu natiuitas multùm ualet, ut tradunt Astrologi in iudicijs natiuitatum.

Appendix 2

The Lost Astronomical Clock of 1583

In the University of Strasbourg Library and the Library of Trinity College, Cambridge, there are two descriptions of an astronomical clock, the first of which is in German and the Cambridge manuscript in Latin, most probably written by David Wolckenstein.

1. The Incipit of the Strasbourg manuscript (BNUS, MS 1562 [Alsat. 997];[1] in the following cited as *S*) reads: "*Beschreybung des Astrolabiums und des gantzen Werckhs Nutz und gebruch uff das Kürtzest*" (Description of the astrolabe and the entire utility of the work and usage in the briefest form). The manuscript represents a user's manual for the options for reading an astronomical clock. The author is unknown and the manuscript is not dated; however, its origin can be assigned to 1583 with some certainty, as an example is given for determining the planet governing the hour: "*Freitag, den 21 tag Brachmonat As 1583*" (Friday, the 21st day of the fallow month [June] of the year 1583) (*S*: fol. 5^v) and the tables attached for positions of the planets and the ascending lunar nodes (*S*: fols. 23^v–28^r) start with January 1584.

2. In 1590 David Wolckenstein wrote a description of an astronomical clock (Library of Trinity College, Cambridge, MS 0.2.32;[2] in the following cited as *C*) which differs from *S* in as much as it does not include any tables, and places more emphasis on the astronomical background and definitions of terms. Otherwise the content of this manuscript parallels that of the Strasbourg manuscript to a great extent. Here, as well, an example is given for determining the planetary hour for the same date: "*Exemplum tale esto: Anno 1583 Die XXI Junij …*" (*C*: fol. 9^r).

Théodore Ungerer is credited with recognizing that the two manuscripts belong together and that they do not refer to the Cathedral clock, but rather to the lost astronomical clock that was commissioned by Hans Fugger, built by Isaac Habrecht in 1583, and was probably set up in Stettenfels Castle near Heilbronn.[3] On 5 October 1583, Isaac Habrecht approached the Council of XV because he wanted to have his brother-in-law from Schaffhausen execute paintings on the wooden clock housings, as the local painters—Wendel Dietterlin, Isaak Knoderer and Pangraz Englisch—were either employed elsewhere or they were not up to the demanding task: "*dieweyl der herr Fugger*

1 Manuscript sources, VIII.A.3; transcription in the Wagner estate, AMS, 90 Z 31, 11, 1–11, 56.

2 Manuscript sources, VIII.A.8. The manuscript was seriously damaged by moisture; a copy from a later date is attached to it. Collated text in the Wagner estate: AMS, 90 Z 31, 13, 1–13, 30. The late Director of the Museum of the History of Science in Oxford, Francis Maddison, generously provided the author with a microfilm copy in 1989; my sincere thanks go to the Master and Fellows of Trinity College.

3 Ungerer 1930, 101–102.

sich umb die astronomia wohl verstehe" (because Mr. Fugger was well acquainted with astronomy).[4]

The appearance of the three tiered clock can be reconstructed relatively accurately from the two manuscripts.

a. Bottom Tier

An astrolabe was located on the left side of the lowest tier (*C*: fol. 6v; on the right side according to *S*: fol. 4r), the outer blue ring of which had golden numerals from 1 to 24 for the hours. On the rete—probably on the outer part of the ecliptic ring—the 28 stations of the Moon (*S*: fol. 4r; *C*: fol. 7r)[5] were painted with numbering and the signs of the zodiac on the inside. The names of the stations of the Moon are given in the German description and their meanings explained in detail in verse (*S*: fols. 11r–17v). The astrolabe was equipped with Sun, Moon, and dragon pointers for the movements of the lunar nodes;[6] its copper tympanum plate with the usual system of lines: almucantars (black lines with the indication of the altitude perpendicular to the center of the plate), yellow lines for the temporal hours, horizon and twilight arcs (in gold and red, respectively), white boundary lines for the twelve loci (*S*: fols. 4v–6r, *C*: fols. 8v–10v). This network of lines was painted on a stereographically projected map of the northern hemisphere. The face of the astrolabe probably had a similar appearance to that of the tympanum of the clock in the Münster Cathedral[7] (*S*: fol. 4v; *C*: fol. 8r); the geographical regions, cities, rivers, etc. that were listed on it were numbered in red and the legend was in a table that was held up by the statue of a girl (*S*: fol. 6v; *C*: fol. 10v). There is no mention of azimuth arcs in either of the manuscripts, but their inclusion was very likely.

The sculpture was located between the astrolabe and the calendar plate behind the globe; in one hand it held an arrow that functioned as the index for the calendar plate.

---

4 AMS, 2 R 293, fols. 84r–84v; see also Rott 1933/38, v. III (Sources, v. I), 240–241.

5 These were the endpoints of the signs of the zodiac through which the Moon passes during approximately one day; the ecliptic is divided into 27 or 28 sections, each under the protection of a planet, in the order of the signs starting with spring. As 4 × 7 = 28, each planet (arranged according to the orbital time) dominates four lunar stations. Christian astrologers assigned angels as guardians to them, which is also the case in the manuscript from the Strasbourg University Library. The stations are named after individual stars or constellations close to the orbit of the Moon, because its brightness made stars in its immediate vicinity imperceptible. Their origin is disputed. The knowledge of the stations of the Moon and their astrological importance came to Europe via Arabian sources; Savage-Smith 1985, 119–132; Ackermann 2004; Ackermann 2005, 76–79.

6 Neither of the manuscripts mentions a pointer for the planets on the astrolabe, so that the fragments of an astrolabe found in Castle Babenhausen, near Augsburg, cannot have been part of this clock, as assumed by Ungerer 1930, 106, 108.

7 Wieschebrink 1983, pls. 40–42.

On the outer ring of the calendar plate on the right side (left side according to *S*: fol. 6^{v}) were the letters for the Sundays, days of the months with the saints' names associated with them, sunrise in hours and minutes, length of the days in hours and minutes, the names of the months. A stationary disk with an opening and vertical slit was mounted in the center. The four seasons could be seen on triumphal wagons in the arch-shaped opening (*S*: fol. 7^{r}), the monthly labors and the sign of the zodiac with their astrological properties, as well; the latter were set off in colors in groups of three (*S*: fol. 7^{r}).[8] The outermost part of the central disk was divided into weeks; an index rotating with the outer ring indicated the weeks that had passed since New Year's Day. The entries on the inner calendar plate could be read through the vertical slit; at the end of the year this was shifted by the outermost ring and registered the letters for the Sundays in normal years and the leap years, the *claves terminorum*, and the golden number (*S*: fols. 7^{v}–9^{r}; *C*: fols. 10^{r}–12^{v}).

In the spandrels around the astrolabe and the calendar plate there were four paintings each, representing the four elements in the form of Biblical scenes: the flood represented water; Sodom and Gomorrah fire; Korah, Dathan, and Abiram, who were swallowed up by the earth because they rebelled against Moses (Num. 16), represented earth. The scene in which Christ calmed the storm on the Sea of Galilee (Mt 8: 26; Mk 4: 39; Lk 8: 24) represents the element air (*S*: fol. 10^{v}). On both sides of the housing, verses explaining the clock were on the columns and Latin inscriptions at the very bottom which named figures from the Old Testament in the first part who prophesied Christ; the other part depicted the fulfillment of the prophecies in the New Testament (*S*: fols. 10^{v}–11^{r}).

A round table with the numbers for the years on an outer white ring was above the sculpture in the middle of the bottom tier. On this there were depictions of the eclipses of the Sun and the Moon, and in the middle the dates, hours, and the extent of each eclipse. In the middle a painting didactically presented the cause of the phases of the Moon (*S*: fols. 9^{r}–10^{r}). The minute dial was located above this (*C*: fol. 12^{v}).

In front of the clock there was a copper celestial globe with paintings of 48 constellations and a total of 1,022 stars, as well as iron rings with copper and gilded images of the Sun and Moon. The symbols of the five planets were mounted on the zodiac so they could move and had to be positioned by hand using ephemerides (*S*: fol. 10^{r}; *C*: fols. 1^{v}–6^{v}). A pointer was attached to the polar axis that made it possible to read the sidereal time from a numbered ring above the north pole (*C*: fol. 6^{v}).

8 This corresponds to dividing the twelve signs of the zodiac into four trigons or triplicities according to the four elements (fire triplicity: Aries, Leo, Sagittarius; earth triplicity: Taurus, Virgo, Capricorn; air triplicity: Gemini, Libra, Aquarius; water triplicity: Cancer, Scorpio, Pisces); Ptolemy 1980, I.18, 83–87.

b. Middle Tier

The middle tier had three dials flanked on the sides by wooden sculptures of the astronomers Ptolemy and Pontano (*S*: fol. 10^{v}; *C*: fol. 13^{v}).[9]

The following was depicted on the dials:

Left: Display of the regency of the day; the planets were presented as personifications on wagons, the respective day and night houses of which were displayed on the wheels (*S*: fols. 3^{v}–4^{r}; *C*: fols. 13^{r}–13^{v}). A pointer with a star indicated the astrological properties of the individual planet, the opposite end pointed to its name.

Right: Quarter hour dial; in an opening at the bottom one could see the age of man on a triumphal wagon corresponding to the quarter hour (*S*: fol. 3^{r}; *C*: fol. 13^{r}).

Middle: Display of the synodic month and the phases of the Moon; a scale divided into 29½ sections at the outer edge. Four astronomers on wagons could be seen in an opening, who were holding the corresponding symbol for the Moon phase in their hands. A table in the middle of the dial indicated the points for bloodletting corresponding to the Moon phases (*S*: fols. 3^{r}–3^{v}; *C*: fol. 13^{r}).

c. Top Tier

There were two openings in the top tier of the housing, out of which a herald appeared at the full hour, followed by a number of noble persons showing their respects to a statue of Mary in the middle. During the procession, Mary lowered her scepter. To the right, a soldier with a hammer in his hand rang out the quarter hours; to the left, death rang out the full hours with a bone. Two angels sat at the feet of these two automata, the right one of which rang the quarter hours with a scepter, after which the left one turned over an hour glass (*S*: fol. 3^{r}; *C*: fols. 13^{v}–14^{r}).

In other openings in the housing, two angels stood to the left and right, who raised their trumpets and turned to face each other when the ten-bell carillon in the top of the tower started to play one of three melodies. At the end of the bell tune, a cock mounted at the very top crowed and flapped its wings. At the full hour a watchman sticking his head out of an opening in the top of the tower moved his jaw to the beat (*S*: fol. 3^{r}; *C*: fol. 14^{r}).

The similarity of this complex clock with the work of Dasypodius is obvious and the plans for it probably came from David Wolckenstein, who was acquainted with the precise plans of his colleague. An approximate idea of what this clock looked like can be taken from the automata clock built by Isaak Habrecht in 1594, which is preserved in Rosenborg Castle, Copenhagen.

9 The Italian humanist Giovanni Gioviano Pontano (1426–1503) translated, among other things, Ptolemy's Geography (Basel, 1541). He wrote a commentary on *Centiloquium* as well as the works *De rebus coelestibus Libri XIII* (Naples, 1512; Basel, 1530) and *Urania sive de stellis libri quinque* (Venice, 1505); Zedler, v. 28, cols. 1460–1462.

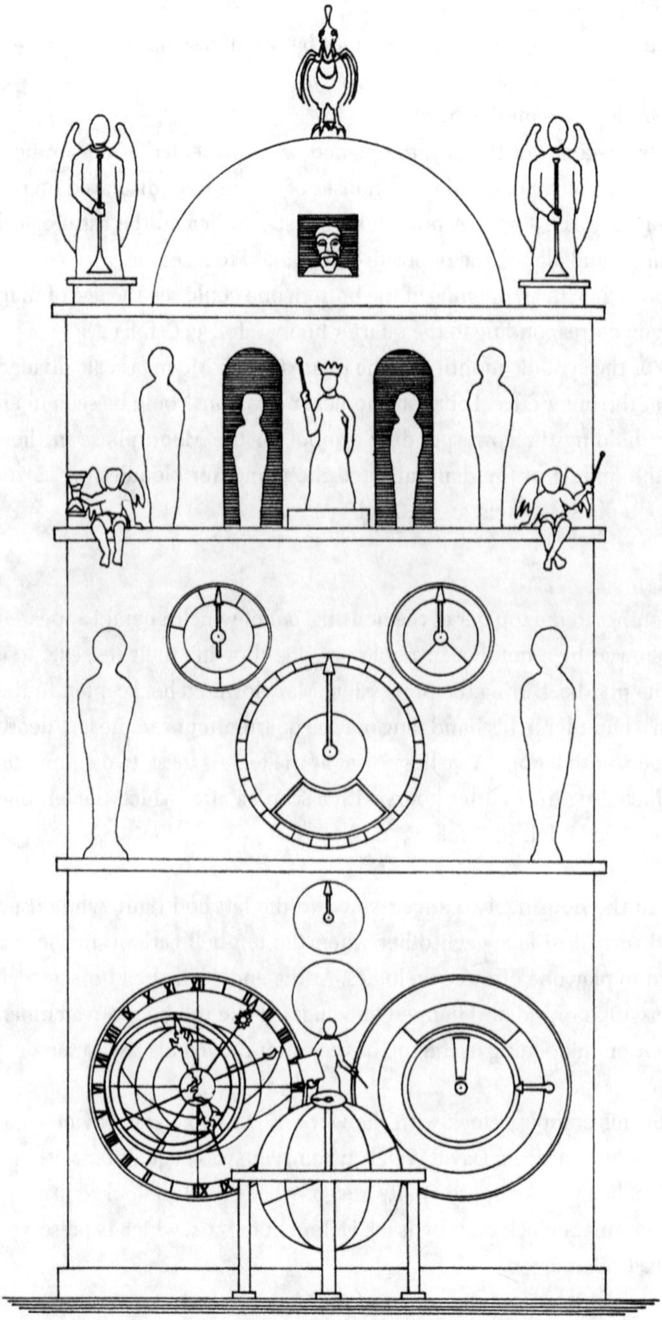

FIGURE 15.1 The lost astronomical clock of 1583

Appendix 3

Excerpt from Conrad Dasypodius, *Protheoria mathematica, in qua non solum disciplinae Mathematica omnes, ordine convenienti enumerantur; verum etiam universalia Mathematica praecepta; explicantur*, Strasbourg, 1593.

20

Volumen III.

PANDECTARVM Mathematicarum est: in quatuor Tomos, & lib: sedecim diuisum : quod omnia Mathemata; plenius & distinctius explicat.

TOMVS. I.

Continet, Libris quinq.

1. Generalem scientiarum Mathematicarum, explicationem.
2. Arithmeticam.
3. Logisticam vulgarem, & Astronomicam.
4. Logisticam analyticam, aut Algebraicam.
5. Musicam Theoreticam, & Practicam.

TOMVS II.

Continet, Libris quinq.

1. Geometriam.
2. Geodæsiam.
3. Opticam.
4. Catoptricam.
5. Scenographiam.

TOMVS

21

TOMVS III.

Continet, Lib: tribus.

1. Astronomiam.
2. Astrologiam.
3. Geographiam:

TOMVS. IIII.

Continet Lib: tribus.

1. Mechanicam Logicam.
2. Mechanicam Cheirurgicam.
3. Architecturæ, quæ omnium scientiarum regina est: elementa.

Atq; hæ sunt præcipuæ INSTITVTIONVM MATHEMATICARVM Propositiones. PANDECTARVM vero, hoc in loco recensere: nimis operosum fuisset: præsertim cum hic nihil aliud, quam specimen aliquod tantum totius CORPORIS MATHEMATICI, non autem integra & plenaria doctrina exhibeatur. hæcquoq; doctrina PANDECTARIA, sese in maximam rerum Mathematicarum copiam, diffundat. deniq; facile ex INSTITVTIONVM Propositionibus, quales PANDECTÆ futuræ sint: qui iudicio valent: percipere pos-

H sint.

22

Adhæc sciendum est, me præsentes, quas prælo subieci INSTITVTIONES MATHEMATICAS, more scholarum per quæstiones formasse: vt studiosis faciliorem discēdi rationem præberem. At in altero TOTIVS CORPORIS MATHEMATICI opere: alio vtar orationis genere, veteribus Mathematicis proprio: qui res suas, non quæstionibus: sed propositionibus exposuerunt. Quæ vero & quales illæ sint: in Protheoria, explicatum fuit.

Tandem vt omnes literati, me ab eo tempore non fuisse otiosum intelligant: quo Cæs: a Maiest: as MAXIMILIANVS II. Sanctisi: æ & honorificentiss: æ memoriæ Imperator pijssimus, & optimus, me meosq; conatus: omnibus Imperij, & extra Imperium, Statibus clementissime commendauit: recensebo, Græcorum manu scriptorum autorum nomina: quos magnis meis sumptibus, ex varijs collectos prouincijs, in latinum sermonem transtuli, correxi, commentarijsq; illustraui.

1. Pappi

23

1. *Pappi Alexandrini, collectanea Mathematica.*
2. *Theonis Smyrnæi, Mathematica.*
3. *Isaaci Monachi, Mathematica.*
4. *Pselli, Mathematica.*
5. *Nicomachi Arithmetica.*
6. *Diophanti Arithmetica.*
7. *Barlaami Arithmeticæ demonstrationes lib: 2. Elem.*
8. *Euclidis Elementa, ex varijs & diuersis græcis exemplaribus, collecta, & emendata, Scholijsq; aucta.*
9. *Procli deprauatissima commentaria, ex varijs græcis exemplaribus correcta.*
10. *Archimedis, Opera.*
11. *Apollonij, Conica.*

H 2 12. *Sere*

24

12. *Sereni, Cylindrica.*
13. *Euclidis, Data.*
14. *Heronis, Geometrica.*
15. *Heronis, Geodæsia.*
16. *Ioannis, Pediasimi, Geodæsia.*
17. *Euclidis, Optica.*
18. *Eiusdem, Catoptrica.*
19. *Scenographica noua.*
20. *Damiani Larißæi, hypotheses Opticæ.*
21. *Euclidis, Phænomena.*
22. *Theodosij, de Sphæra.*
23. *Eiusdem, de habitationibus.*
24. *Eiusdem, de diebus & noctibus.*
25. *Autolyci, de Sphæra mobili.*
26. *Eiusdem, de ortu & occasu Stellarum.*
27. *Aristarchi, de distantia & magnitudine Solis, & Lunæ.*

28. The

25

28. *Theonis, Ephodus in canones Astronomicos.*
29. *Gemini, quædam in Phænomena.*
30. *Hipsiclis, Anaphoricus.*
31. *Barlaami Monachi, Logistica Astronomica.*
32. *Eiusdem, de Eclipsibus.*
33. *Eiusdem, de decemnouenali canone Paschatis, non abyciendo, aut emendando.*
34. *Menelai, Sphærica, diu multum q; desyderata.*
35. *Plurima Arabum, scripta Astrologica.*
36. *Ptolomæi, Apotelesmatica.*
37. *Pauli Alexandrini, Astrologica.*
38. *Ptolomæi, aliorum græcorum, & recentioru, scripta Geographica.*

H 3 39. Pto-

39. *Ptolemæi, Musica.*
40. *Barlaami, quædam in Musicam Ptolemæi.*
41. *Aristoxenis, Musica.*
42. *Quintiliani Aristidis, Musica.*
43. *Alypij, Musica.*
44. *Bachæi Senioris, Eisagoge Musicæ.*
45. *Brennij, Harmonica.*
46. *Euclidis, Harmonica.*
47. *Heronis, Belopœtica.*
48. *Heronis, Dioptrica.*
49. *Eiusdem, Pneumatica.*
50. *Eiusdem, Clasmatica.*
51. *Eiusdem, Automatopœtica.*
52. *Ptolemæi Sphæropœtica.*
53. *Gnomonica quædam nondum edita.*

Quæ

Quæ cum maxima ex parte in latinum sermonem a me translata: lucubrationibus, figuris & delineationibus illustrata: prælo deniq; Typographico præter INSTITVTIONES MATHEMATICAS præparata sint: D. O. M. diuino suo auxilio iuuante: Imper: Regibus, Principibus & Magnatibus, Clementissima, & clementi liberalitate & munificentia patrocinantibus: doctis & literatis, maiore librorum copia, & eruditis consilijs adiuuantibus: librarijs quoq; & opificibus, debita fide, & diligentia cooperantibus: in Reipub: literariæ gratiam: & posteritatis commodum: in lucem dare: omnibus meis viribus, conabor.

FINIS.

H 4 PAVLLI

Appendix 4

Excerpt from a Letter by Paul Virdung to Kepler, 14 February 1604. ÖNB, Cod. Vin. 10703, fol. 258.

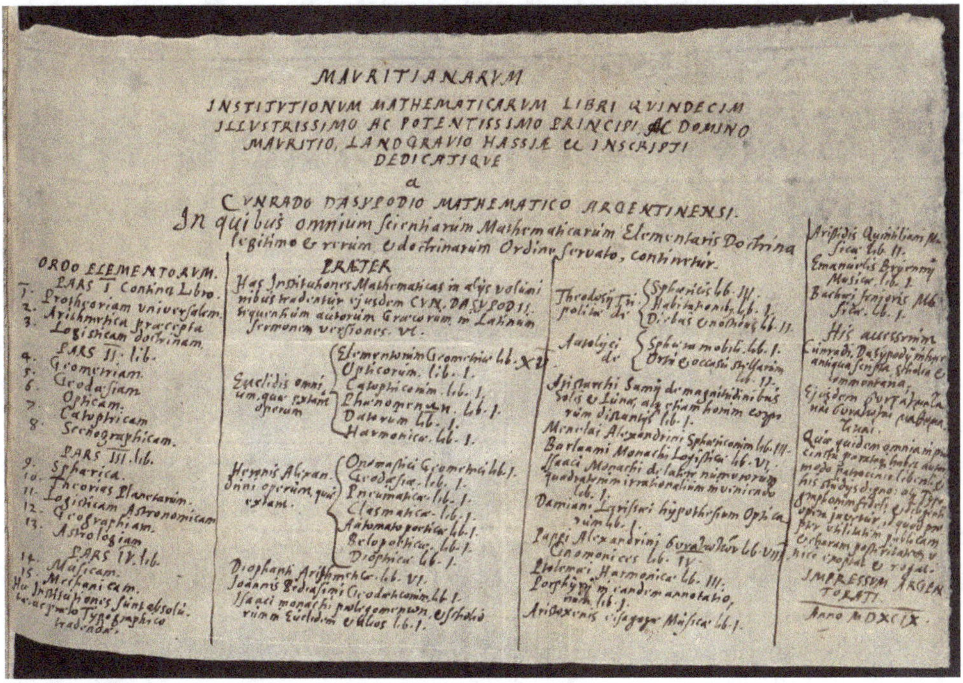

MAVRITIANARVM
INSTITVTIONVM MATHEMATICARVM LIBRI QVINDECIM
ILLVSTRISSIMO AC POTENTISSIMO PRINCIPI AC DOMINO
MAVRITIO, LANDGRAVIO HASSIÆ &c. INSCRIPTI
DEDICATIQUE
a
CVNRADO DASYPODIO MATHEMATICO ARGENTINENSI.
In quibus omnium scientiarum Mathematicarum Elementaris Doctrina legitimo & rerum & doctrinarum Ordine servato, continetur.

ORDO ELEMENTORVM.
PARS I. Continet Libro.
1. Prothеoriam universalem.
2. Arithmetica præcepta
3. Logisticam doctrinam.
PARS II. lib.
4. Geometriam.
5. Geodæsiam
6. Opticam.
7. Catoptricam
8. Scenographicam.
PARS III. lib.
9. Sphærica.
10. Theorias Planetarum.
11. Logisticam Astronomicam
12. Geographiam.
13. Astrologiam
PARS IV. lib.
14. Musicam.
15. Mechanicam.
Hæ Institutiones sunt absolutæ, ac prælo Typographico tradendæ.

PRÆTER
Has Institutiones Mathematicas in alijs voluminibus traduntur ejusdem CVN. DASYPODII, sequentium autorum Græcorum in Latinum sermonem versiones. VI.

Euclidis omnium, quæ extant operum: Elementorum Geometriæ lib. XV. Opticorum. lib. I. Catoptricorum. lib. I. Phænomenωn. lib. I. Datorum lib. I. Harmonica. lib. I.

Heronis Alexandrini operum, quæ extant: Onomastici Geometrici lib. I. Geodæsiæ. lib. I. Pneumatica. lib. I. Clasmatica. lib. I. Automatopoetica lib. I. Belopoetica. lib. I. Dioptrica lib. I.

Diophanti Arithmetica. lib. VI.
Joannis Pediasimi Geodæticorum lib. I.
Isaaci monachi prolegomenωn, & scholiorum in Euclidem & alios lib. I.

Theodosii Tripolitæ de: Sphæricis lib. III. Habitationibus lib. I. Diebus & noctibus lib. II.

Autolyci de: Sphæra mobili. lib. I. Ortu & occasu stellarum lib. II.

Aristarchi Samij de magnitudinibus Solis & Lunæ, ac etiam horum corporum distantijs lib. I.
Menelai Alexandrini Sphæricorum lib. III.
Barlaami Monachi Logistica lib. VI.
Isaaci Monachi de latere numerorum quadratorum irrationalium inveniendo lib. I.
Damiani Larissæi hypothesium Opticarum lib. I.
Pappi Alexandrini συναγωγῶν lib. VIII.
Gnomonices lib. IV.
Ptolemæi Harmonica lib. III.
Porphyrij in eandem annotationum lib. I.
Aristoxenis eisagoge Musica lib. I.

Aristidis Quintiliani Musica lib. II.
Emanuelis Bryennij Musica. lib. I.
Bacchei senioris Musica. lib. I.
His accesserunt Cunradi Dasypodij [illegible] antiqua scripta, scholia & commentaria.
Ejusdem [illegible] [illegible] [illegible].
Quæ quidem omnia in [illegible] [illegible] modo patrocinio liberali his studijs digno: ac [illegible] [illegible] [illegible] opera juvetur, & quod per utilitatem publicam [illegible] posteritatemque [illegible] hic exposita & rogat.
IMPRESSVM ARGENTORATI
Anno MDXCIX.

Appendix 5

The Divisions of Mathematics according to Geminos.
Mathematics in its entirety falls into two main fields:

One of these fields comprises things that deal with the intelligible items, that is, which can be comprehended with the soul. Among these are:

Arithmetic (today one would say: number theory); here delimited from the art of computation (logistics) consisting of:

- Simple (linear) numbers,
- Surface (2D) numbers,
- Body (3D) numbers;

Geometry, consisting of

- Planimetry,
- Stereometry.

The other field deals with things with material perceptions:
Geodesy,
Logistics (the art of computation),
Optics, consisting of

- Optics in the proper sense,
- Catoptrics (mirror optics; the study of the reflection of light),
- Theatre perspectives (scenography)

Canonics (harmonics);
Mechanics, consisting of

- Manufacture of war machines,
- Wonder-working (automata),
- Science of equilibrium and center of gravity,
- Manufacture of celestial models (planetariums);

Astronomy, consisting of

- Celestial movements (size, form, illumination, distance from the Earth of celestial bodies),
- Gnomonics,
- Meteoroscopy (investigating the height and the separation of the fixed stars),
- Dioptrics (determining the separation between the five planets).

Evans/Berggren 2006, 246–249 (English translation from the critical, annotated edition by Gottfried Friedlein, *Procli Diadochi primum Euclides elementorum librum commentarii*, Leipzig 1873, 38–42).

[p. 246] This, then, is the position of the Pythagoreans and their division of the four sciences. But others, such as Geminos, think it proper to divide mathematics according to another scheme. They make one branch concerned with mental things only, and one concerned with perceptible things or touching on them. They call mental, of course, those objects of contemplation that the soul itself calls up, separating itself from material forms. In the branch engaged with mental things, they place arithmetic[10] and geometry as the two first and most important parts. In the branch that operates with perceptible things they place six parts: mechanics, astronomy, optics, geodesy, canonics,[11] and logistics.[12]

On the other hand, they do not think it right to call tactics one of the parts of mathematics, as others do. Rather, they hold that it uses sometimes logistics, as in the tallying up of the companies, and sometimes geodesy, as the division and measurement of camps. Still less do they consider history or medicine to be part of mathematics, even if those who write histories often use mathematical theorems in writing on the positions of the *klimata* or in inferring the sizes, diameters, and perimeters of cities, and the physicians often clarify their own discussions by such methods: for Hippokratēs and all those who have said anything on seasons and places make clear the utility of astronomy to medicine. In the same way, then, the tactician will use the theorems of mathematics, although he is not a mathematician, if, when he wishes to show the smallest size he should form the camp in a circle or to show the largest, in a square, pentagon, or some other polygon.

These, then, are the species of mathematics as a whole. Geometry is in turn divided [p. 247]
into the theory of the plane and stereometry.[13] There is no branch of study specially devoted to points and lines, inasmuch as no figure would arise from them without planes or solid bodies: and, indeed, it is always the task of geometry, both plane and solid, to construct [figures] or to compare or to divide those that have already been constructed.

Similarly, there is a division of arithmetic into the study of linear numbers, plane numbers, and solid numbers:[14] for it examines the classes of numbers in themselves, as they proceed from the unit, and the generation of the plane numbers, both similar and dissimilar, and the progression to the third dimension.

Geodesy and logistics are analogous to these [i.e. to geometry and arithmetic], since they do not discourse about mental numbers or figures, but rather about perceptible ones. For the task of geodesy is to measure not a cylinder or a cone, but rather heaps as cones and pits as cylinders; and to measure not by means of mental straight lines, but

---

10 *Arithmētike* is the 'arithmetical art', the science of pure numbers; not to be confused with the elementary computation taught to school children.

11 *Kanonikē* is the mathematical theory of musical scales.

12 *Logistikē*, the art of practical computation.

13 The 'measurement of solids', which Geminos uses here to mean 'solid geometry'. But Hero of Alexandria wrote a work under the title *Stereometrica* that was concerned more with practical applications, i.e. the measurement of solids.

14 Similar and dissimilar plane numbers. Two plane numbers are similar (*homoioi*) if the factors composing them stand in the same ratio; Euclid, *Elements* VII, def. 21.

by means of perceptible ones, sometimes more precise ones, such as rays of sunlight, and sometimes coarser ones, such as ropes or a carpenter's rule. Nor, again, does the calculator [*logistikos*] consider the very properties of numbers in themselves, but rather as present in perceptible things, which is why he takes the names for them from the things being counted, calling some apples and others bowls. And he does not concede
[p. 248] that something is least, as does the arithmetician, since he assumes his least in relation to a particular class: one man, as unit, is his measure for a crowd.

Again, optics and canonics are offspring of geometry and arithmetic. The former uses visual lines and the angles formed from them. It is divided into: a part called optics proper, which explains the cause of false appearances due to the distances of the things observed, such as the meeting of parallels or the perception of squares as circles; a part called general catoptrics, which is concerned with all the varieties of reflections and involves the study of images; and the part called scene-painting,[15] which explains how one may represent appearances in pictures without disproportion or distortion when the things drawn are at a distance or raised to a height. Canonics, in turn, treats the apparent ratios of musical notes by discovering the divisions of the *kanōn*,[16] relying always on sense-perception and, as Plato says,[17] putting the ears before the mind.

Besides these sciences there is the one called mechanics, which is a part of the study of perceptible and material objects. Under this comes the construction of instruments useful in war, such as the engines for defense, that Archimēdēs is said to have built during the siege of Syracuse, and also the science of wonder-working,[18] which works
[p. 249] its contrivances by means of air, as both Ktēsibios and Hero describe, or by means of weights whose disequilibrium is the cause of motion and whose equilibrium is the cause of rest, as the *Timaeus* has established,[19] or, finally, by means of cords and ropes mimicking the tugs and movements of living creatures. Also under mechanics come the general science of things in equilibrium and the determination of what are called

15 *Skēnographikē*, scenography, is the art of applied perspective, useful for the realistic representation of buildings in theatrical sets.

16 *Kanōn* is here used as a synonym for *monochordos*, the monochord, the ancient instrument with a single taut string and a movable bridge.

17 Plato (*Republic* VII, 531a–c) disapproves of those who rely upon experiment to determine the rules governing musical harmony, rather than inquiring into the number itself to determine which numbers are inherently concordant and why. Geminos himself does not seem to object to putting the ears before the mind.

18 *Thaumatopoiïkē*, the art of devising automata and other gadgets, often operated by means of fluid or air pressure.

19 According to Plato (*Timaeus* 57d–58a), the cause of motion is always to be found in nonuniformity or unevenness (*anōmalotēs*), by which he means a nonuniformity in the arrangement of bodies. If all was uniform, it would be impossible for one body to move and for another to be moved. Rest is therefore associated with uniformity.

centers of gravity, as well as sphere-makingin imitation of the celestial motions,[20] such as Archimēdēs practiced, and, generally, all that is concerned with matter in motion.

There remains astronomy, which treats the cosmic motions, the sizes and shapes of the heavenly bodies, their illuminations and their distances from the Earth, and all such questions. It benefits greatly from sense perception but also has a good deal in common with physical theory. Its parts are: gnomonics, which is engaged with the measurement of the hours through the placement of gnomons; meteoroscopy,[21] which discovers the different altitudes [of the pole] and the distances of the stars and teaches many other complex matters from astronomical theory; and dioptrics, which examines the positions of the Sun, Moon, and the other stars by means of just such instruments [ie. dioptras].

Such are the writings on the parts of mathematics that we have received from the ancients.

---

20 *Sphairopoiïa* is the science of constructing models of the heavens, including celestial globes, armillary spheres, and planetaria.

21 *Meteōroskopikē* is most likely concerned with the armillary astrolabe ("astrolabon organon") as a specialized instrument of observation.

Sources and Bibliography

A. Manuscripts

Strasbourg, Archives Municipales

3 OND 589

I: Jean-Baptiste Schwilgué, *Description du mécanisme et des Fonctions de l'horloge Astronomique de la Cathédrale de Strasbourg* (11 pages, numbered, dated 16 May 1836).

II.1: *Projet pour la restauration de l'horloge Astronomique de la Cathédrale de Strasbourg.*

II.2: Ditto, *Devis.*

III: Ditto, *Detail Estimatif.*

IV: Ditto, *Bordereau Analytique.*

Jean-Baptiste Schwilgué, *Supplément du Bordereau Analytique des Prix des Travaux faits à l'horloge astronomique de la Cathédrale de Strasbourg en sus du Devis.*

Paper, 12 pp., the last left blank, not numbered, dated 10 March 1843.

Johann Philipp Maybaum and Son, *Bericht über die dermalige Lage der grossen Kunst-Uhr im Münster.*

Paper, 8 pp., the last two left blank, not numbered, undated [1788].

1 OND 15

Verding des Urwercks im Münster Alhie.

Paper, 8 pp., the last left blank, not numbered, dated 26 July 1571.

Inventaire des Archives de la Ville de Strasbourg, Sér. II, 46, No. 15 Ordnung der Pfleger, des Schaffners und der Gesinde des Frauenwerkes [Rules for caretakers, supervisors, and the workers of the 'Frauenwerk'].

Paper, 28 pp., not numbered. The rules for chaplains is dated 1464; among other things, the rules for Cathedral laborers are included in the same document. The manuscript contains rules for diverse occupations from a later author as well as regulations for celebrating the mass from 1481.

Inventaire des Archives de la Ville de Strasbourg, Sér. II, 45a/4 Invoices for the years 1382–1383 and 1385 (?).

Paper, 12 pp., the last three left blank, reprinted by Fuchs 1990.

1 OND 43
Invoices for the 'Frauenwerk'.
The following is a list of the extant invoice books up until the end of the sixteenth century. The Roman numerals following the years are the year halves. After the second half of 1432, the figures were no longer listed for the respective week—as was previously the case—but according to categories, and starting in 1437–1438 the books include invoices for an entire year, beginning with *Sungichten* (summer solstice or 24 June; the Feast Day of Saint John the Baptist). There are no records between 1568–1569 and 1650–1651.

1414 (II); 1415 (II); 1416 (I); 1418 (I); 1419 (I, II); 1420 (II); 1422 (I); 1423 (I); 1424 (I); 1425 (II); 1426 (II); 1427 (II); 1428 (II); 1429 (I); 1430 (II); 1431 (I); 1432 (I, II); 1433 (II); 1434 (II); 1435 (II); 1436 (II); 1437–1438; 1438–1439; 1439–1440; 1440–1441; 1443–1444 or 1444–1445 (text loss by tear on the cover page); 1445–1446; 1447–1448; 1448–1449; 1449–1450; 1450–1451; 1451–1452; 1459–1460; 1460–1461; 1461–1462; 1462–1463; 1475–1476; 1492; 1500–1501; 1501–1502; 1502–1503; 1504–1505; 1505–1506; 1506–1507; 1508–1509; 1512–1513; 1531–1532; 1539–1540; 1543–1544; 1548–1549; 1568–1569.

1 OND 8
Liber contractuum fabricae ecclesiae Argentinae (1368–1389).

1 OND 4
Statbuch.

KS 17
Kontraktsprotokolle des Frauenwerkes (*Kontraktstube*), v. XVII (1524–1526).

4 R 103, 105
Bürgerbuch I, III

Inventaire des Archives de la Ville de Strasbourg, Sér. II, 46, no. 35 Letter to the Strasbourg city council written by Conrad Dasyposius.
Paper, 2 pp., dated 5 June 1577.
Reports of the caretakers of the 'Frauenwerk'.
Paper, 8 pp., dated 5/19 October 1577.
Inventaire des Archives de la Ville de Strasbourg, Sér. II, 46, no. 32 Letter of the Rottweil city council.
Paper, 1 p., dated 1398; reprinted in Kraus 1876, 385.

1 R 51, 58, 81, 117, 149
Einundzwanzigerprotokolle, v. 51 (1571); 58 (1577); 81 (1600); 117 (1636); 149 (1669).

2 R 293
Fünfzehnerprotokolle, v. 293 (1583).

1 OND 15
Excerpts from the minutes concerning winding up the weights for the clock on the tower platform.
Paper, 2 pp., dated 21 June 1786.
Inventaire des Archives de la Ville de Strasbourg, Sér. II, 46, no. 34.
Note on the Cathedral servants (*Münsterknechte*) who were negligent about winding up the weights of the tower clock.
Paper, 1 p., undated (probably from the fifteenth century).

Estate of Charles-Eugène Wagner.
90 Z 15
Orgeln von St. Thomas (1524–1681); therein, material on David Wolckenstein's biography and his musical activities, 1576–1589.
90 Z 18
Musiklehrer an der Schule (1538–1681), 1195–1316.
90 Z 22
Urkundliches über die Straßburger Münster-Uhr von 1571–1574 und ihre Urheber: Konrad Dasypodius, David Wolckenstein, Isaac und Josias Habrecht, Tobias Stimmer u.s.w.
90 Z 23
7, 1–14 Transcription of *Cunradi Dasypodii, Scholia et Resolutiones seu Tabulae in Lib. IIII. Apotelesmaticos Cl. Ptolemaei* […], Basel, 1578.
90 Z 24 *Urkundliches über die Straßburger Münster-Uhr von 1571–74 und ihre Urheber: Konrad Dasypodius, David Wolckenstein, Isaac und Josias Habrecht, Tobias Stimmer u.s.w.* fols. 201–245; created in the years 1929–1936.
90 Z 31
11,1–11,56 Transcription of *Beschreybung des Astrolabiums und des gantzen Werckhs Nutz und gebruch uff das Kürtzest*; BNUS, MS 1562 (Alsat. 997).
13,1–13,30 Transcription of David Wolckenstein, *Descriptio Horologii* (Cambridge, Trinity College) in two copies.
Inventaire des Archives de la Ville de Strasbourg, Sér. V, 14, no. 188.
Privileges granted to Conrad Dasypodius by Emperor Maximilian II (first supplement to the petition by Dasypodius of 14 April 1571).
Paper, 2 pp., dated 30 September 1570.
Inventaires des Archives de la Ville de Strasbourg, Sér. V, 14, no. 5.
Missive from Maximilian II, Holy Roman Emperor, to the city council of Strasbourg for the benefit of Conrad Dasypodius (second supplement to the petition by Dasypodius of 14 April 1571).
Paper, 2 pp., dated 8 November 1570.

Inventaires des Archives de la Ville de Strasbourg, Sér. V, 14, no. 5.
Petition of Conrad Dasypodius to the city council of Strasbourg.
Paper, 2 pp., dated 14 April 1571.
Inventaires des Archives de la Ville de Strasbourg, Sér. V, 14, no. 189.
Petition of Conrad Dasypodius to the city council of Strasbourg.
Paper, 6 pp., dated 3 November 1574.
58 NOT 21
Inventory of the estate of Conrad Dasypodius, dated: 30 July–4 September 1601.
1 AST 353/10, 185–187
Petition of the rector and deans of the University of Strasbourg to the city council, requesting support for buying the library of Matthias Bernegger.
Paper, 3 pp., dated 3 August 1636.
1 AST 353/10, 183–184
Matthias Bernegger's letter to Johann Georg Dorsch concerning the sale of his library and scientific instruments.
Paper, 2 pp., dated 12 September 1635.

Strasbourg, Musée de l'Œuvre Notre-Dame

Box 43, no. 2
Johann Philipp Maybaum and Son, *Bericht über die dermalige Lage der großen Kunstuhr im Münster.*
Paper, 5 partially numbered pp., dated 1788.
Box 43, no. 18–20 (combined file and page numbering)
Johann Martin Franz Theodor Lobstein (?), *Recherches sur le Carillon de l'horloge astronomique de la Cathédrale de Strasbourg.*
Paper, 4 pp., undated (probably before 1839).
Box 43, no. 78
Brief note on the astronomical clock in the Cathedral and the new construction by Schwilgué.
Paper, 1 p., undated.
Inv. no. 1108/I
Jean-Baptiste Schwilgué, *Description de l'Horloge Astronomique de la Cathédrale de Strasbourg construite par J.B. Schwilgué 1838–1842* (tome I).
Paper, bound in leather, 13-pp. preface plus 304 pp. (numbering of the main text starting with p. 3).
Inv. no. 1108/II
Jean-Baptiste Schwilgué, *Résultats des Calculs, servant a la composition de l'Horloge Astronomique de la Cathédrale de Strasbourg. Anciennes Horloges—Devis et divers travaux* (tome II).
Paper, bound in leather, 280 numbered pp. plus 13 pls.

Strasbourg, Bibliotheque Nationale et Universitaire

MS 1562 (Alsat. 997)

Paper, bound in leather with embossed cover, 31 pp., two of which being the endpapers front and back; pages numbered later in a different handwriting with endpapers counted in the numbering. Undated (1583). The file includes:

Beschreybung des Astrolabiums und des gantzen Werckhs Nutz und gebruch uff das Kürtzest, fols. 3ʳ–11ʳ.

Der acht und zweintzig Mansion. des Mons Im Lauff der zwolff himlischen Zeichen (Moon stations), fols. 11ʳ–17ᵛ.

Zusammen. Komung (Conjunctions), fols. 17ᵛ–22ᵛ.

Caracter der 12 himelischen Zeichen. Caracter der 7. planetten, fol. 23ʳ.

Tables for the position of the planets and the ascending lunar nodes for the year 1582–88, fols. 23ᵛ–28ʳ.

List of the saints' days, fols. 29ʳ–30ᵛ.

In Wickersheimer 1923, 340, this file is erroneously listed as a description of the astrolabe of the Cathedral clock. See Baillie 1978, 24.

MS 1563 (Alsat. 998)

Verding des Uhrwercks Im Münster alhie.

Paper, 8 pp., the last left blank, dated 26 July 1571; notes in Isaac Habrecht's handwriting on fol. 1ᵛ.

Reprinted in Edel 1777, 196–204; Edel 1843, 16–21; Maurice 1976, v. II, 7–9. See also Baillie 1978, 22.

MS 1373 (Alsat. 704)

Auszug Eines Manuskripts des Herrn Hecklers Werkmeisters des Münsters zu Straßburg, dessen Bemerkungen enthaltend über die Beschreibung des besagten Münsters welche Herr Israel Mürschel gewesener Pfarrer zu Bishheim und Hönheim zusamen getragen und dessen Sohn Johann Friderich Mürschel Pfarrer ebendaselbst heraus zu geben gedacht hatte; obige Bemerkungen verfaßt ungefehr um 1,666.

Paper, 18 pp. plus cover page, dated at bottom of fol. 1ʳ "*Copié de l'original* [...] *1817*". The original manuscript was burned in 1870. Mürschel's excerpts are reprinted in Kraus 1876, 693–703.

MS 1566 (Alsat. 1001)

Genealogia. Das ist Geburt Büchlin, und Register unserer VorEltern. Zur Glückwünschung Eines newen glückseligen Fried- und Frewdenreichen Jahrs, Herren Hans Sellingern, Seinem Freundlich Schwagern, zusamen getrag. Auff den Newen Jahrstag 1621.

Paper, 7 pp., the last left blank; dated New Year's Day 1621. At bottom of fol. 7ʳ: "*Alles trewlich ausgeschriben, Collationirt und verglich. aus geschribnen Zedlen, und büchlin. Ao 1620. von Is. Habrecht*".

The manuscript was written by Isaac II Habrecht (1589–1633), a son of Isaac Habrecht. Excerpts in Edel 1777, 193–195; Edel 1843, 23–26.

MS 1565 (Alsat. 1000)

Invoice for numerous items produced by Isaac Habrecht later for the Cathedral clock Paper, 1 page, undated (1574?).

This is a reprint from the eighteenth century. The original belonged to Matthäus Edel, who published it in Edel 1777, 204–206. The text was later reprinted in Edel 1843, 22–23; Maurice 1976, v. II, 9. What has happened to the original is unknown.

MS 4847

Jean-Baptiste Schwilgué, *Description de l'horloge astronomique de la Cathédrale de Strasbourg, construite par J.B. Schwilgué. 1838–1842* (tome I).

Paper, half bound in leather, 13-pp. preface plus 318 numbered pp., the last left blank.

MS 4848

Jean-Baptiste Schwilgué, *Résultats des Calculs servant à la composition de l'horloge astronomique de la Cathédrale de Strasbourg. Anciennes horloges—Devis et divers travaux* (tome II).

Paper, half bound in leather, 192 numbered pp., the last left blank. The last section *Anciennes horloges*, is missing; see the second copy of the manuscript in the Musée de l'Œuvre Notre-Dame.

Paris, Bibliothèque Nationale

MS fr. 2756, no. 8

Actes et memoires servans à l'histoire d'Allemagne, "*De l'orloge de la ville de Strazburg*", fols. 230^{r}–231^{v}.

Catalogue des manuscrits français: Ancien fonds, v. I.

Brief description of the Cathedral clock; probably a reprint of an older manuscript from the eighteenth century.

Paris, Archives Nationales

Fonds de l'administration des Cultes, F/19/7892, nos. 65–84

Official documents on the restoration of the Cathedral clock from 1836–1844.

Schaffhausen, Stadtarchiv

City invoices from 1540–1541 (AII 05/208) and 1566–1567 (AII 05/247)

Schaffhausen, Staatsarchiv

Protokolle der Schulherren, 1581–1603, v. II, fol. 2^{v} (1 April 1581)

Id., 1554–1791, v. XVI, IV. *Stück*, fol. 1^{r} (6 March 1567)

Villingen-Schwenningen, Stadtarchiv

MS GG 4a

Documents of the schultheiss, mayor, and city council of Villingen for the clockmaker Claus Gutsch concerning the construction of an astronomical clock.

Parchment, dated 6 May 1401.

Reprinted in Roder 1880, 291–292; Kraus 1890, 124; Fuchs 1970/71, 59–62.

Cambridge, Library of Trinity College

MS 0.2.32

Incipit, "*Descriptio Astronomici Horologii etc; Horologium hoc αὐτώματον est*"

Explicit, "*Dauid Wolckenstein Vratislaviensis* [...] *describebat A.D. 1590 XIIX Februarii*"

Parchment, 16 partially numbered pp., two of which being the unnumbered endpapers. Attached is a later transcription of the manuscript in another handwriting; paper, 11 pp.

Erroneously referred to as a description of the Cathedral clock by James 1900/04, v. III, 131–132, no. 1136. See Baillie 1978, 27.

Olomouc, Státní okresní archiv

Archiv města Olomouc, Zlomky registratur 1426–1786 (Archive of the City of Olomouc, fragments of the depositories 1426–1786), box 161

Inv. no. 4471

Letter of Recommendation for Conrad Dasypodius from Emperor Maximilian II. Paper, 1 p., dated 30 September 1570.

Inv. no. 4481

Letter to the Olomouc city council written by Conrad Dasyposius. Paper, 1 p., dated 20 September 1574.

Osnabrück, Niedersächsisches Staatsarchiv

Rep. 100, Abschnitt 332, no. 6

Incipit, "*Instructio quaedam brevissima. Ein kurtzer unterricht und einleytung des new erfundenen astromischen* [sic!] *uhrwerks zu Oßnabrug im Dhome, durch Josten Bodeker von Wartbergh, Vicarius daselbst im Dhome, erfunden, verordtnet und eingestellt, nun aber dieses lauffenden jars entlich verfertigt. 1587*".

Explicit, "*Hiermit auch die gantze Contrafactur in seinem wesen gentzlich uberliebern thun*".

Paper, 24 pp., six of which left blank. Added later in a different handwriting on fol. 1 (cover):

"*N.24.P.4. Capitularia. Nachricht wegen des gewesenen astronomischen Uhrwercks im Thumb zu Osnabrück. 1587*".

Baillie 1978, 24–26; reprinted in Veltman 1890, 241–263.

Schleswig, Schleswig-Holsteinisches Landesarchiv

Breitenburg Estate Archives, manuscripts, 293 (depository)

Five letters from Conrad Dasypodius to Heinrich Rantzau

1. 5 April 1573, pp. 55–67
2. 4 September 1578, pp. 161–162
3. 1 August 1583, pp. 185–187
4. 20 October 1583, pp. 213–217
5. 7 December 1587, pp. 395–396

A typed index is included in the collection of letters, which is printed in Neumann 1968.

Hamburg, Staats- und Universitätsbibliothek

Supellex Epistolica Uffenbachii et Wolfiorum, 4°, 31

1. Draft of a letter from Matthias Bernegger to Daniel Heinsius dated 3/13 April 1625 (Old and New Style), pp. 169–173. New page numbering in catalog by Krüger 1978, v. I, 68: fols. 85^r–87^r. Excerpts reprinted in Heumann 1722/32, v. II, Book 2, 226–229.
2. Draft of a letter from Matthias Bernegger to Daniel Heinsius dated 31 March 1627, p. 308. In Krüger 1978, vol. I, 69: fol. 154^v.

Supellex Epistolica Uffenbachii et Wolfiorum, 2°

Four letters from Conrad Dasypodius to Matthias Ritter:

1. 7 March [1558], fols. 13^v–14^v, in Latin;
2. 1558, fols. 14^v–15^v, in Latin;
3. 1 September 1558, fols. 15^v–17^v, in French;
4. 1 September 1559, fols. 18^r–19^r, in Latin.

These are eighteenth century copies that Zacharias Konrad von Uffenbach had made for his collection of letters.

Vienna, Österreichische Nationalbibliothek

Cod. Vin. 9737 z14–18

25 letters from Conrad Dasypodius to Hugo Blotius:

1. 23 April 1570, Cod. 9737 z14/I,9;
2. 6 August 1570, Cod. 9737 z14/I,16;
3. 13 February 1571, Cod. 9737 z14/I,27;
4. 22 February 1571, Cod. 9737 z14/I,28;
5. 17 March 1571, Cod. 9737 z14/I,35;
6. 14 May 1571, Cod. 9737 z14/I,45;
7. 1 June 1571, Cod. 9737 z14/I,48;
8. 14 August 1571, Cod. 9737 z14/I,56;
9. 11 December 1571, Cod. 9737 z14/I,73;
10. 1 January 1572, Cod. 9737 z14/I,81;

11. 14 January 1572, Cod. 9737 z14/I,86;
12. 25 February 1572, Cod. 9737 z14/I,97;
13. 13 April 1572, Cod. 9737 z14/I,114;
14. 24 September 1573, Cod. 9737 z14/I,271;
15. 11 July 1574, Cod. 9737 z14/I,399;
16. 17 May 1575, Cod. 9737 z15/II,58;
17. 28 June 1575, Cod. 9737 z15/II,89;
18. 1 August 1577, Cod. 9737 z15/II,295;
19. 19 January 1592, Cod. 9737 z17/IV,179;
20. 31 Oct. 1593 (Old Style), Cod. 9737 z17/IV,280;
21. 6 March 1594, Cod. 9737 z17/IV,302;
22. 1 Oct. 1597 (New Style), Cod. 9737 z18/V,61;
23. 15 August 1599, (Old Style) Cod. 9737 z18/V,105;
24. Undated (23 April 1570?), Cod. 9737 z.18/V,335;
25. Undated (1570–1571), Cod. 9737 z18/V,336.

Three letters from Conrad Dasypodius to Heinrich Rantzau:

1. 7 June 1587, Cod. 9737 l, fols. 240^{r}–241^{r};
2. 5 September 1587, Cod. 9737 l, fols. 277^{r}–278^{v};
3. 22 September 1587, Cod. 9737 l, fols. 284^{r}–284^{v}.

Letter from Heinrich Rantzau to Conrad Dasypodius dated 26 July 1588, Cod. 9737 l, fol. 262^{r}.

Cod. Vin. 10702

Letter from Conrad Dasypodius to Johannes Kepler dated 31 December 1599; fol. 177. Reprinted in KGW, v. 14, 105.

Cod. Vin. 10703

Letter from Paul Virdung to Johannes Kepler dated 14 February 1604; fols. 256–260. Reprinted in KGW, vol. 15, 32–38.

Cod. Vin. 11449

Latin manuscript with 321 pages of astrological content.

A brief summary of the content of the manuscript can be found in *Tabulae codicum Vindobonensi*, v. VI, 329. For a more comprehensive table of contents, see Oestmann 2004, 140–145. According to Houzeau/Lancaster 1882/89, v. I.1, 788, no. 4965, this is the manuscript of Heinrich Rantzau's *Tractatus astrologicus* published in Frankfurt am Main in 1593, which is erroneous. On fols. 97^{r}–98^{r} a copy of the horoscope cast by Dasypodius for Heinrich Rantzau and his wife.

Series nova, 362

Letter from Hugo Blotius to Conrad Dasypodius from 25 January 1589; fols. 65^{r}–66^{v}.

Series nova, 363

Two draft letters from Hugo Blotius to Conrad Dasypodius:

1. 6 December 1590, fols. 150^{r}–156^{v};

2. 16 December 1600; fols. 183^r–184^v. On fols. 185^r–186^r, a later copy of the same letter in a different handwriting.

Letter from Hugo Blotius to Federico Commandino dated 7 July 1571, fol. 7^r.

Nancy, Archives Départementales de Meurthe et Moselle

B 1332

"*Somme payée à Isaac Habrecht, horloger à Strasbourg, pour une horloge faite comme celle du temple de Strasbourg, pour être placée dans la galerie des peintures du Palais Ducal*", 29 October 1611.

B 1398

"*Sommes payées à Abraham Habrecht, envoyé à Paris pour enseigner comment il faut gouverner une grand horloge qu'il a faite et laquelle le duc a envoyée à Paris*", 14 July 1618.

München, Bayerische Staatsbibliothek

Cgm 4189

"*Deß künstreichen Astronomischen, züerlichen und nützlichen newen Calender und Uhrwerckhs so unlangst in unser lieben Frawen Thumbstifft zu Augspurg aufgericht worden, Beschreibung und Erclärung, allen Liebhabern und erkundigern Himlischer Astronomischer verenderungen und würckungen zu guetem gestelt Durch Thobiam Klieber Mitbürgern zu Augspurg*".

Paper, 61 pp., blind embossed leather binding, 1603.

Ungerer 1931, 193–195; Bobinger 1966, 90–93; Baillie 1978, 31. Reproduction of the text in Bassermann-Jordan 1929b.

B. Printed Sources and Secondary Literature

Ackermann 2004 – Silke Ackermann, "The Path of the Moon Engraved: Lunar Mansions on European and Islamic Instruments", *Micrologus: Natura, Scienze e Societè Medievali* 12, 2004, 135–164.

Ackermann 2005 – Silke Ackermann, "Astrological Scales on the National Maritime Museum Astrolabes", in Koenraad Van Cleempoel, *Astrolabes at Greenwich: A Catalogue of the Astrolabes in the National Maritime Museum Greenwich*, Oxford, 2005, 73–89.

Adam 1615 – Melchior Adam, *Vitae Germanorum Superiori et quod excurrit, seculo philosophicis et humanioribus literis clarorum; sive Literati in Germania seculi*, Frankfurt am Main, 1615.

Alexander 1943 – Anneli Alexander, "Straßburgische Bibliographie", *Deutsches Archiv für Landes- und Volksforschung* 7, 1943, 164–226.

Anon. 1840 – Anonymus (P.F.), “L’horloge de Strasbourg”, *Revue d’Austrasie, new ser.* 2, 1840, 180–194.

Anz 1905 – Heinrich Anz, *Die lateinischen Magierspiele: Untersuchungen und Texte zur Vorgeschichte des deutschen Weihnachtsspiels*, Leipzig, 1905.

Aristoteles 1955 – Aristoteles, *Meteorologica* (tr. by Erwin W. Webster), in *The Works of Aristotle*, v. III, Oxford, 1955³.

Arlt 1888 – Otto von Arlt, “Die Rathausuhr zu Ulm”, *Jahreshefte des Vereines für Mathematik und Naturwissenschaften in Ulm* 1, 1888, 11–20.

Armstrong 1974 – Joe Edwin Armstrong, *The Waterclocks of Ktesibius: Their Origin and Reconstructions*, Master’s thesis, San José State University, 1974.

Aschbach 1865/88 – Joseph Ritter von Aschbach, *Geschichte der Wiener Universität*, Vienna, 1865–1888.

Aujac 1970 – Germaine Aujac, “La sphéropée, ou la mécanique au service de la découverte du monde”, *Revue d’histoire des sciences et de leurs applications* 23, 1970, 93–107.

Avelar/Ribeiro 2010 – Helena Avelar, Luis Ribeiro, *On the Heavenly Spheres: A Treatise on Traditional Astrology*, Tempe (AZ), 2010.

Bach 1978 – Henri Bach, “Einige interessante alte astronomische Uhren und besondere Räderwerke aus solchen: Die zwei ersten [sic!] Strassburger Münsteruhren”, *Schriften der Freunde alter Uhren* 17, 1978, 23–46.

Bach 1979 – Henri Bach, “Einige interessante alte astronomische Uhren und besondere Räderwerke aus solchen: Der Globus des Dasypodius”, *Schriften der Freunde alter Uhren* 18, 1979, 19–36.

Bach 1982/86 – Henri Bach, “Les trois horloges astronomiques”, in *Encyclopédie de l’Alsace*, Strasbourg, 1982–1986; v. III, 1292–1346, s.v. “Cathédrale”.

Bach/Rieb 1992 – Henri Bach, Jean-Pierre Rieb, *Die drei astronomischen Uhren des Straßburger Münsters* (tr. by Robert Wilhelm), Strasbourg, 1992.

Baeschlin 1879/80 – Johann Heinrich Baeschlin, *Schaffhauser Glasmaler des XVI. und XVII. Jahrhunderts, I, II*, “Neujahrsblatt des Kunst-Vereins in Schaffhausen für 1879; 1880”, Schaffhausen, 1879–1880.

Baillie 1931 – Granville Hugh Baillie, “An Old Book about an Elaborate Clock”, *The Watch and Clock Maker*, 1931, 244–245.

Baillie 1948 – Granville Hugh Baillie, “Sixteenth Century Public Clocks and earlier Public Clocks: A List of Public Clocks erected before 1600”, *Horological Journal* 90, 1948, 735–747.

Baillie 1978 – Granville Hugh Baillie, *Clocks and Watches: An Historical Bibliography*, v. I., London, 1978².

Balmer 1971 – Heinz Balmer, “Keplers Beziehungen zu Jost Bürgi und anderen Schweizern”, in *Ausstellungskatalog Johannes Kepler: Werk und Leistung*, Linz, 1971, 115–126.

Barack 1895 – Karl August Barack, *Katalog der Kaiserlichen Universitäts- und Landesbibliothek in Strassburg: Elsass-Lothringische Handschriften und Handzeichnungen*, Strasbourg, 1895.

Baranowski 1958/73 – Henryk Baranowski, *Bibliografia Kopernikowska 1509–1555; 1956–1971*, Warszawa, 1958; 1973.

Baranowski 1977 – Henryk Baranowski, "Copernican Bibliography: Selected Materials for the Years 1972–1975", *Studia Copernicana* 17, 1977, 179–201.

Barker 2005 – Peter Barker, "Lutheran Contributions to the Astronomical Revolution", in John Hedley Brooke, Ekmeleddin İhsanoğlu (eds.), *Religious Values and The Rise of Science in Europe*, "Series of Studies and Sources on History of Science 13", Istanbul, 2005, 31–62.

Barker 2009 – Peter Barker, "The Hypotyposes orbium coelestium (Strasbourg, 1568)", in Miguel Ángel Granada, Édouard Mehl (eds.), *Nouveau ciel, nouvelle terre: La révolution copernicienne dans l'Allemagne de la Réforme (1530–1630)*, Paris, 2009, 85–108.

Barnass 1930 – Margarete Barnass, "Tobias Stimmer und Tintoretto", *Oberrheinische Kunst: Jahrbuch der oberrheinischen Museen* 6, 1930, 163–170.

Barnass 1932 – Margarete Barnass, *Die Bibelillustration Tobias Stimmers: Ein Beitrag zur Bibelillustration des 16. Jahrhunderts*, PhD thesis, University of Heidelberg, 1932.

Bassermann-Jordan 1921 – Ernst von Bassermann-Jordan, *Die Geschichte der Räderuhr unter besonderer Berücksichtigung der Uhren des Bayrischen Nationalmuseums*, Frankfurt am Main, 1921[2].

Bassermann-Jordan 1929a – Ernst von Bassermann-Jordan, "Die einstige Monumentaluhr im Dome zu Frankfurt am Main", *Die Uhrmacher-Woche: Handelszeitung für den Uhren- und Bijouteriewaren-Markt* 48, 1929, 440–442.

Bassermann-Jordan 1929b – Ernst von Bassermann-Jordan, "Die verlorene Monumentaluhr im Dome zu Augsburg", *Die Uhrmacherkunst* 48, 1929, 179–184, 200–203 and 220–225.

Batowski 1942 – Zygmunt Batowski, *Coppernicusbildnisse* (tr. by Alfons Triller), Berlin, 1942; (Typoscript of the original Polish publication *Wizerunki Kopernika*, Toruń, 1933; Leipzig, Deutsche Nationalbibliothek: 1943 B 407).

Baumann/Köster/Kuhl 2017 – Kirsten Baumann, Constanze Köster, Uta Kuhl (eds.). *Adam Olearius: Neugier als Methode* (proceedings of a conference at Gottorf Castle, Schleswig, 24–27 June 2015), Petersberg, 2017.

Beck 1909 – Theodor Beck, "Herons des Älteren Automatentheater", *Beiträge zur Geschichte der Technik und Industrie: Jahrbuch des Vereines Deutscher Ingenieure* 1, 1909, 182–199.

Beer 1952 – Ellen Judith Beer, *Die Rose der Kathedrale von Lausanne*, "Berner Schriften zur Kunst, VI", Bern, 1952 (PhD dissertation, University of Bern).

Behling 1957 – Lottlisa Behling, *Die Pflanze in der mittelalterlichen Tafelmalerei*, Weimar, 1957.

Behr 1747 – Georg Heinrich Behr, *Strasburger Münster- und Thurn-Büchlein; oder Kurtzer Begriff der merkwürdigsten Sachen, so im Münster und dasigem Thurn zu finden; Mit neuen beygefügten Kupfern*, Strasbourg, 1747; French ed. in Boehm 1743.

Bendel 1925/26 – Max Bendel, "Tobias Stimmer und die venezianische Malerei", *Oberrheinische Kunst: Vierteljahresberichte der oberrheinischen Museen* 1, 1925–1926, 126–135.

Bendel 1940 – Max Bendel, *Tobias Stimmer: Leben und Werke*, Zürich – Berlin, 1940.

Bengel 2011 – Sabine Bengel, *Das Straßburger Münster: Seine Ostteile und die Südquerhauswerkstatt*, "Studien zur internationalen Architektur- und Kunstgeschichte, 84", Petersberg, 2011 (PhD dissertation, Technische Universität Berlin, 2007).

Benzing 1958 – Josef Benzing, *Walther H. Ryff und sein literarisches Werk: Eine Bibliographie*, Hamburg, 1959 (Offprint from *Philobiblon: Eine Vierteljahrsschrift für Buch- und Graphik-Sammler* 2, 1958, 126–154 and 203–226).

Berg 1943 – Alexander Berg, "Der Arzt Nikolaus Kopernikus und die Medizin des ausgehenden Mittelalters", in Johannes Papritz, Hans Schumauch (eds.), *Kopernikus-Forschungen*, "Deutschland und der Osten: Quellen und Forschungen zur Geschichte ihrer Beziehungen 22", Leipzig, 1943, 172–201.

Berns 1989 – Jörg Jochen Berns, "'Prinz aller hohen Türm': Notizen zur literarischen Wahrnehmung des Straßburger Münsters in der Frühen Neuzeiit", *Marburger Jahrbuch für Kunstwissenschaft* 22, 1989, 83–102.

Bertele 1961 – Hans von Bertele, *Globes and Spheres*, Lausanne, 1961.

Beschreibung Heilbronn 1865 – *Beschreibung des Oberamts Heilbronn*, Stuttgart, 1865.

Beyer 1955 – Victor Beyer, *La sculpture Strasbourgeoise au quatorzième siècle*, Strasbourg, 1955.

Beyer 1973 – Victor Beyer, "L'horloge astronomique", in Victor Beyer, Roland Recht, Jean-Daniel Ludmann, Jean-Richard Haeusser, *La cathédrale de Strasbourg*, Strasbourg, 1973, 399–410.

Beyer/Bach/Muller 1960 – Victor Beyer, Henri Bach, Ernest Muller, "Le globe céleste de Dasypodius: I. Étude historique par V. Beyer, II. Étude astronomique et horlogère par Henri Bach, III. Étude hymnologique par E. Muller", *Bulletin de la Société des Amis de la Cathédrale de Strasbourg*, 2nd Ser. 7, 1960, 103–139.

Białostocki 1989 – Jan Białostocki, "Vivitur ingenio", in Stephan Füssel, Joachim Knape (eds.), *Poesis et pictura: Studien zum Verhältnis von Text und Bild in Handschriften und alten Drucken. Festschrift für Dieter Wuttke zum 60. Geburtstag*, "Saecula Spiritalia, Sonderband", Baden – Baden, 1989, 223–234.

Bilfinger 1886 – Gustav Bilfinger, "Die Zeitmesser der antiken Völker", in *Festschrift zur Jubelfeier des Eberhard-Ludwigs-Gymnasiums in Stuttgart: Zugleich Programm zum Schlusse des Schuljahrs 1885/86*, Stuttgart, 1886, 1–78.

Bilfinger 1892 – Gustav Bilfinger, *Die Mittelalterlichen Horen und die Modernen Stunden: Ein Beitrag zur Kulturgeschichte*, Stuttgart, 1892; reprint Vaduz, 1985.

Birkenmajer 1960 – Aleksander Birkenmajer, "Le commentaire inédit d'Erasme Reinhold sur le De Revolutionibus de Nicolas Copernicus", in *La science au seizième siècle: Colloque international, Royaumont 1–4 Juillet 1957*, Paris 1960, 169–177; also in *Studia Copernicana*, 4, 1972, 761–766.

Bīrūnī 1934 – Abū Rayḥān Muḥammad ibn Aḥmad al-Bīrūnī, *The Book of Instruction in the Elements of the Art of Astrology* (ed. and tr. by Ramsay Wright), London, 1934.

Blume 2000 – Dieter Blume, *Regenten des Himmels: Astrologische Bilder in Mittelalter und Renaissance*, "Studien aus dem Warburg-Haus, 3", Berlin, 2000 (enlarged version of Post PhD dissertation, Ludwig-Maximilians-Universität, München, 1990).

Blume 2004 – Dieter Blume, "Children of the Planets: The Popularization of Astrology in the 15th Century", *Micrologus: Natura, Scienze e Societè Medievali* 12, 2004, 549–563.

Blumhof 1796 – Johann Georg Ludolph Blumhof, *Vom alten Mathematiker Conrad Dasypodius*, Göttingen, 1796.

Boas 1949 – Marie Boas, "Hero's Pneumatica: A Study of its Transmission and Influence", *Isis: An International Review Devoted to the History of Science and its Cultural Influences* 40, 1949, 38–48.

Bobinger 1966 – Maximilian Bobinger, *Alt-Augsburger Kompaßmacher: Sonnen-, Mond- und Sternuhren. Astronomische und mathematische Geräte. Räderuhren*, "Abhandlungen zur Geschichte der Stadt Augsburg: Schriftenreihe des Stadtarchivs Augsburg 16", Augsburg, 1966.

Boehm 1743 – François Joseph Boehm, *Description nouvelle de la Cathedrale de Strasbourg et ses fameuse tour*, Strasbourg, 1743 (tr. of Behr 1732).

Boer/Strohmaier 1979 – Emilie Boer, Gotthard Strohmaier, "Mittellateinische Inschriften auf astronomischen Uhren des 14. Jhs. in Stralsund und Bad Doberan", *Philologus: Zeitschrift für klassische Philologie* 123, 1979, 108–114; also in Gotthard Strohmaier, *Hellas im Islam: Interdisziplinäre Studien zur Ikonographie, Wissenschaft und Religionsgeschichte*, "Diskurse der Arabistik 6", Wiesbaden, 2003, 220–224.

Boll 1894 – Franz Boll, *Studien über Claudius Ptolemäus: Ein Beitrag zur Geschichte der griechischen Philosophie und Astrologie*, "*Jahrbücher für classische Philologie 21*, suppl.", Leipzig, 1894.

Boll 1899 – Franz Boll, "Beiträge zur Überlieferungsgeschichte der griechischen Astrologie und Astronomie", *Sitzungsberichte der Bayerischen Akademie der Wissenschaften, Philosophisch-philologische und historische Klasse* 1, 1899, 77–140.

Boll 1913 – Franz Boll, *Die Lebensalter: Ein Beitrag zur antiken Ethologie und zur Geschichte der Zahlen: Mit einem Anhang über die Schrift von der Siebenzahl*, Leipzig – Berlin, 1913 (offprint from *Neue Jahrbücher für das klassische Altertum, Geschichte und Deutsche Literatur* 16, 1913, 89–145).

Boll/Bezold 1926 – Franz Boll, Carl Bezold, *Sternglaube und Sterndeutung: Die Geschichte und das Wesen der Astrologie*, Berlin, 1926³.

Bopp 1959 – Marie-Joseph Bopp, *Die evangelischen Geistlichen und Theologen in Elsaß und Lothringen von der Reformation bis zur Gegenwart*, "Genealogie und Landesgeschichte 1", Neustadt an der Aisch, 1959.

Bornkamm 1943 – Heinrich Bornkamm, "Kopernikus im Urteil der Reformatoren", *Archiv für Reformationsgeschichte: Forschungen zur Geschichte des Protestantismus und seiner Weltwirkungen* 40, 1943, 171–183.

Bouché-Leclercq 1899 – Auguste Bouché-Leclercq, *L'Astrologie grecque*, Paris, 1899; reprint Aalen, 1979.

Boyle 1772 – Robert Boyle, *The Works* (ed. by Thomas Birch), London, 1772; reprint Hildesheim, 1966.

Brachvogel 1919 – Eugen Brachvogel, "Die Bildnisse der ermländischen Bischöfe", *Zeitschrift für die Geschichte und Altertumskunde Ermlands* 20, 1919, 516–601.

Brachvogel 1939 – Eugen Brachvogel, "Zur Kunde der Coppernicus-Bildnisse", *Zeitschrift für die Geschichte und Altertumskunde Ermlands* 27, 1939, 281–286.

Brandt 1927/28 – Paul Brandt, *Schaffende Arbeit und bildende Kunst*, Leipzig, 1927–1928.

Brennan 2017 – Christopher W. Brennan, *Hellenistic Astrology: The Study of Fate and Fortune*, Denver (CO), 2017.

Brosseder 2004 – Claudia Brosseder, *Im Bann der Sterne: Caspar Peucer, Philipp Melanchthon und andere Wittenberger Astrologen*, Berlin, 2004 (PhD dissertation, Ludwig-Maximilians-Universität München, 2001).

Brown 1932 – Basil J.W. Brown, *Astronomical Atlases, Maps and Charts: An Historical and General Guide*, London, 1932; reprinted London, 1968.

Brummel 1965 – Leendert Brummel, "Hugo Blotius in Straßburg", in *Buch und Welt: Festschrift Gustav Hofmann*, Wiesbaden, 1965, 211–217.

Brun 1905/17 – Carl Brun (ed.), *Schweizerisches Künstler-Lexikon*, Frauenfeld, 1905–1917.

Brunn 1937 – Walter Albert Leopold von Brunn, "Geschichtliches vom Maiglöckchen", in W.A.L. von Brunn, Lothar Klotz, *Das Maiglöckchen eine alte, vielgerühmte Heilpflanze: Zum 50jährigen Bestehen der Firma Dr. Degen & Kuth/Düren-Rheinland*, Düren, 1937, 5–55.

Bucher 1968 – François Bucher, "Design in Gothic Architecture: A Preliminary Assessment", *Journal of the Society of Architectural Historians* 27, 1968, 49–71.

Buczkowski 1989 – Malgorzata Buczkowski, *Beitrag zum gegenwärtigen Stand der Forschung über die ärztliche Tätigkeit des Nicolaus Copernicus*, PhD dissertation, Ludwig-Maximilians-Universität München, 1989.

Büeler 1918 – Gustav Büeler, "Der Familienname des Petrus Dasypodius", *Thurgauische Beiträge zur vaterländischen Geschichte* 57–58, 1918, 172–176.

Bund 1986 – Konrad Bund, "Eine Chronik des Frankfurter Domgeläutes", in K. Bund (ed.), *Frankfurter Glockenbuch*, "Mitteilungen aus dem Frankfurter Stadtarchiv IV", Frankfurt am Main, 1986, 228–314.

Bünger 1893 – Carl Bünger, *Matthias Bernegger: Ein Bild aus dem geistigen Leben Strassburgs zur Zeit des Dreißigjährigen Krieges*, Strasbourg, 1893.

Bunte 1877 – Bernhard Bunte, "Über Archimedes, mit besonderer Berücksichtigung der Lebens- und Zeitverhältnisse, sowie zweier von demselben herrührenden mechanischen Kunstwerke", in *Programm der Realschule zu Leer*, Leer, 1877, 3–21.

Burmeister 1978 – Karl Heinz Burmeister, "Der Kartograph Heinrich Zell (1518–1564)", *Studia Copernicana* 16, 1978, 427–442.

Burmeister 2015 – Karl Heinz Burmeister, *Magister Rheticus und seine Schulgesellen: Das Ringen um Kenntnis und Durchsetzung des heliozentrischen Weltsystems um 1540/50*, "Forschungen zur Geschichte Vorarlbergs, *new ser.* 11", Konstanz – München, 2015.

Cantamessa 2011 – Leandro Cantamessa, *Astrologia: Ins & Outs; Opere a Stampa (1468–1930)*, Mailand, 2011 (enlarged and revised edition of L. Cantamessa, *Astrologia Opere a Stampa (1472–1900)*, "Biblioteca di bibliografia italiana 187", Florence, 2007).

Cantor 1907 – Moritz Cantor, *Vorlesungen über Geschichte der Mathematik*, v. I., Leipzig 1907^3; reprinted New York, 1965.

Cardano 1578 – Girolamo Cardano, *In Cl. Ptolemaei De astrorum iudiciis, aut (ut vulgo appellant) Quadripartitae Constructionis Lib. IIII. Commentaria, ab Autore postremum castigata, et locupletata.* [...] *Item, Cunradi Dasypodii, Mathematici Argent. Scholia et Resolutiones seu Tabulae in Lib. IIII. Apotelesmaticos Cl. Ptolemaei: Una cum Aphorismis eorundem librorum. Denique brevis explicatio Astronomici Horologii Argentoratensis, ad veri et exacti temporis investigationem extructi*, Basel, 1578.

Châtelet-Lange 2010 – Liliane Châtelet-Lange, "Architecture et humanisme à l'horloge astronomique de la cathédrale de Strasbourg", *Bulletin de la Cathédrale de Strasbourg* 29, 2010, 9–28.

Chojecka 1967 – Ewa Chojecka, *Astronomische und astrologische Darstellungen und Deutungen bei kunsthistorischen Betrachtungen alter wissenschaftlicher Illustrationen des XV. bis XVIII. Jahrhunderts*, "Veröffentlichungen des Staatlich Mathematisch-Physikalischen Salons Dresden 4", East Berlin, 1967.

Chrisman 1982 – Miriam Usher Chrisman, *Lay Culture, Learned Culture: Books and Social Change in Strasbourg 1480–1599*, New Haven – London, 1982.

Christianson 1973 – John Robert Christianson, "Copernicus and the Lutherans", *The Sixteenth Century Journal* 4 (2), 1973, 1–10.

Christianson 2000 – John Robert Christianson, *On Tycho's Island: Tycho Brahe and His Assistants, 1570–1601*, Cambridge, 2000.

Cicero 1945 – Marcus Tullius Cicero, *Tusculan Disputations* (tr. by John E. King), "Loeb Classical Library 141", Cambridge (MA), 1945.

Cicero 1970 – Marcus Tullius Cicero, *De re Publica* (*On the Republic*), *De Legibus* (*On the Laws*) (tr. by Clinton Walker Keyes), "Loeb Classical Library 213", Cambridge (MA) – London, 1970^{8}.

Claudian 1922 – Claudius Claudianus, *The Shorter Poems* (tr. by Maurice Platnauer), "Loeb Classical Library 136", Cambridge (MA), 1922.

Colson 1926 – Francis Henry Colson, *The Week: An Essay on the Origin and Development of the Seven-Day Cycle*, Cambridge, 1926.

Cordez 2015 – Philippe Cordez, *Schatz, Gedächtnis, Wunder: Die Objekte der Kirchen im Mittelalter*, "Quellen und Studien zur Geschichte und Kunst im Bistum Hildesheim, 10", Regensburg, 2015 (PhD dissertation, École des Hautes Études en Sciences Sociales, Paris, 2010).

Cordonnier 2012 – Rémy Cordonnier, "L'iconographie du coq à la cloche", in Fabienne Pomel (ed.), *Cloches et horloges dans les textes médiévaux: Mesurer et maîtriser le temps*, Rennes, 2012, 79–108.

Coryat 1611 – Thomas Coryat, *Coryats Crudities. Hastily gobled up in five Moneths travells in France, Savoy, Italy, Rhetia commonly called the Grisons country, Helvetia alias Switzerland, some parts of high Germany, and the Netherlands; Newly digested in the hungry aire of Odcombe in the County of Somerset, and now dispersed to the nourishment of the trauelling Members of this Kingdome*, London 1611; reprinted Glasgow, 1905.

Crombach 1654 – Hermann Crombach, *Primitiae gentium seu historia SS. Trium Regum magorum evangelicorum*, Cologne, 1654.

Curtze 1874a – Maximilian Curtze, "Reliquiae Copernicanae", *Zeitschrift für Mathematik und Physik* 19, 1874, 76–82, 432–458; 20, 1875, 221–248.

Curtze 1874b – Maximilian Curtze, "Fünf ungedruckte Briefe von Gemma Frisius", *Archiv der Mathematik und Physik* 56, 1874, 313–325.

Dacheux 1887 – Léon Dacheux (ed.), *La petite chronique de la cathédrale. La chronique strasbourgeoise de Sébald Büheler*, "Fragments des anciennes chroniques d'Alsace 1", Strasbourg, 1887.

DaCosta Kaufmann 1993 – Thomas DaCosta Kaufmann, "Astronomy, Technology, Humanism and Art at the Entry of Rudolf II into Vienna, 1577: The Role of Paulus Fabricius", in T. DaCosta Kaufmann, *The Mastery of Nature: Aspects of Art, Science, and Humanism in the Renaissance*, Princeton, 1993, 136–150.

Dann 1989 – Ingrid Dann, "Walther Ryff", in Hubertus Günther (ed.), *Deutsche Architekturtheorie zwischen Gotik und Renaissance*, Darmstadt, 1988, 79–84.

Dasypodius 1572 – Conrad Dasypodius, *Sphaericae Doctrinae Propositiones Graecae et Latinae*, Strasbourg, 1572.

Dasypodius 1578a – Conrad Dasypodius, *Warhafftige Außlegung des Astronomischen Uhrwercks zu Straßburg beschriben Durch M. Cunradum Dasypodium der solches Astronomische Uhrwerck anfenglichs erfunden und angeben*, Strasbourg, 1578.

Dasypodius 1578b – Conrad Dasypodius, *Von Cometen und ihrer würckung, Durch M. Cunradum Dasypodium beschrieben*, Strasbourg, 1578.

Dasypodius 1578c – Conrad Dasypodius, *Brevis doctrina de Cometis et Cometarum effectibus*, Strasbourg, 1578.

Dasypodius 1579 – Conrad Dasypodius, *Oratio Cunradi Dasypodii de disciplinis Mathematicis: Ad Fridericum II. Sereniss. Regem Daniae etc. Eiusdem Hieronis Alexandrini Nomenclaturae Vocabulorum Geometricorum translatio. Eiusdem. Lexicon Mathematicum, ex diversis collectum antiquis scriptis*, Strasbourg, 1579.

Dasypodius 1580a – Conrad Dasypodius, *Warhafftige Außlegung und Beschreybung des Astronomischen Uhrwercks zu Straßburg welches er Anfaenglichs Erfunden und angeben hat. Auch Ein Altes Lied vom dem Kampff und streyt so entstanden Zwischen dem Roraffen (welcher under der Orglen im Muenster zu Straßburg ist) und dem Hanen so auff der Alten Uhren war vor 200. Jharen Gedicht*, Strasbourg, 1580.

Dasypodius 1580b – Conrad Dasypodius, *Heron Mechanicus: Seu De Mechanicis artibus, atque disciplinis. Eiusdem Horologii astronomici, Argentorati in summo Templo erecti, descriptio*, Strasbourg, 1580.

Dasypodius 1593a – Conrad Dasypodius, *Institutionum Mathematicarum Voluminis primi Erotemata. Logisticae. Geometriae. Sphaerae. Geographiae*, Strasbourg, 1593.

Dasypodius 1593b – Conrad Dasypodius, *Protheoria mathematica, in qua non solum disciplinae Mathematica omnes, ordine convenienti enumerantur; verum etiam universalia Mathematica praecepta; explicantur*, Strasbourg, 1593.

Dasypodius 1596 – Conrad Dasypodius, *Institutionum Mathematicarum Voluminis primi Erotematum. Appendix. Elementorum. Arithmeticae. Geodaesiae. Opticae. Catoptricae. Scenographiae. Theoriae Planetarum. Logisticae. Astronomicae Astrologiae. Musicae. Mechanicae*, Strasbourg, 1596.

Dasypodius 2008 – Conrad Dasypodius, *Heron mechanicus: Seu de Mechanicis artibus, atque disciplinis. Eiusdem Horologii astronomici, Argentorati in summo Templo erecti, descriptio; Argentorati 1580* (ed. by Günther Oestmann, tr. by Bernard Aratowsky), "Algorismus: Studien zur Geschichte der Mathematik und der Naturwissenschaften 68", Augsburg, 2008.

Debauvais/Befort 2002 – Francis Debauvais, Paul-André Befort, *Cueillir les étoiles: Autour des astrolabes de Strasbourg*, Strasbourg, 2002.

Delambre 1817 – Jean Baptiste Joseph Delambre, *Histoire de l'Astronomie Ancienne*, Paris, 1817.

Delambre 1821 – Jean Baptiste Joseph Delambre, *Histoire de l'Astronomie Moderne*, Paris, 1821.

Deluz 2019 – Vincent Deluz, "Création mécanique: Le coq automate de la prémière horologe astronomique de Strasbourg", in: Florian Besson, Viviane Griveau-Genest, Julie Pilorget (eds.), *Créer: Créateurs, créations, créatures au Moyen Âge*, Paris 2019, 215–236.

Dengler 2011 – Michael Dengler, *Zeitmaschinen, Sakralautomaten, Frömmigkeitsapparate: Die Produktion sakraler Zeiten im Kirchenraum der Vormoderne*, PhD dissertation, Universität Konstanz, 2011 (http://nbn-resolving.de/urn:nbn:de:bsz:352-263349).

Deplasne 1921 – L. Deplasne [Léopold Reverchon], "La bibliographie de l'Horloge de Strasbourg", *Hora: Gazette mensuelle pour l'horloger, le mécanicien, le bijoutier et l'orfèvre* 64, 1921, 153–157.

Deutsch 2017 – Kristina Deutsch, "Mensch und Maschine im Mittelalter", *Kunstchronik* 70, 2017, 36–44 (book review of Elly Rachel Truitt, *Medieval Robots: Mechanism, Magic, Nature, and Art*, Philadelphia, 2015).

D'Hollander 1993 – Raymond D'Hollander, *L'Astrolabe: Les Astrolabes du Musée Paul Dupuy*, Toulouse, 1993.

Dicks 1953/54 – David Reginald Dicks, "Ancient Astronomical Instruments", *Journal of the British Astronomical Association* 64, 1953–1954, 77–85.

Diels 1917 – Hermann Diels, *Über die von Prokop beschriebene Kunstuhr von Gaza: Mit einem Anhang enthaltend Text und Übersetzung der ΕΚΦΡΑΣΙΣ ΩΡΟΛΟΓΙΟΥ des Prokopius von Gaza*, "Abhandlungen der Königl. Preußischen Akademie der Wissenschaften, Philosophisch-Historische Klasse", Berlin, 1917.

Diels 1924 – Hermann Diels, *Antike Technik*, Leipzig, 1924[3]; reprint Osnabrück, 1965.

Diemar 1909 – Hermann Diemar (ed.), *Die Chroniken des Wigand Gerstenberg von Frankenberg*, "Veröffentlichungen der Historischen Kommission für Hessen und Waldeck: Chroniken von Hessen und Waldeck I", Marburg, 1909.

Dijksterhuis 1956 – Eduard Jan Dijksterhuis, *Archimedes*, "Acta Historica Scientiarum Naturalium et Medicinalium XII", Copenhagen, 1956.

Dohrn-van Rossum 1996 – Gerhard Dohrn-van Rossum, *History of the Hour: Clocks and Modern Temporal Orders* (Tr. by Thomas Dunlap), Chicago – London, 1996.

Drachmann 1948 – Aage Gerhardt Drachmann, *Ktesibios, Philon and Heron: A Study in Ancient Pneumatics*, "Acta Historica Scientiarum Naturalium et Medicinalium IV", Copenhagen, 1948 (PhD dissertation, University of Copenhagen, 1947).

Drachmann 1963 – Aage Gerhardt Drachmann, *The Mechanical Technology of Greek and Roman Antiquity: A Study of the Literary Sources*, "Acta Historica Scientiarum Naturalium et Medicinalium XVII", Copenhagen, 1963.

Drachmann 1969 – Aage Gerhardt Drachmann, *Heron von Alexandria*, "Veröffentlichungen des Forschungsinstituts des Deutschen Museums für die Geschichte der Naturwissenschaften und der Technik B-2", Munich, 1969.

Drechsler 1855 – Adolph Drechsler, *Astrologische Vorträge zur Einführung in das Verständniß des Systems und der Geschichte der Astrologie gehalten zu Dresden im Winter 1854/55*, Dresden, 1855.

Drecker 1909 – Joseph Drecker, *Gnomone und Sonnenuhren*, "Wissenschaftliche Beilage zum Jahresbericht der Oberrealschule mit Reform-Realgymnasium zu Aachen 672", Aachen, 1909.

Drecker 1927 – Joseph Drecker, "Das Planisphaerium des Claudius Ptolemaeus", *Isis: An International Review Devoted to the History of Science and its Cultural Influences* 9, 1927, 255–278.

Drecker 1931 – Joseph Drecker, "Hermannus Contractus: Über das Astrolab", *Isis: An International Review Devoted to the History of Science and its Cultural Influences* 16, 1931, 200–219.

Drescher 1993 – Georg Drescher, "Frühe Copernicus-Porträts", in *450 Jahre Copernicus 'De revolutionibus': Astronomische und mathematische Bücher aus Schweinfurter Bibliotheken*, "Veröffentlichungen des Stadtarchivs Schweinfurt 9", Schweinfurt, 1993, 23–28.

Dreyer 1890 – John Louis Emil Dreyer, *Tycho Brahe: A Picture of Scientific Life and Work in the Sixteenth Century*, Edinburgh, 1890.

Drost 1963 – Willi Drost, *Die Marienkirche in Danzig und ihre Kunstschätze*, "Bau- und Kunstdenkmäler des deutschen Ostens 4", Stuttgart, 1963.

Eade 1984 – John Christopher Eade, *The Forgotten Sky: A Guide to Astrology in English Literature*, Oxford, 1984.

Ebhardt 1918 – Bodo Ebhardt, *Die zehn Bücher der Architektur des Vitruv und ihre Herausgeber seit 1484*, Berlin, 1918.

Edel 1777 – Matthäus Edel, "Nachricht von den Uhrenmachern des Uhrwerks im Münster: Aus Originalien abgedruckt", *Der Bürgerfreund: Eine Straßburgische Wochenschrift* 2 (1), 1777, XIII. Stück, 193–207.

Edel 1843 – Friedrich Wilhelm Edel, *Die astronomische Münsteruhr in Straßburg: Mittheilungen zur Kenntniß ihrer ältern und neuern Geschichte*, Strasbourg, 1843.

Eis 1952 – Gerhard Eis, "Zu den medizinischen Aufzeichnungen des Nicolaus Copernicus", *Lychnos: Lärdomshistoriska Samfundets Årsbok*, 1952, 186–209.

Erler 1954 – Adalbert Erler, *Das Straßburger Münster im Rechtsleben des Mittelalters*, "Frankfurter wissenschaftliche Beiträge: Rechts- und wirtschaftswissenschaftliche Reihe 9", Frankfurt am Main, 1954.

Euw 1975 – Anton von Euw, "Imago mundi", in *Monumenta Annonis: Weltbild und Kunst im hohen Mittelalter* (catalog of the exhibition at the Schnütgen-Museum in St. Cäcilien of Cologne, 30 April–27 July 1975), Cologne, 1975, 89–99.

Evans 1999 – James Evans, "The Material Culture of Greek Astronomy", *Journal for the History of Astronomy* 30, 1999, 237–307.

Evans/Berggren 2006 – James Evans, John Lennart Berggren (eds. and tr.), *Geminos's Introduction to the Phenomena: A Translation and Study of a Hellenistic Survey of Astronomy*, Princeton – Oxford, 2006.

Fallet-Scheurer 1917 – Marius Fallet-Scheurer, *Geschichte der Uhrmacherkunst in Basel 1370–1874: Ein Beitrag zur Entwicklungsgeschichte der Uhrmacherkunst im allgemeinen, sowie zur Wirtschafts- und Kulturgeschichte Basels*, PhD dissertation, University of Bern, 1917.

Ficker/Winckelmann 1902/05 – Johannes Ficker, Otto Winckelmann (eds.), *Handschriftenproben des sechzehnten Jahrhunderts nach Straßburger Originalen*, Strasbourg, 1902–1905.

Firmicus Maternus 1897/1913 – Firmicus Maternus, *Iulii Firmici Materni Matheseos Libri VIII* (ed. by Wilhelm Kroll, Franz Skutsch, Konrat Ziegler), Leipzig, 1897–1913; reprint Stuttgart, 1968.

Firmicus Maternus 1975 – Firmicus Maternus, *Ancient Astrology: Theory and Practice* (tr. by Jean Rhys Bram), New Jersey, 1975 (English tr. of Firmicus Maternus 1897/1913).

Fischart 1977 – Johann Fischart, *Das Glückhafft Schiff von Zürich* (ed. by Alois Haas), Stuttgart, 1977.

Fix 1792 – [Christian Gotthelf Fix], *Geschichte der Astronomie von den ältesten bis auf gegenwärtige Zeiten, in zwey Bänden*, Chemnitz, 1792, v. 1 (the only one published).

Flik-Fizek 1992 – M. Flik-Fizek, "Najstarsze wizerunki Mikolaja Kopernika w malarstwie i grafice z XVI i XVII wieku" [The oldest pictures of Nicolaus Copernicus in the paintings and graphics of the 16th and 17th centuries], *Rocznik Muzeum w Toruniu* 9, 1992, 166–181.

Forssman 1956 – Erik Forssman, *Säule und Ornament: Studien zum Problem des Manierismus in den nordischen Säulenbüchern und Vorlageblättern des 16. und 17. Jahrhunderts*, Stockholm, 1956.

Forstreuter 1973 – Kurt Forstreuter, "Bemerkungen zu den ältesten Bildern und Biographien von Copernicus", *Preußenland* 11, 1973, 18–32.

Frauenfelder 1951 – Reinhard Frauenfelder, *Die Kunstdenkmäler des Kantons Schaffhausen, Bd. I: Die Stadt Schaffhausen*, "Die Kunstdenkmäler der Schweiz 26", Basel, 1951.

Friedensburg 1917 – Walter Friedensburg, *Geschichte der Universität Wittenberg*, Halle, 1917.

Friese 1792 – Johannes Friese, *Neue Vaterländische Geschichte der Stadt Straßburg und des ehemaligen Elsaßes. Von den ältesten Zeiten bis auf das Jahr 1791*, Strasbourg, 1792.

Frischlin 1575 – Nikodemus Frischlin, *Carmen de Astronomico Horologio Argentoratensi, scriptum à M. Nicodemo Frischlino Balingensi, Academiae Tubingensis Professore. Item de eodem Schediasma Guilielmi Xylandri Augustani*, Strasbourg, 1575.

Fröde 2010 – Tino Fröde, "Paulus Fabricius—ein universaler Humanist aus Lauban: Sein Leben, seine Schriften und seine Beziehungen zur Oberlausitz", *Neues Lausitzisches Magazin: Zeitschrift der Oberlausitzischen Gesellschaft der Wissenschaften, new ser.* 13, 2010, 55–70.

Frommhold 2004 – Katrin Frommhold, *Bedeutung und Berechnung der Empfängnis in der Astrologie der Antike*, "Orbis Antiquus 38", Münster, 2004 (PhD dissertation, Westfälische Wilhelms-Universität Münster, 2013–2014).

Fry 1845 – Henrietta Joan Fry, *Hymns of the Reformation by Dr. Martin Luther and others from the German*, London, 1845.

Fuchs 1970/71 – Joseph Fuchs, "Die Villinger Münsteruhr", *Schriften der Freunde alter Uhren* 10, 1970–1971, 59–62.

Fuchs 1990 – François Joseph Fuchs, "Les comptes de l'Œuvre Notre Dame des années 1382/83 et 1385 (?)", *Bulletin de la Cathédrale de Strasbourg* 19, 1990, 41–52.

Gall/Szaivert 1967 – Franz Gall, Willy Szaivert (eds.), *Die Matrikel der Universität Wien*. Graz – Vienna – Cologne, 1967.

Gass 1926 – Joseph Gass, "Le portrait de Copernic à Strasbourg", *Revue catholique d'Alsace*, 41, 1926, 75–78.

Gassendi 1658 – Pierre Gassendi, *Nicolai Copernici Warmiensis Canonici Astronomi illustris vita*, "Opera omnia v", Lyon, 1658; reprint Stuttgart – Bad Cannstatt, 1964.

Geiger 1983 – Angelika Geiger, *Wallensteins Astrologie: Eine kritische Überprüfung der Überlieferung nach dem gegenwärtigen Quellenbestand*, Graz, 1983 (PhD dissertation, Bonn 1982).

Geisberg 1937 – Max Geisberg, *Die Stadt Münster*, "Die Bau- und Kunstdenkmäler von Westfalen 41, pt. 5", Münster, 1937.

Gelenius 1645 – Aegidius Gelenius, *De admiranda, sacra, et civili magnitudine Coloniae Claudiae Agrippinensis Augustae Ubiorum Urbis Libri IV*, Cologne, 1645.

Gélis 1949 – Edouard Gélis, *L'horlogérie ancienne: Histoire, Décor et Technique*, Paris, 1949.

Gérard 1872/63 – Charles Gérard, *Les artistes de l'Alsace pendant le moyen-age*, Colmar – Paris, 1872–1873.

Gerhardt 1871 – Carl Immanuel Gerhardt (ed. and tr.), *Der Sammlung des Pappus von Alexandrien siebentes und achtes Buch*, Halle, 1871.

Gericke 1955 – Helmuth Gericke, *Zur Geschichte der Mathematik an der Universität Freiburg*, "Beiträge zur Freiburger Wissenschafts- und Universitätsgeschichte 7", Freiburg im Breisgau, 1955.

Gerstinger 1926 – Hans Gerstinger, "Johannes Sambucus als Handschriftensammler", in *Festschrift der Nationalbibliothek in Wien*, Vienna, 1926, 251–400.

Gerstinger 1968 – Hans Gerstinger, "Die Briefe des Johannes Sambucus (Zsamboky) 1554–1584", *Sitzungsberichte der Österreich. Akademie der Wissenschaften, Philosophisch-Historische Klasse* 255, 1968.

Gibbs 1976 – Sharon Louise Gibbs, *Greek and Roman Sundials*, New Haven – London, 1976.

Gigon 1973 – Olof Gigon, "Posidoniana—Ciceroniana—Lactantiana", in Willem den Boer *et al.* (eds.), *Romanitas et Christianitas: Studia Iano Henrico Waszink A.D. VI Kal. Nov. A. MCMLXXIII XIII lustra complenti oblata*, Amsterdam – London, 1973, 145–180.

Gingerich 1973 – Owen Gingerich, "The Role of Erasmus Reinhold and the Prutenic Tables in the Dissemination of Copernican Theory", *Studia Copernicana* 6, 1973, 43–62; 123–125.

Gingerich 2002 – Owen Gingerich, *An Annotated Census of Copernicus' De Revolutionibus Nuremberg, 1543, and Basel, 1566*, "Studia Copernicana 2", Leiden – Boston – Cologne, 2002.

Good 1985 – Richard Good, "Isaac Habrechts Uhr aus dem Jahr 1589 mit Automaten und Glockenspiel", *Alte Uhren* 8 (1), 1985, 9–20.

Gormans 1999 – Andreas Gormans, *Geometria et Ars memorativa: Studien zur Bedeutung von Kreis und Quadrat als Bestandteile mittelalterlicher Mnemonik und ihrer Wirkungsgeschichte an ausgewählten Beispielen*, PhD dissertation, Rheinisch-Westfälische Technische Hochschule Aachen, 1999 (http://publications.rwth-aachen.de/record/61189/files/Gormans_Andreas.pdf).

Grandidier 1782 – Philippe André Grandidier, *Essais historiques et topographiques sur l'église cathédrale de Strasbourg*, Strasbourg, 1782.

Graux 1889 – Charles Graux, "Notices sommaires des manuscrits grecs de Suède", *Archives des missions scientifiques et littéraires: Choix de rapports et instructions publié sous les auspices du Ministère de l'instruction publique et des cultes, 3rd ser.* 15, 1889, 293–370.

Gundel 1926 – Wilhelm Gundel, "Textkritische und exegetische Bemerkungen zu Manilius", *Philologus: Zeitschrift für das klassische Altertum und sein Nachleben* 81 (*new ser.* 35), 1926, 168–191; 309–338.

Gundel 1934 – Wilhelm Gundel, "Astronomie, Astralreligion, Astralmythologie und Astrologie: Darstellung und Literaturbericht 1907–1933", *Jahresbericht über die Fortschritte der klassischen Altertumswissenschaft* 243, 1934, 1–162.

Gundel 1936 – Wilhelm Gundel, "Neue astrologische Texte des Hermes Trismegistos: Funde und Forschungen auf dem Gebiet der antiken Astronomie und Astrologie", *Abhandlungen der Bayerischen Akademie der Wissenschaften. Philosophisch-Historische Klasse, new ser.* 12, 1936.

Gundel/Gundel 1966 – Wilhelm Gundel, Hans-Georg Gundel, *Astrologumena: Die astrologische Literatur in der Antike und ihre Geschichte*, "Sudhoffs Archiv für Geschichte der Medizin und der Naturwissenschaften 6", Wiesbaden, 1966.

Gunella/Lamprey 2007 – Alessandro Gunella, John Lamprey (eds. and tr.), *Stoeffler's Elucidatio—The Construction and Use of the Astrolabe*, Cheyenne (WY), 2007.

Gunnoe 2010 – Charles D. Gunnoe Jr., "German Protestantism and Astrology: The Debate between Thomas Erastus and the Melanchthon Circle", in: Kaspar v. Greyerz (ed.), *Religion und Naturwissenschaften im 16. und 17. Jahrhundert: Eine Einführung*, "Schriften des Vereins für Reformationsgeschichte 210", Gütersloh, 2010, 86–101.

Günther 1988 – Hubertus Günther (ed.), *Deutsche Architekturtheorie zwischen Gotik und Renaissance*, Darmstadt, 1988.

Haber 1975 – Francis C. Haber, "The Cathedral Clock and the Cosmological Clock Metaphor", in Julius Thomas Fraser, Nathaniel Lawrence (eds.), *The Study of Time II: Proceedings of the International Society for the Study of Time*, Berlin – Heidelberg – New York, 1975, 399–416.

Hahnloser 1972 – Hans R. Hahnloser, *Villard de Honnecourt: Kritische Gesamtausgabe des Bauhüttenbuches ms. fr. 19093 der Pariser Nationalbibliothek*, Graz, 1972².

Hain 1948 – Ludwig Hain, *Repertorium bibliographicum, in quo libri omnes ab arte typographica inventa usque ad annum MD. typis expressi …*, Stuttgart, 1837; reprint Mailand, 1948.

Hamel 1995 – Jürgen Hamel, "Georg Spalatin contra Nicolaus Copernicus? Eine Betrachtung über ein angebliches Copernicus-Portrait", *Die Sterne: Monatsschrift über alle Gebiete der Himmelskunde* 71, 1995, 110–113.

Hamm 2011 – Elizabeth Anne Hamm, *Ptolemy's Planetary Theory: An English Translation of Book One, Part A of the Planetary Hypotheses with Introduction and Commentary*, PhD dissertation, University of Toronto 2011, https://tspace.library.utoronto.ca/handle/1807/32069.

Hamm 2016 – Elizabeth Anne Hamm, "Modeling the Heavens: *Sphairopoiia* and Ptolemy's *Planetary Hypotheses*", *Perspectives on Science* 24, 2016, 416–424.

Hanauer 1876/78 – Auguste Charles Hanauer, *Études économiques sur l'Alsace ancienne et moderne publiées sous les auspices de la Société Industrielle de Mulhouse*, Paris – Strasbourg, 1876–1878.

Hanfmann 1951 – George Maxim Anossov Hanfmann, *The Season Sarcophagus in Dumbarton Oaks*, "Dumbarton Oaks Studies 2", Cambridge (MA), 1951.

Harbison 1976 – Craig Harbison, *The Last Judgment in Sixteenth Century Northern Europe: A Study of the Relation Between Art and the Reformation*, New York – London, 1976.

Harms/Kemp 1987 – Wolfgang Harms, Cornelia Kemp (eds.), *Die Sammlungen der Hessischen Landes- und Hochschulbibliothek in Darmstadt*, "Deutsche illustrierte Flugblätter IV", Tübingen, 1987.

Hartig 1925 – Otto Hartig, "Mitteilungen aus Münchner Archiven über Uhrmacher und Uhren", *Die Uhrmacher-Woche: Handelszeitung für den Uhren- und Bijouteriewaren-Markt* 15, 1925, 246–247.

Hartner 1938 – Willy Hartner, "The Pseudoplanetary Nodes of the Moon's Orbit in Hindu and Islamic Iconographies: A Contribution to the History of Ancient and Medieval Astrology", *Ars Islamica* 5, 1938, 113–154 (reprinted as Hartner 1968c).

Hartner 1964 – Willy Hartner, "Medieval Views on Cosmic Dimensions and Ptolemy's Kitāb al-Manshūrāt", in *Mélanges Alexandre Koyré publiés à l'occasion de son soixantedixième Anniversaire*, Paris, 1964, 254–282.

Hartner 1968a – Willy Hartner, "The Principle and Use of the Astrolabe", in W. Hartner, *Oriens-Occidens: Ausgewählte Schriften zur Wissenschafts- und Kulturgeschichte. Festschrift zum 60. Geburtstag*, "Collectanea 3", Hildesheim, 1968, 287–311.

Hartner 1968b – Willy Hartner, "Asṭurlāb", in W. Hartner, *Oriens-Occidens: Ausgewählte Schriften zur Wissenschafts- und Kulturgeschichte. Festschrift zum 60. Geburtstag*, "Collectanea 3", Hildesheim, 1968, 312–318 (reprint from Hamilton Alexander Rosskeen Gibb, Evariste Lévi-Provençal, Joseph Schacht (eds.), *Encyclopedia of Islam: New Edition*, Leiden, 1960–, v. I, 722–728).

Hartner 1968c – Willy Hartner, "The Pseudoplanetary Nodes of the Moon's Orbit in Hindu and Islamic Iconographies: A Contribution to the History of Ancient and Medieval Astrology", in W. Hartner, *Oriens-Occidens: Ausgewählte Schriften zur Wissenschafts- und Kulturgeschichte. Festschrift zum 60. Geburtstag*, "Collectanea 3", Hildesheim, 1968, 349–404.

Hass 1927 – P. Hass, "Das Planisphaerium der astronomischen Uhr in der Marienkirche zu Lübeck", *Mitteilungen der Mathematischen Gesellschaft in Hamburg* 6, 1927, 273–290.

Hasse 1984 – Max Hasse, "Baldungs Flügel einer astronomischen Uhr", *Kunstchronik* 37, 1984, 183–185.

Hasse/Wartenberg 2005 – Hans-Peter Hasse, Günther Wartenberg (eds.), *Caspar Peucer (1525–1602): Wissenschaft, Glaube und Politik im konfessionellen Zeitalter*, Leipzig, 2005.

Hauber 1916 – Anton Hauber, *Planetenkinder und Sternbilder: Zur Geschichte des menschlichen Glaubens und Irrens*, "Studien zur deutschen Kunstgeschichte 194", Strasbourg, 1916.

Hauffen 1896 – Adolf Hauffen, "Fischart-Studien II: Die Beschreibung des Astronomischen Uhrwercks", *Euphorion: Zeitschrift für Literaturgeschichte* 3, 1896, 705–710.

Haug 1933 – Hans Haug, "Contributions à l'histoire de la sculpture strasbourgeoise (1350–1550)", *Archives alsaciennes d'histoire de l'art* 12, 1933, 1–49.

Hayard 2004 – Michel Hayard, *Chefs d'œuvre de l'horlogerie ancienne: Collection du musée Paul-Dupuy de Toulouse*, Paris, 2004.

Heath 1921 – Thomas Little Heath, *A History of Greek Mathematics*, Oxford, 1921.

Hegel 1870/71 – Carl Hegel (ed.), *Die Chroniken der oberrheinischen Städte: Straßburg*, "Die Chroniken der deutschen Städte vom 14. bis ins 16. Jh., VIII/IX", Leipzig, 1870–1871.

Heiberg 1879 – Johan Ludvig Heiberg, *Quaestiones Archimedeae*, Dissertation, University of Copenhagen, 1879.

Heiberg 1886 – Johan Ludvig Heiberg, "Nogle Eftervirkninger af græsk Mechanik", *Det Kongelige Danske Videnskabernes Selskab: Oversigt over det Kongelige Danske Selskabs Forhandlinger og dets Medlemmers Arbejder*, 1886, 1–14.

Heilen 2010 – Stephan Heilen, "Ptolemy's Doctrine of the Terms and Its Reception", in Alexander Jones (ed.), *Ptolemy in Perspective: Use and Criticism of his Work from Antiquity to the Nineteenth Century*, "Archimedes: New Studies in the History and Philosophy of Science and Technology 23", Dordrecht – Heidelberg – London – New York, 2010, 45–93.

Helfrich-Dörner 1957 – Alma Helfrich-Dörner, "Die Heilbronner Rathausuhr, ihre Erbauer und ihre Baugeschichte: Ein Bericht über die astronomische Kunstuhr am Rathaus zu Heilbronn", *Die Uhr* 11–1, 1957, 9–13.

Hellman 1944 – Clarisse Doris Hellman, *The Comet of 1577: Its Place in the History of Astronomy*, "Studies in History, Economics and Public Law Edited by the Faculty of Political Science of Columbia University 510", New York, 1944.

Heninger 1977 – Simeon Kahn Heninger, *The Cosmographical Glass: Renaissance Diagrams of the Universe*, San Marino (CA), 1977.

Henkel/Schöne 1967 – Arthur Henkel, Albrecht Schöne (eds), *Emblemata: Handbuch zur Sinnbildkunst des XVI. und XVII. Jahrhunderts*, Stuttgart, 1967.

Heron 1899 – Heron of Alexandria, *Druckwerke und Automatentheater* (ed. by Wilhelm Schmidt), "Heronis Alexandrini opera quae supersunt omnia 1", Leipzig, 1899; reprint Stuttgart, 1976.

Heron 1971 – Heron of Alexandria, *The Pneumatics of Hero of Alexandria: A Facsimile of the 1851 Woodcroft Edition* (tr. by Bennet Woodcroft, ed. by Maria Boas Hall), London – New York, 1971.

Herrera 2012 – Breanne Herrera, *The Children of the Planets: Freedom, Necessity, and the Impact of the Stars—The Iconographic Dimensions of a Pan-European Early Modern Discourse*, PhD dissertation, Central European University Budapest, 2012, www.etd.ceu.hu/2012/herrera_breanne.pdf.

Herzog August Bibliothek 1984 – *Architekt und Ingenieur: Baumeister in Krieg und Frieden* (catalog of the exhibition on the Herzog August Bibliothek Wolfenbüttel, 5 May–18 November 1984), Braunschweig, 1984.

Heumann 1722/32 – Christoph August Heumann, *Poecile sive epistolae miscellaneae ad literatissimos aevi nostri viros. Accedit appendix exhibens dissertationes argumenti rarioris*, Halle, 1722–1732.

Heydon 1977 – Christopher Heydon, *A Defence of Iudiciall Astrologie*, "The English Experience: Its Record in Early Printed Books Published in Facsimile 873", Amsterdam – Norwood (NJ), 1977.

Hipler 1875 – Franz Hipler, "Die Porträts des Nikolaus Kopernikus", *Mittheilungen des Ermländischen Kunstvereins* 3, 1875, 73–152.

Hipp 1974 – Hermann Hipp, *Studien zur 'Nachgotik' des 16. und 17. Jahrhunderts in Deutschland, Böhmen, Österreich und der Schweiz*, PhD dissertation, University of Tübingen, 1974.

Hipp 2008 – Hermann Hipp, "Die 'Nachgotik' in Deutschland—kein Stil und ohne Stil", in Stephan Hoppe, Norbert Nussbaum, Matthias Müller (eds.), *Stil als Bedeutung in der nordalpinen Renaissance: Wiederentdeckung einer methodischen Nachbarschaft*, Regensburg, 2008, 14–46.

Hofmann 1975 – Hans Hofmann, *Die Heiligen Drei Könige: Zur Heiligenverehrung im kirchlichen, gesellschaftlichen und politischen Leben des Mittelalters*, "Rheinisches Archiv: Veröffentlichungen des Instituts für geschichtliche Landeskunde der Rheinlande an der Universität Bonn 94", Bonn, 1975 (PhD dissertation, Universität Göttingen, 1972).

Holden 1977 – Ralph William Holden, *The Elements of House Division*, London, 1977; German ed.: *Astrologische Häusersysteme: Entstehung, Berechnung, Bewertung* (tr. by Reinhardt Stiehle), Mössingen, 1998.

Holländer 2000 – Hans Holländer, "Maschinen- und Labyrinthmetaphern als Topoi neuzeitlicher Weltbeschreibung", in H. Holländer (ed.), *Erkenntnis, Erfindung, Konstruktion: Studien zur Bildgeschichte von Naturwissenschaften und Technik vom 16. bis zum 19. Jahrhundert*, Berlin, 2000, 577–586.

Hon/Goldstein 2008 – Giora Hon, Bernard R. Goldstein, *From Summetria to Symmetry: The Making of a Revolutionary Scientific Concept*, "Archimedes: New Studies in the History of Sciences and Technology 20", Dordrecht, 2008.

Horn-Oncken 1967 – Alste Horn-Oncken, "Über das Schickliche: Studien zur Geschichte der Architekturtheorie I", *Abhandlungen der Akademie der Wissenschaften in Göttingen, Philosophisch-Historische Klasse, III. ser.* 70, 1967.

Horsky 1967 – Zdeněk Horsky, "Astronomy and the Art of Clockmaking in the Fourteenth, Fifteenth and Sixteenth Centuries", *Vistas in Astronomy* 9, 1967, 25–34.

Horský/Šimková 1985 – Zdeněk Horský, Anežka Šimkovà, *Olomoucky orloj*, Olomouc, 1985.

Höss 1996 – Irmgard Höss, "Das Jugendbildnis Georg Spalatins von Lucas Cranach d. Ä. 1509", *Archiv für Reformationsgeschichte: Internationale Zeitschrift zur Erforschung der Reformation und ihrer Weltwirkungen* 87, 1996, 400–404.

Houzeau/Lancaster 1882/89 – Jean-Charles Houzeau, Albert Lancaster, *Bibliographie Générale de l'Astronomie*, Brüssel 1882–1889.

Hultsch 1877 – Friedrich Hultsch, "Über den Himmelsglobus des Archimedes", *Zeitschrift für Mathematik und Physik: Historisch-literarische Abtheilung* 22, 1877, 106–107.

Hüttenhain 1936 – Erich Hüttenhain, "Zur Berechnung der Zahnräder des astronomischen Werkes an der Domuhr zu Münster in Westfalen", *Die Uhrmacher-Woche: Handelszeitung für den Uhren- und Bijouteriewaren-Markt* 43 (12–13), 1936, 152–154, 166–167; extract reprinted in Wieschebrink 1983, 72–79.

Ideler 1825/26 – Ludwig Ideler, *Handbuch der mathematischen und technischen Chronologie*, Berlin, 1825–1826.

Jachmann 2006 – Julian Jachmann, *Die Architekturbücher des Walter Hermann Ryff: Vitruvrezeption im Kontext mathematischer Wissenschaften*, "Cultural and Interdisciplinary Studies in Art 1", Stuttgart, 2006.

James 1900/04 – Montague Rhodes James, *The Western Manuscripts in the Library of Trinity College, Cambridge: A Descriptive Catalogue*, Cambridge, 1900–1904.

Janson 1937 – Horst W. Janson, "The Putto with the Death's Head", *The Art Bulletin* 19, 1937, 423–449.

Jenzen 1989 – Igor A. Jenzen (ed.), *Uhrzeiten: Die Geschichte der Uhr und ihres Gebrauches*, "Kleine Schriften des Historischen Museums Frankfurt a. M. 42", Marburg, 1989.

Jolles 1906 – Johannes André Jolles, *Vitruvs Aesthetik*, PhD dissertation, University of Freiburg im Breisgau, 1906.

Jordan 2012 – Benoît Jordan, "Hommage à François-Joseph Fuchs (1921–2012)", *Bulletin de Cathédrale de Strasbourg* 30, 2012, 243–244.

Jud 2008 – Roger Jud, *Konrad Dasypodius: Ein Mathematiker des 16. Jahrhunderts*, Diplomarbeit Hochschule St. Gallen, http://archive.is/eXu1v.

Kaltenböck 1981 – Frederick Kaltenböck, *Die Sammlung von Kaiser Rudolf II. unter besonderer Berücksichtigung der Uhren*, Dissertation, University of Salzburg, 1981.

Karlsen 1986 – Helge B.J. Karlsen, "Jost Bürgi's Rock Crystal Clock from the Period 1622–1627", *Horological Journal* 128 (11), 1986, 8–14.

Kästner 1796/1800 – Abraham Gotthelf Kästner, *Geschichte der Mathematik seit der Wiederherstellung der Wissenschaften bis an das Ende des achtzehnten Jahrhunderts*, Göttingen, 1796–1800.

Kaufmann 1930 – Arnold Kaufmann, "Die Kunstuhr des Zeitglockenturms in Solothurn", *Jahrbuch für Solothurnische Geschichte* 3, 1930, 245–349.

Kejlbo 1995 – Ib Rønne Kejlbo, *Rare Globes: A Cultural-Historical Exposition of Selected Terrestrial and Celestial Globes Made before 1850—Especially connected with Denmark*, Copenhagen, 1995.

Kennedy 1994 – Edward Stewart Kennedy, "Ibn'Muadh on the Astrological Houses", *Zeitschrift für Geschichte der Arabisch-Islamischen Wissenschaften* 9, 1994, 153–160.

Kennedy 1996 – Edward Stewart Kennedy, "The Astrological Houses as Defined by Medieval Islamic Astronomers", in Josep Casulleras, Julio Samsó (eds.), *From*

Baghdad to Barcelona: Studies in the Islamic Exact Sciences in Honour of Prof. Juan Vernet, Barcelona, 1996, v. II, 535–578.

Kern 1982 – Hermann Kern, *Labyrinthe: Erscheinungsform und Deutungen, 5000 Jahre Gegenwart eines Urbilds*, Munich, 1982.

Ketz/Ketz 1978 – Hans-Jürgen Ketz, Klaudia Ketz, "Das älteste deutsche Buch über Räderuhren: Zum 400. Jahrestag seines Erscheinens", *Alte Uhren* 4, 1978, 328–331.

Keussen 1907 – Hermann Keussen, "Die drei Reisen des Utrechters Arnoldus Buchelius nach Deutschland, insbesondere sein Kölner Aufenthalt", *Annalen des Historischen Vereins für den Niederrhein insbesondere die alte Erzdiözese Köln* 84, 1907, 1–102.

Khautz 1755 – Franz Constantin Florian von Khautz, *Versuch einer Geschichte der Oesterreichischen Gelehrten*, Frankfurt am Main – Leipzig, 1755.

King/Millburn 1978 – Henry C. King, John R. Millburn, *Geared to the Stars: The Evolution of Orreries and Astronomical Clocks*, Toronto – Buffalo – London, 1978.

Klein 1971 – Ulrich Klein, "Johannes Keplers Bemühungen um die Harmonieschriften des Ptolemaios und Porphyrios", in *Ausstellungskatalog Johannes Kepler: Werk und Leistung*, Linz, 1971, 51–60.

Klein 2008 – Bruno Klein, "Das Straßburger Münster als Ort kommunaler Repräsentation", in Jörg Oberste (ed.), *Repräsentationen der mittelalterlichen Stadt*, "Forum Mittelalter: Studien 4", Regensburg, 2008, 83–93.

Kleinert 2003 – Andreas Kleinert, "'Eine handgreifliche Geschichtslüge': Wie Martin Luther zum Gegner des copernicanischen Weltsystems gemacht wurde", *Berichte zur Wissenschaftsgeschichte* 26, 2003, 101–111.

Klibansky/Panofsky/Saxl 1964 – Raymond Klibansky, Erwin Panofsky, Fritz Saxl, *Saturn and Melancholy: Studies in the History of Natural Philosophy Religion and Art*, London, 1964; reprint Nendeln, 1979.

Knell 1985 – Heiner Knell, *Vitruvs Architekturtheorie: Versuch einer Interpretation*, Darmstadt, 1985.

Knod 1895 – Gustav Knod, "Hugo Blotius in seinen Beziehungen zu Straßburg", *Centralblatt für Bibliothekswesen* 12, 1895, 266–275.

Koch 1908 – Ernst Koch, "Magister Erasmus Reinhold aus Saalfeld", *Saalfelder Weihnachtsbüchlein* 55, 1908, 3–16.

Koch/Knappich 1959/60 – Walter Koch, Wilhelm Knappich, *Horoskop und Himmelshäuser, I. Teil: Grundlagen und Altertum*, Göppingen, 1959.

Koseleff 1934 – Olga Koseleff, *Die Monatsdarstellungen der französischen Plastik des 12. Jahrhunderts*, PhD dissertation, University of Marburg, 1934.

Kraai 2003 – Jesse Kraai, *Rheticus' Heliocentric Providence: A Study Concerning the Astrology, Astronomy of the Sixteenth Century*, PhD dissertation, Ruprecht -Karls-Universität Heidelberg, 2003, http://www.ub.uni-heidelberg.de/archiv/3254.

Kraus 1876 – Franz Xaver Kraus, *Kunst und Alterthum im Unter-Elsass: Beschreibende Statistik*, "Kunst und Alterthum in Elsass-Lothringen I", Strasbourg, 1876.

Kraus 1890 – Franz Xaver Kraus (ed.), *Die Kunstdenkmäler des Kreises Villingen*, "Die Kunstdenkmäler des Großherzogthums Baden: Beschreibende Statistik II", Freiburg im Breisgau, 1890.

Krebs 1968 – Manfred Krebs (ed.), *Die Protokolle des Speyerer Domkapitels, Bd: I: 1500–1517*, "Veröffentlichungen der Kommission für geschichtliche Landeskunde in Baden-Württemberg, Ser. A, 17", Stuttgart, 1968.

Kren 1985 – Claudia Kern, *Medieval Science and Technology: A Selected, Annotated Bibliography*, "Bibliographies of the History of Science and Technology 11", New York – London, 1985.

Kris/Kurz 1924 – Ernst Kris, Otto Kurz, *Die Legende vom Künstler: Ein geschichtlicher Versuch*, Vienna, 1934; reprint Frankfurt am Main, 1980.

Krüger 1978 – Nilüfer Krüger, *Supellex Epistolica Uffenbachii et Wolfiorum: Katalog der Uffenbach-Wolfschen Briefsammlung*, Hamburg, 1978.

Kuehn 2011 – Sara Kuehn, *The Dragon in Medieval East Christian and Islamic Art*, "Islamic History and Civilization: Studies and Texts 86", Leiden – Boston, 2011.

Kühlmann 2016 – Wilhelm Kühlmann, "Poesie und Mechanik als Weltmodell: Zu Faktur und ideengeschichtlichem Gehalt von Nicodemus Frischlins Lehrgedicht (1575) über die Straßburger Münsteruhr", *Scientia poetica: Jahrbuch für Geschichte der Literatur und der Wissenschaften* 20, 2016, 1–26.

Kuhn 1971 – Werner Kuhn, *Die Studenten der Universität Tübingen zwischen 1477 und 1534: Ihr Studium und ihre spätere Lebensstellung*, "Göppinger akademische Beiträge 37–38", Göppingen, 1971.

Kühne 2001 – Andreas Kühne, "Die Bedeutung von 'Perspectiva communis' und 'Perspectiva artificialis' im Kanon der wissenschaftlichen Bildung um 1500 am Beispiel von Nicolaus Copernicus", in Menso Folkerts, Stefan Kirschner, Theodor Schmidt-Kaler (eds.), *Florilegium Astronomicum: Festschrift für Felix Schmeidler*, "Algorismus: Studien zur Geschichte der Mathematik und der Naturwissenschaften 37", Munich, 2001, 231–248.

Kühne/Kirschner 2004 – Andreas Kühne, Stefan Kirschner (eds.), *Biographia Copernicana: Die Copernicus-Biographien des 16. bis 18. Jahrhunderts: Texte und Übersetzungen. Mit einem Katalog der frühen Copernicus-Porträts von Gudula Metze*, "Nicolaus Copernicus-Gesamtausgabe IX", Berlin, 2004.

Kunstmuseum Basel 1984 – *Spätrenaissance am Oberrhein: Tobias Stimmer 1539–1584* (catalog for the exhibition at the Kunstmuseum, Basel, 23 September–9 December 1984), Basel, 1984.

Kux/Kreß 1904 – Hans Kux, Max Kress, *Das Rathaus zu Olmütz: Ein Gedenkblatt zu seiner Wiederherstellung*, Olmütz, 1904.

Lactantius 1964 – Lucius Caecilius Firmianus Lactantius, *The Divine Institutes, Books I–VII* (tr. by Mary Francis McDonald), "The Fathers of the Church: A New Translation 49", Washington (DC), 1964.

Ladendorf 1963 – Heinz Ladendorf, "Das Labyrinth in Antike und neuerer Zeit", *Archäologischer Anzeiger: Beiblatt zum Jahrbuch des Deutschen Archäologischen Instituts*, 1963, cols. 761–796.

Lang 1903 – Robert Lang, "Die Beaufsichtigung der Schaffhauser Stipendiaten in der Fremde", *Mitteilungen der Gesellschaft für deutsche Erziehungs- und Schulgeschichte* 13, 1903, 146–155.

Lang 1932 – Robert Lang, *Geschichte des Stipendiatenwesens in Schaffhausen: Festschrift zum 75jährigen Bestehen des historisch-antiquarischen Vereins des Kantons Schaffhausen*, "Beiträge zur Vaterländischen Geschichte 12", Schaffhausen, 1932.

Lange 1888 – Carl Lange, "Ein Dreikönigsspiel aus Straßburg", *Zeitschrift für deutsches Alterthum und deutsche Litteratur* 32 (new ser. 20), 1888, 412–415.

Laube 2011 – Stefan Laube, *Von der Reliquie zum Ding: Heiliger Ort—Wunderkammer—Museum*, Berlin, 2011 (Postdoctoral dissertation, Humboldt University, Berlin).

Laugel 1888 – [Auguste Laugel], "Légendes, Moeurs et Coutumes d'Alsace I: Le Singe et le Coq de la Cathédrale de Strasbourg. (Der Rohraff und der Guller.)", *Revue alsacienne: Organe des Intérèts Alsaciens-Lorrains* 11, 1888, 557–565.

Lecocq 1867 – Adolphe Lecocq, "Notice historique et archéologique sur les horloges de l'église Notre-Dame de Chartres", *Mémoires de la société archéologique d'Eure et Loir* 4, 1867, 284–340.

Legin 1986 – Philippe Legin, *Die Astronomische Uhr in Bildern*, Strasbourg, 1986.

Legros 1914 – Henri-Marie Legros, "Vieilles horloges du Maine", *La Province du Maine* 22, 1914, 81–90.

Lehr 1981 – André Lehr, *De Geschiedenis van het Astronomisch Kunstuurwerk: Zijn techniek en muziek*, Den Haag, 1981.

Leichtle 1940 – Christian Leichtle, "Kunstuhren in Württemberg", *Schwäbisches Heimatbuch*, 1940, 81–93.

Leitner 1968 – Gertraud Leitner, *Hugo Blotius und der Straßburger Freundeskreis: Studien zur Korrespondenz des ersten Bibliothekars der Österreichischen Nationalbibliothek*, PhD dissertation, University of Vienna, 1968.

Lemay 1977/78 – Richard Lemay, "Origin and Success of the Kitāb Thamara of Abū Ja'far ibn Yūsuf ibn Ibrāhīm", in *Proceedings of the First International Symposium for the History of Arabic Science 5–12 April 1976*, Aleppo, 1977–1978, v. II, 91–107.

Leovitius 1558 – Cyprianus Leovitius, *Brevis et perspicua ratio iudicandi genituras, ex physicis causis et vera experientia extructa ... Praefixa est Admonitio de vero et licito Astrologiae usu: per Hieronymum Wolfium ... in Dialogo conscripta*, London, 1558.

Leupold 1725 – Jacob Leupold, *Theatrum machinarum oder: Schau-Platz der Heb-Zeuge ...*, Leipzig, 1725.

Linnard/Robey 2019a – William Linnard, John A. Robey, "Three Medieval Clockmakers: Children of Mercury", in *Antiquarian Horology and the Proceedings of the Antiquarian Horological Society* 40, 2019, 102–108.

Linnard/Robey 2019b – William Linnard, John A. Robey, "Children of Mercury: An Addendum", in *Antiquarian Horology and the Proceedings of the Antiquarian Horological Society* 40, 2019, 545–551.

Lippmann 1895 – Friedrich Lippmann, *Die sieben Planeten*, Berlin, 1895.

Lobstein 1839a – Johann Martin Franz Theodor Lobstein, "Notice historique sur le carillon de l'ancienne horloge astronomique de la cathédrale de Strasbourg", *Album Alsacien: Revue de l'Alsace littéraire, historique et artistique* 2 (17), 1839, 417–420; also in *Revue et Gazette Musicale de Paris* 6 (49), 1839, 390.

Lobstein 1839b – Johann Martin Franz Theodor Lobstein, "Musikalische Topografie von Straßburg", *Allgemeine Musikalische Zeitung* 46, 1839, cols. 885–886, 891, 893–895.

Lobstein 1840 – Johann Martin Franz Theodor Lobstein, *Beiträge zur Geschichte der Musik im Elsass und besonders in Straßburg von der ältesten bis auf die neueste Zeit*, Strasbourg, 1840.

Loeske 1926 – Michael Loeske, "Die Uhrmacherfamilie der Habrecht", *Deutsche Uhrmacherzeitung* 50, 1926, 552–554 (review of Ungerer 1925).

Lorch 1980 – Richard P. Lorch, "Al-Khāzinī's 'Sphere That Rotates by Itself'", *Journal for the History of Arabic Science* 4, 1980, 287–329.

Lorenz 1973 – Dona Marie Lorenz, *Tools of Astrology—Houses*, Topanga (CA), 1973.

Lühning 1975 – Arnold Lühning, *Gemessene Zeit: Uhren in der Kulturgeschichte Schleswig-Holsteins* (catalog of the exhibition at the Schleswig-Holsteinisches Landesmuseum, Schleswig, 29 June–19 October 1975), Schleswig, 1975.

Lühning 1997 – Felix Lühning, *Der Gottorfer Globus und das Globushaus im 'Newen Werk': Dokumentation und Rekonstruktion eines frühbarocken Welttheaters* (catalogue of the exhibition "Gottorf im Glanz des Barock: Kunst und Kultur am Schleswiger Hof 1544–1713"), v. IV, Schleswig, 1997.

Lutz 1902/12 – Jules Lutz, "Les réformateurs de Mulhouse: Nicolas Prugner", *Bulletin du Musée historique de Mulhouse* 26, 1902, 32–68 (pt. I); 27, 1903, 10–68 (pt. II); 35, 1911, 35–60 (pt. III); 36, 1912, 31–51 (pt. IV).

Maass 1898 – Ernst Maass (ed.), *Commentariorum in Aratum reliquiae*, Berlin, 1898.

Mackensen 1978a – Ludolf von Mackensen, "Neue Ergebnisse zur ägyptischen Zeitmessung: Die Inbetriebnahme und Berechnung der ältesten erhaltenen Wasseruhr", *Alte Uhren* 1 (1), 1978, 13–18.

Mackensen 1978b – Ludolf von Mackensen, "Der silberne Copernicus von Kassel: Eine astronomische Tafeluhr von Jost Bürgi und Hans Jakob Emck von 1591 als frühestes künstlerisches Bekenntnis zum neuen Weltsystem", *Pantheon: Internationale Zeitschrift für Kunst* 36, 1978, 132–138.

Mackensen 1995 – Ludolf von Mackensen, "Der Astronom im Porträt: Nicolaus Prugner, ein Zeitgenosse Apians und sein identifiziertes Gelehrtenbildnis", in Karl Röttel (ed.), *Peter Apian: Astronomie, Kosmographie und Mathematik am Beginn der Neuzeit*, Eichstätt, 1995, 127–132.

Malleolos 1628 – Isaac Malleolos (ed.), *Quaestiones in quatuor primarias mathematicarum disciplinarum partes, Arithmeticam, Geometriam, Astronomiam, et Geographiam, ex Conradi Dasypodii Argentinensis, et Davidis Wolckenstenii Vratislaviensis lectionibus publicis, nec non aliis authoribus, iussu Superiorum, in usum Gymnasii Argentoratensis, nec non Philosophia Candidatorum collecta*, Strasbourg, 1628.

Manilius 1977 – Manilius, *Astronomica* (tr. by George P. Goold), "Loeb Classical Library 469", Cambridge (MA) – London, 1977.

Manitius 1888 – Karl Manitius, *Des Hypsikles Schrift Anaphorikos nach Überlieferung und Inhalt kritisch behandelt* (Gymnasium program at Heiligen Kreuz, Dresden), Dresden, 1888.

Manitius 1904/05 – Karl Manitius, "Fixsternbeobachtungen des Altertums", *Das Weltall* 5, 1904–1905, 14–18, 23–26, 399–407.

Marckwald/Mentz/Wilhelm 1908/11 – Ernst Marckwald, Ferdinand Mentz, Ludwig Wilhelm, *Katalog der Elsass-Lothringischen Abteilung der Kaiserlichen Universitäts- und Landesbibliothek Strassburg*, v. 1, Strasbourg, 1908–1911.

Martin 1854 – Thomas Henri Martin, "Recherches sur la vie et les ouvrages d'Héron d'Alexandrie disciple de Ctésibius, et sur tous les ouvrages mathématiques grecs, conservés ou perdus, publiés ou inédits, qui ont été attribués à un auteur nommé Héron", in *Mémoires présentés par divers savants à l'Académie des inscriptions*, ser. 1, v. 4, Paris, 1854, 1–488.

Martini 1777 – Georg Heinrich Martini, *Abhandlung von den Sonnenuhren der Alten*, Leipzig, 1777.

Maurer 1964 – Wilhelm Maurer, "Melanchthon und die Naturwissenschaft seiner Zeit", in W. Maurer, *Melanchthon-Studien*, "Schriften des Vereins für Reformationsgeschichte 181", Gütersloh, 1964, 39–66.

Maurice 1976 – Klaus Maurice, *Die deutsche Räderuhr: Zur Kunst und Technik des mechanischen Zeitmessers im deutschen Sprachraum*, Munich, 1976.

Maurice/Mayr 1980 – Klaus Maurice, Otto Mayr (eds.), *The Clockwork Universe: German Clocks and Automata 1550–1650* (catalog of the exhibition at the Bayerisches Nationalmuseum München, 15 April–30 September 1980 and at the Smithsonian Institution, Washington (DC), 13 November 1980–15 February 1981), New York, 1980.

Maurmann 1976 – Barbara Maurmann, *Die Himmelsrichtungen im Weltbild des Mittelalters: Hildegard von Bingen, Honorius Augustodunensis und andere Autoren*, "Münstersche Mittelalter-Schriften 33", Munich, 1976.

Mayr 1974 – Otto Mayr, "Automatenlegenden in der Spätrenaissance", *Technikgeschichte* 41, 1974, 20–32.

Mayr 1986 – Otto Mayr, *Authority, Liberty and Automatic Machinery in Early Modern Europe*, "Johns Hopkins Studies in the History of Technology, new ser. 8", Baltimore – London, 1986.

Melanchthon 1999 – Philipp Melanchthon, *Orations on Philosophy and Education* (ed. by Sachiko Kusukawa, tr. by Christine F. Salazar), Cambridge, 1999.

Melanchthon 2008 – Philipp Melanchthon, *Initia Doctrinae Physicae: Dictata in Academia Vuitebergensi/Die Anfänge der physikalischen Lehre, vorgetragen an der Universität Wittenberg* (tr. by Walther Ludwig), "Subsidia Classica 11", Rahden, 2008.

Menhardt 1959 – Hermann Menhardt, "Hugo Blotius (1533–1608): Beiträge zu seiner Lebens-geschichte bis zur Adelsverleihung", *Senftenegger Monatsblatt für Genealogie und Heraldik* 1, 1959, cols. 129–142.

Menke-Glückert 1912 – Emil Menke-Glückert, *Die Geschichtsschreibung der Reformation und Gegenreformation: Bodin und die Begründung der Geschichtsmethodologie durch Bartholomäus Keckermann*, Leipzig, 1912.

Metze 2004 – Gudula Metze, *Die Entwicklung des Copernicus-Porträts vom 16. Jahrhundert bis zum 18. Jahrhundert*, PhD dissertation, Ludwig-Maximilians-Universität of Munich, 2004, https://edoc.ub.uni-muenchen.de/6796/.

Meyer 1954 – Heinrich Meyer, "More on Copernicus and Luther", *Isis: An International Review Devoted to the History of Science and its Cultural Influences* 45, 1954, 99.

Meyer 1975 – Heinz Meyer, *Die Zahlenallegorese im Mittelalter: Methode und Gebrauch*, "Münstersche Mittelalter-Schriften 25", Munich, 1975.

Meyer/Suntrup 1987 – Heinz Meyer, Rudolf Suntrup, *Lexikon der mittelalterlichen Zahlenbedeutungen*, "Münstersche Mittelalter-Schriften 56", Munich, 1987.

Michalik 1943 – Rudolf Michalik, *Die Kunstuhr*, "Das schöne Olmütz 1", Olmütz, 1943.

Michel 1947 – Henri Michel, *Traité de l'Astrolabe*, Paris, 1947; reprint Paris, 1976.

Mieg 1929 – Philippe Mieg, "Quelques détails nouveaux sur Nicolas Prugner, réformateur de Mulhouse", *Bulletin du Musée historique de Mulhouse* 49, 1929, 47–55.

Mieg 1939 – Philippe Mieg, "Notes sur les Réformateurs de Mulhouse", *Bulletin du Musée historique de Mulhouse* 58, 1939, 33–51.

Mieg 1957 – Philippe Mieg, "Une médaille à l'effigie de Nicolas Prugner, réformateur de Mulhouse", *Bulletin du Musée historique de Mulhouse* 65, 1957, 73–75.

Mieg 1970 – Philippe Mieg, "L'Epitaphe et l'Ex Libris de Nicolas Prugner", *Bulletin du Musée historique de Mulhouse* 78, 1970, 33–44.

Mills 1982/83 – Allan Alfred Mills, "Newton's Water Clocks and the Fluid Mechanics of Clepsydrae", *Notes and Records of the Royal Society of London* 37, 1982–1983, 35–61.

Moran 1973 – Bruce T. Moran, "The Universe of Philip Melanchthon: Criticism and Use of the Copernican Theory", *Comitatus: A Journal of Medieval and Renaissance Studies* 4, 1973, 1–23.

Morrison 2009 – James E. Morrison, *The Astrolabe*, Rehoboth Beach (DE), 2009.

Mosel 1835 – Ignaz Franz Edler von Mosel, *Geschichte der kaiserl. königl. Hofbibliothek zu Wien*, Vienna, 1835.

Müller 1914 – Walter L. Müller, "Zur Geschichte der Uhrmacherkunst in Straßburg", in *XIV. Verbandstag Verband Deutscher Juweliere, Gold- & Silberschmiede*

Straßburg, vom 7.–11. August 1914 verbunden mit einer Ausstellung alter & neuer Goldschmiedearbeiten & Uhren, [Strasbourg, 1914], 99–135.

Müller 1963/64 – Konrad Müller, "Ph. Melanchthon und das kopernikanische Weltsystem", *Centaurus* 9, 1963–1964, 16–28.

Müller 1993 – Uwe Müller (ed.), *450 Jahre Copernicus 'De revolutionibus': Astronomische und mathematische Bücher aus Schweinfurter Bibliotheken* (catalog of the exhibition at the State Archive Schweinfurt in collaboration with the Library Otto Schäfer, 21 November 1993–19 June 1994), "Veröffentlichungen des Stadtarchivs Schweinfurt 9", Schweinfurt, 1993.

Müller 2008 – Kathrin Müller, *Visuelle Weltaneignung: Astronomische und kosmologische Diagramme in Handschriften des Mittelalters*, "Historische Semantik 11", Göttingen, 2008 (PhD dissertation, University of Hamburg, 2006).

Müller-Jahncke 1999 – Wolf-Dieter Müller-Jahncke, "Michael Herr, Arzt in Straßburg: Ein Vorbericht", in W.-D. Müller-Jahncke, Wilhelm Kühlmann (eds.), *Iliaster: Literatur und Naturkunde in der frühen Neuzeit. Festgabe für Joachim Telle zum 60. Geburtstag*, Heidelberg, 1999, 201–209.

Murdoch 1984 – John Emery Murdoch, *Album of Science: Antiquity and the Middle Ages*, New York, 1984.

Murphy 1995 – Susan Murphy, "Heron of Alexandria's *On Automaton Making*", *History of Technology* 17, 1995, 1–44.

Murschel 1995 – Andrea Murschel, "The Structure and Function of Ptolemy's Physical Hypotheses of Planetary Motion", *Journal for the History of Astronomy* 26, 1995, 33–61.

Musée Dupuy 1978 – *Musée Paul Dupuy: Horlogérie et instruments de mesure du temps passé* (catalog for the exhibition), [Toulouse], 1978.

Neddermayer 1988 – Uwe Neddermayer, *Das Mittelalter in der deutschen Historiographie vom 15. bis zum 18. Jahrhundert: Geschichtsgliederung und Epochenverständnis in der frühen Neuzeit*, "Kölner historische Abhandlungen 34", Cologne – Vienna, 1988 (PhD dissertation, University of Cologne, 1986).

Neugebauer 1975 – Otto Neugebauer, *A History of Ancient Mathematical Astronomy*, Berlin – Heidelberg – New York, 1975.

Neumann 1968 – Otto Neumann, "Verzeichnis zu dem Briefband des Heinrich Rantzau 1570–1594", *Die Heimat: Monatsschrift des Vereins zur Pflege der Natur- und Landeskunde in Schleswig-Holstein und Hamburg* 75, 1968, 248–250.

Nolte 1922 – Friedrich Nolte, *Die Armillarsphäre*, "Abhandlungen zur Geschichte der Natur- wissenschaften und der Medizin 2", Erlangen, 1922.

Norlind 1953 – Wilhelm Norlind, "Copernicus and Luther: A Critical Study", *Isis: An International Review Devoted to the History of Science and its Cultural Influences* 44, 1953, 273–276.

Norman 1929 – John Norman, "A Habrecht Clock in London", *The Practical Watch and Clock Maker* 2, 1929, 358–360, 406–409, 452–454.

North 1974 – John David North, "The Astrolabe", *Scientific American* 230 (1), 1974, 96–106.

North 1986 – John David North, *Horoscopes and History*, "Warburg Institute: Surveys and Texts 13", London, 1986.

North 2004 – John David North, "Diagram and Thought in Medieval Science", in Marie-Thérèse Zenner (ed.), *Villard's Legacy: Studies in Medieval Technology, Science and Art in Memory of Jean Gimpel*, "AVISTA—The Association Villard de Honnecourt for Interdisciplinary Study of Medieval Technology, Science and Art, 2", Aldershot, 2004, 265–287.

Nye 2019 – James Nye, "Andrew James (1954–2019)", in *Antiquarian Horology and the Proceedings of the Antiquarian Horological Society* 41, 2020, 114–116.

Oestmann 1989 – Günther Oestmann, *Die zweite Uhr des Straßburger Münsters: Funktion und Bedeutung*, Master dissertation, Department of Cultural History and Cultural Studies, University of Hamburg, 1990.

Oestmann 1993/2000 – Günther Oestmann, *Die Straßburger Münsteruhr: Funktion und Bedeutung eines Kosmos-Modells des 16. Jahrhunderts*, Stuttgart, 1993; Berlin – Diepholz, 2000.

Oestmann 1996 – Günther Oestmann, "Ein unbekanntes Bildnis des Reformators und Astronomen Nikolaus Prugner (1488–1557)", *Bulletin de la Cathédrale de Strasbourg* 22, 1996, 64–70.

Oestmann 1999a – Günther Oestmann, "The Strasbourg Cathedral Clock" (10th Dingwall-Beloe Lecture, British Museum, London, 25 November 1998), *Antiquarian Horology and the Proceedings of the Antiquarian Horological Society* 25 (1), 1999, 50–63 (summary in *Horological Journal* 141, 1999, 46–47).

Oestmann 1999b – Günther Oestmann, "Ein unbekannter Brief des Adam Olearius über den Bau des Gottorfer Globus", *Nordelbingen: Beiträge zur Kunst- und Kulturgeschichte Schleswig-Holsteins* 68, 1999, 39–43.

Oestmann 2003 – Günther Oestmann, "Tycho Brahe's Attitude towards Astrology and his Relations to Heinrich Rantzau", in John Robert Christianson, Alena Hadravová, Petr Hadrava, Martin Šolc (eds.), *Tycho Brahe and Prague: Crossroads of European Science. Proceedings of the International Symposium on the History of Science in the Rudolphine Period, Prague, 22–25 October 2001*, "Acta Historica Astronomiae 16", Frankfurt am Main, 2003, 84–94.

Oestmann 2004 – Günther Oestmann, *Heinrich Rantzau und die Astrologie: Ein Beitrag zur Kulturgeschichte des 16. Jahrhunderts*, "Disquisitiones Historiae Scientiarum: Braunschweiger Beiträge zur Wissenschaftsgeschichte 2", Braunschweig, 2004 (Habilitation thesis, Department of Mathematics, University of Hamburg, 2001).

Oestmann 2005 – Günther Oestmann, "Astrologi und Mechanici im Umkreis Ottheinrichs", in Suzanne Bäumler, Evamaria Brockhoff, Michael Henker (eds.),

Von Kaisers Gnaden: 500 Jahre Pfalz-Neuburg (catalog of the Bavarian State exhibition, Neuburg an der Donau, 3 June–16 October 2005), "Veröffentlichungen zur Bayerischen Geschichte und Kultur 50", Augsburg, 2005, 256–260.

Oestmann 2014a – Günther Oestmann, "Die astronomische Uhr am Rathaus zu Olmütz/Olomouc", in Manfred Schukowski, Uta Jahnke, Wolfgang Fehlberg (eds.), *Mittelalterliche astronomische Großuhren: Internationales Symposium in Rostock 25. bis 28. Oktober 2012*, "Acta Historica Astronomiae 49", Leipzig, 2014, 125–154.

Oestmann 2014b – Günther Oestmann, *Geschichte, Konstruktion und Anwendung des Astrolabiums bei Zifferblättern astronomischer Uhren*, Oberhausen, 2014.

Oestmann 2017 – Günther Oestmann, "Changing the Angle of Vision: Astrolabe Dials on Astronomical Clocks", *Medieval Encounters: Jewish, Christian and Muslim Culture in Confluence and Dialogue* 23, 2017, 404–420.

Oestmann 2018 – Günther Oestmann, "Displaying Astrological Knowledge through Tabulation: Some Notes Pertaining to Particular Arrangements on Instruments", in Richard Dunn, Silke Ackermann, Giorgio Strano (eds.), *Heaven and Earth United: Instruments in Astrological Contexts*, Leiden, 2018, 86–103.

Oestmann/Schramm 1992 – Günther Oestmann, Matthias Schramm, "Die astronomische Uhr des Straßburger Münsters von 1571/74", *Bulletin de la Cathédrale de Strasbourg* 20, 1992, 82–99.

Oliger 1918 – Livarius Oliger, "Die Leidensuhr eines Straßburger Franziskaners aus dem 15. Jahrhundert: Zugleich ein Beitrag zur Geschichte der Betrachtung des bitteren Leidens", *Der Katholik: Zeitschrift für katholische Wissenschaft und kirchliches Leben* 98 (4th ser., 21), 1918, 99–112, 158–175.

Omodeo/Tupikova 2013 – Pietro Omodeo, Irina Tupikova, "The Post-Copernican Reception of Ptolemy: Erasmus Reinhold's Commented Edition of the *Almagest*, Book One (Wittenberg, 1549)", *Journal for the History of Astronomy* 44, 2013, 235–256.

Osten 1983 – Gert von der Osten, *Hans Baldung Grien: Gemälde und Dokumente*, Berlin, 1983.

Ovid 1989 – Publius Ovidius Naso, *Fasti* (tr. by James George Frazer), "Loeb Classical Library 253", Cambridge (MA) – London, 1989[6].

Panofsky/Saxl 1923 – Erwin Panofsky, Fritz Saxl, *Dürers* Melencolia I*: Eine quellen- und typengeschichtliche Untersuchung*, "Studien der Bibliothek Warburg 2", Leipzig – Berlin, 1923.

Pantaleon 1565/66 – Heinrich Pantaleon, *Prosopographiae Heroum atque illustrium virorum totius Germaniae*, Basel, 1565–1566.

Pappus 1875/78 – Pappus Alexandrinus, *Pappi Alexandrini Collectio quae supersunt* (ed. by Friedrich Hultsch), Berlin, 1875–1878.

Pariset 1932 – François-Georges Pariset, "Dessins de Tobias Stimmer", *Archives alsaciennes d'histoire de l'art* 11, 1932, 123–138.

Pariset 1933 – François-Georges Pariset, "Tobias Stimmer et l'horloge de la Cathédrale de Strasbourg", in *XIIIe Congrès International d'Histoire de l'Art: Résumés des communications présentées au Congrès*, Stockholm, 1933, 154–155.

Pauli 1901 – Gustav Pauli, *Hans Sebald Beham: Ein kritisches Verzeichnis seiner Kupferstiche, Radierungen und Holzschnitte*, "Studien zur deutschen Kunstgeschichte 33", Strasbourg, 1901.

Peter 2016 – Claus Peter, "Zum astronomischen Zeigerwerk und einigen Sondereinrichtungen der astronomischen Uhr im Dom zu Münster (Westfalen)", in Wolfgang R. Dick, Jürgen Hamel (eds.), *Beiträge zur Astronomiegeschichte* 13, "Acta Historica Astronomiae 58", Leipzig, 139–163.

Peucer 1568 – [Caspar Peucer], *Hypotyposes orbium Coelestium, quas appellant Theoricas Planetarum: congruentes cum Tabulis Alphonsinis et Copernici, seu etiam tabulis Prutenicis: in usum Scholarum publicatae*, Strasbourg, [1568].

Peucer 1571 – Caspar Peucer, *Hypotheses Astronomicae, seu Theoriae Planetarum. Ex Ptolemaei et aliorum veterum doctrina ad observationes Nicolai Copernici, et canones motuum ab eo conditos accommodatae*, Wittenberg, 1571.

Peucer 1573 – [Caspar Peucer], *Absolutissimae orbium coelestium Hypotyposes, quas Planetarum Theoricas vocant: congruentes cum Tabulis Alphonsinis et Copernici, seu etiam tabulis Prutenicis*, Cologne, 1573.

Piccolomini 1962 – Enea Silvio Piccolomini, *Deutschland: Der Brieftraktat an Martin Mayer und Jakob Wimpfelings 'Antworten und Einwendungen gegen Enea Silvio'* (tr. by Adolf Schmidt), "Die Geschichtsschreiber der deutschen Vorzeit 3", Cologne – Graz, 1962.

Podestà 1869 – Bartolomeo Podestà, "I primi oriuoli pubblici in Bologna nei secoli XIV e XV", *Atti e Memorie delle R. Deputazione di Storia Patria per la provincie Romagna* 8, 1869, 141–165.

Poulle 1955 – Emmanuel Poulle, "La fabrication des astrolabes au moyen âge", *Techniques et Civilisations* 4, 1955, 117–128.

Poulle 1974 – Emmanuel Poulle, "Les mécanisations de l'astronomie des épicycles: l'horloge d'Oronce Fine", *Comptes rendus des séances de l'Académie des Inscriptions et Belles Lettres* 118 (1), 1974, 59–79.

Poulle 1980 – Emmanuel Poulle, *Les instruments de la théorie des planètes selon Ptolémée: Équatoires et horlogerie planétaire du XIIIe au XVIe siècle*, "Collection des travaux de l'Académie Internationale d'Histoire des Sciences 27", Genf, 1980.

Poulle 1983 – Emmanuel Poulle, "L'horloge planétaire de la cathédrale de Strasbourg", *Journal for the History of Astronomy* 14, 1983, 33–46.

Prager 1973 – Frank D. Prager, "Kepler als Erfinder", in Fritz Krafft, Karl Meyer, Bernhard Sticker (eds.), *Internationales Kepler-Symposium Weil der Stadt 1971*, "Arbor Scientiarum: Beiträge zur Wissenschaftsgeschichte, Ser. A, 1", Hildesheim, 1973, 385–405.

Proklus 1945 – Proklus Diadochus, *Euklid-Kommentar* (ed. by Max Steck, tr. by Leander Schönberger), Halle, 1945.

Prowe 1883/84 – Leopold Prowe, *Nicolaus Coppernicus*, Berlin, 1883–1884.

Prugner 1551 – Nicolaus Prugner (Ed.), *Iulii Firmici Materni Iunioris Siculi V.C. ad Mavortium Lollianum, Astronomicon Libri VIII, per Nicolaum Prucknerum Astrologum nuper ab innumeris mendis vindicati* ..., Basel, 1551.

Ptolemy 1961 – Claudius Ptolemy, *ΚΑΡΠΟΣ: Pseudo-Ptolemaei Fructus sive Centiloquium* (ed. by Emilie Boer), "Claudii Ptolemaei opera quae exstant omnia III.2", Leipzig, 1961[2].

Ptolemy 1963 – Claudius Ptolemy, *Handbuch der Astronomie* (tr. by Karl Manitius), Leipzig, 1963.

Ptolemy 1980 – Claudius Ptolemy, *Tetrabiblos* (ed. and tr. by Frank Eggleston Robbins), "Loeb Classical Library 435", Cambridge (MA) – London, 1980[6].

Ptolemy 1996 – Claudius Ptolemy, *Tetrabiblos Book III*, (tr. by Robert Schmidt, ed. by Robert Hand), "Project Hindsight, Greek Track, XII", Berkeley Springs, 1996.

Ræder/Strömgren/Strömgen 1946 – Hans Ræder, Elis Strömgren, Bengt Strömgren (ed. and tr.), *Tycho Brahe's Description of his Instruments and Scientific Work as given in Astronomiae Instauratae Mechanica* (*Wandesburgi 1598*), Copenhagen, 1946.

Ramus 1569 – Petrus Ramus [Pierre de La Ramée], *Scholarum mathematicarum, libri unus et triginta*, Basel, 1569.

Rathgeber 1876 – Julius Rathgeber, *Die handschriftlichen Schätze der früheren Straßburger Stadtbibliothek: Ein Beitrag zur elsässischen Bibliographie*, Gütersloh, 1876.

Rathgeber 1888 – Julius Rathgeber, "Die Schicksale einer Straßburger Bibliothek", *Jahrbuch für Geschichte, Sprache und Litteratur Elsass-Lothringens* 4, 1888, 63–69.

Rau 1963 – Reinhold Rau, "Nicolaus Brucknerus mathematicus: Ein Lebensbild aus der Reformationszeit", *Tübinger Blätter* 50, 1963, 10–14.

Rau 1964 – Reinhold Rau, "Über eine Sammlung von Inschriften des 16. Jahrhunderts", *Zeitschrift für Württembergische Landesgeschichte* 23, 1964, 418–438.

Rauch 1929 – Moriz von Rauch, "Über die Heilbronner Rathausuhr", *Bericht des Historischen Vereins Heilbronn* 16 (1925–1928), 1929, 42–48.

Rave 1959 – Paul Ortwin Rave, "Paolo Giovio und die Bildnisvitenbücher der Renaissance", *Jahrbuch der Berliner Museen* 1, 1959, 119–154.

Recht 1974 – Roland Recht, *L'Alsace gothique de 1300 à 1365: Etude d'architecture religieuse*, Colmar, 1974.

Recht 1989 – Roland Recht (ed.), *Les bâtisseurs des cathédrales gothiques* (catalog of the exhibition at the Musées de la Ville de Strasbourg, 3 September–26 November 1989), Strasbourg, 1989.

Regiomontanus 1490 – Johannes Regiomontanus, *Tabule directionum profectionumque famosissimi viri Magistri Joannis Germani de Regiomonte in nativitatibus multum utiles*, Augsburg, 1490.

Reifferscheid 1889 – Alexander Reifferscheid (ed.), *Briefe G.M. Lingelsheims, M. Berneggers und ihrer Freunde*, "Quellen zur Geschichte des geistigen Lebens in Deutschland während des siebzehnten Jahrhunderts I", Heilbronn, 1889.

Reinhardt 1935 – Hans Reinhardt, "Der Rohraff an der Straßburger Münsterorgel", *Schweizer Volkskunde: Korrespondenzblatt der Schweizerischen Gesellschaft für Volkskunde* 25, 1935, 41–48.

Reinhardt 1939 – Hans Reinhardt, "Deux Mécanismes du Moyen Age à la Cathédrale", in "La Cathédrale de Strasbourg 1439–1939", *Le Point: Revue artistique et littéraire* (special ed.) 19, 1939, 36–41.

Reinhardt 1960 – Hans Reinhardt, "Les textes relatifs a l'histoire de la Cathédrale de Strasbourg depuis les origines jusqu'a l'année 1522", *Bulletin de la Société des Amis de la Cathédrale de Strasbourg, 2nd ser.* II (7), 1960, 11–29.

Reinhardt 1972 – Hans Reinhardt, *La Cathédrale de Strasbourg*, Grenoble, 1972.

Remmert 2009 – Volker R. Remmert, "On Picturing the Past: Arithmetic and Geometry as Wings of the Mind", *The Mathematical Intelligencer* 31, 2009, 42–47.

Reusens/Wils/Schillings 1903/67 – Edmond Henri Joseph Reusens, Joseph Wils, A. Schillings (eds.), *Matricule de l'université de Louvain*, "Collection de chroniques belges inédites 32", Brussels, 1903–1967.

Reusner 1587 – Nicolaus Reusner, *Icones sive Imagines virorum literis illustrium quorum fide et doctrina religionis et bonarum literarum studia, nostra patrumque memoria, in Germania praesertim, in integrum sunt restituta. Additis eorundem elogiis diversorum auctorum*, Strasbourg, 1587.

Reuss 1871 – Rodolphe Reuss, *Les bibliothèques publiques de Strasbourg incendiées dans la nuit du 24 aout 1870*, Paris, 1871 (Offprint from *Revue critique d'histoire et de littérature* 5 (1), 1871, 160–180).

Reuss 1890 – Rodolphe Reuss (ed.), *Les collectanées de Daniel Specklin, chronique strasbourgeoise du seizième siècle*, "Fragments des anciennes chroniques d'Alsace II", Strasbourg, 1890.

Riccioli 1651 – Giovanni Battista Riccioli, *Almagestum novum astronomiam veterem novamque complectens observationibus aliorum, et propriis novisque theorematibus, problematibus, ac tabulis promotam*, Bologna, 1651.

Richards 1998 – Edward Graham Richards, *Mapping Time: The Calendar and its History*, Oxford – New York – Toronto, 1998.

Rieger 1996 – Theodore Rieger, "Bernard Nonnenmacher, Maitre d'œuvre de la Cathédrale de Strasbourg de 1519 à 1551, et l'introduction du style renaissance dans l'architecture réligieuse alsacienne", *Bulletin de la Cathédrale de Strasbourg* 22, 1996, 85–90.

Riegl 1889 – Alois Riegl, "Die mittelalterliche Kalenderillustration", *Mitteilungen des Instituts für österreich. Geschichtsforschung* 10, 1889, 1–74.

Riemenschneider 1933 – Albert Riemenschneider (ed.), *The Liturgical Year* (*Orgelbüchlein*) *by Johann Sebastian Bach: Forty-Five Organ Chorals*, Bryn Mawr (PA), 1933.

Riess 1892/93 – Ernst Riess, "Nechepsonis et Petosiridis fragmenta magica", *Philologus: Zeitschrift für das classische Alterthum* 6 (supplement), 1892–1893, 325–394.

Ritter 1613 – Franz Ritter, *Astrolabium, Das ist: Gründliche Beschreibung und Unterricht wie solches herrliche und hochnützliche Astronomische Instrument auff allerley Polus Höh so wol auch nach eines jeden selbst gefälligen Größ auffgerissen und verfertigt werden soll. Darnach wie dasselbe vielfältig zu gebrauchen: Mit Kupferstücken verfertiget*, Nuremberg, [1613].

Rivius 1547 – Gualtherus Hermenius Rivius [Walter Hermann Ryff], *Der furnembsten notwendigsten der gantzen Architectur angehoerigen Mathematischen und Mechanischen kuenst eygentlicher bericht und fast klare verstendliche unterrichtung zu rechtem verstandt der lehr Vitruvii in drey furneme Buecher abgetheilet …*, Nuremberg, 1547.

Rivius 1548 – Gualtherus Hermenius Rivius [Walter Hermann Ryff], *Vitruvius Teutsch. Nemlichen des aller namhafftigisten und hocherfarnesten Roemischen Architecti und Kunstreichen Werck oder Baumeisters Marci Vitruvii Pollionis Zehen Buecher von der Architectur und kuenstlichem Bauen …*, Nuremberg, 1548; reprint Hildesheim – New York, 1973.

Rocca 1612 – Angelo Rocca, *De campanis commentarius*, Rome, 1612.

Rocca 1719 – Angelo Rocca, *Opera omnia*, Rome, 1719.

Roder 1880 – Christian Roder, "Das älteste Münsteruhrwerk in Villingen", *Schriften des Vereins für Geschichte und Naturgeschichte der Baar und der angrenzenden Landestheile in Donaueschingen* 3, 1880, 291–292.

Roettinger 1914a – Heinrich Roettinger, *Die Holzschnitte des Georg Pencz*, Leipzig, 1914.

Roettinger 1914b – Heinrich Roettinger, *Die Holzschnitte zur Architektur und zum Vitruvius Teutsch des Walther Rivius*, "Studien zur deutschen Kunstgeschichte 167", Strasbourg, 1914.

Rohde 1923 – Alfred Rhode, *Die Geschichte der wissenschaftlichen Instrumente vom Beginn der Renaissance bis zum Ausgang des 18. Jahrhunderts*, "Monographien des Kunstgewerbes 16", Leipzig, 1923.

Rohr 1971 – René R.J. Rohr, *Les cadrans solaires anciens d'Alsace*, Colmar, 1971.

Röhrich 1855 – Timotheus Wilhelm Röhrich, *Mittheilungen aus der Geschichte der evangelischen Kirche des Elsasses*, Paris – Strasbourg, 1855.

Rome 1927 – Adolphe Rome, "L'Astrolabe et le Météoroscope d'après le commentaire de Pappus sur le 5e livre de l'Almageste", *Annales de la Société Scientifique de Bruxelles, Ser. A: Sciences mathematiques* (*Mémoires*) 47, 1927, 77–102.

Rose 1975 – Paul Lawrence Rose, *The Italian Renaissance of Mathematics: Studies on Humanists and Mathematicians from Petrarch to Galileo*, Genf, 1975.

Rosen 1954 – Edward Rosen, "Did Galileo claim he invented the Telescope?", *Proceedings of the American Philosophical Society* 98, 1954, 304–312.

Rosen 1971 – Edward Rosen (ed. and tr.), *Three Copernican Treatises: The Commentariolus of Copernicus. The Letter against Werner. The Narratio prima of Rheticus*, "Records of Civilization: Sources and Studies 30", New York, 1971[3].

Rosen 1986 – Edward Rosen, *Three Imperial Mathematicians: Kepler Trapped Between Tycho Brahe and Ursus*, New York, 1986.

Rosenberg 1875 – Adolf Rosenberg, *Sebald und Barthel Beham: Zwei Maler der deutschen Renaissance*, Leipzig, 1875.

Rosenberg 1973 – Bernhard-Maria Rosenberg, "Das ärztliche Wirken des Frauenburger Domherrn Nicolaus Copernicus", in Friedrich Kaulbach, Udo Wilhelm Bargenda, Jürgen Blühdorn (eds.), *Nicolaus Copernicus zum 500. Geburtstag*, Cologne – Vienna, 1973, 97–137.

Roth 1877 – Rudolf Roth, *Urkunden zur Geschichte der Universität Tübingen aus den Jahren 1476 bis 1550*, Tübingen, 1877.

Rott 1905 – Hans Rott, *Ott Heinrich und die Kunst*, "Mitteilungen zur Geschichte des Heidelberger Schlosses 5", Heidelberg, 1905, 220–226.

Rott 1917 – Hans Rott, *Kunst und Künstler am Baden-Durlacher Hof bis zur Gründung Karlsruhes*, Karlsruhe, 1917.

Rott 1933/38 – Hans Rott, *Quellen und Forschungen zur südwestdeutschen und schweizerischen Kunstgeschichte im XV. und XVI. Jahrhundert*, Stuttgart, 1933–1938.

Rott 1971 – Jean [Hans Georg] Rott, "Sources et grandes lignes de l'histoire des bibliothèques publiques de Strasbourg détruites en 1870", *Cahiers alsaciens d'archéologie d'art et d'histoire* 15, 1971, 145–180.

Ruelle 1871 – Charles Emile Ruelle, "Notice et variantes d'un manuscrit grec relatif à la musique, qui a péri pendant le bombardement de Strasbourg", *Académie des Inscriptions et Belles-Lettres: Comptes rendus, new ser.* 7, 1871, 335–339.

Sander 1950 – Hjalmar Sander, "Zur Identifizierung zweier Bildnisse von Lucas Cranach d. Ä", *Zeitschrift für Kunstwissenschaft* 4, 1950, 35–48.

Sarton 1927/48 – George Sarton, *Introduction to the History of Science*, "Carnegie Institution of Washington, Publication 376", Baltimore, 1927–1948.

Sauer 1924 – Joseph Sauer, *Symbolik des Kirchengebäudes und seiner Ausstattung in der Auffassung des Mittelalters: Mit Berücksichtigung von Honorius Augustodunensis, Sicardus und Durandus*, Freiburg im Breisgau, 1924[2].

Savage-Smith 1985 – Emilie Savage-Smith, *Islamicate Celestial Globes: Their History, Construction and Use*, "Smithsonian Studies in History and Technology 46", Washington, 1985.

Saxl 1918/19 – Fritz Saxl, "Probleme der Planetenkinderbilder", *Kunstchronik und Kunstmarkt: Wochenschrift für Kenner und Sammler* 54 (*new ser.* 30), 1918–1919, 1013–1021.

Saxl 1936 – Fritz Saxl, "Veritas filia temporis", in Raymond Klibansky, H.J. Paton (eds.), *Philosophy and History: Essays presented to Ernst Cassirer*, Oxford, 1936; reprint Gloucester (MA), 1975, 197–222.

Schadaeus 1617 – Osea Schadaeus, *Summum Argentoratensium Templum: Das ist: Auszführliche und Eigendtliche Beschreibung deß viel Künstlichen sehr Kostbaren und in aller Welt berühmten Münsters zu Straßburg: Auch alles dessen so An und In demselben Denckwürdigs zu sehen: Mit schönen Figuren und beygefügten underschiedlichen Kupfferstücken gezieret: Jetzmahls zum Ersten seinem vilgeliebten Vatterland und Teutscher Nation zu Ehren in Truck verfertiget Durch M. Oseam Schadaeum Argentoratensem, der Kirchen beym Alten S. Peter daselbsten Diaconum*, Strasbourg, 1617.

Schade 1984 – Richard Erich Schade, "Kunst, Literatur und die Straßburger Uhr", in *Spätrenaissance am Oberrhein: Tobias Stimmer 1539–1584* (exhibition catalogue), Basel, 1984, 112–117.

Schaldach 1998 – Karlheinz Schaldach, *Römische Sonnenuhren: Eine Einführung in die antike Gnomonik*, Frankfurt am Main, 1998[2].

Schallenbach 1771 – [Max Wilhelm Schallenbach], *Historische Beschreibung derer stadtköllnischen Kollegiatstiftern, wie solche Erhard von Winheim und Aegidius Gelenius in lateinischer Sprache mitgetheilet, nunmehro aber zum gemeinen Besten ins Teutsche übersetzet, und mit verschiedenen lesenswürdigen Zusätzen reichlich vermehret worden*, Cologne, 1771.

Scheibel 1769/87 – Johann Ephraim Scheibel, *Einleitung zur mathematischen Bücherkenntniß*, Breslau, 1769–1787.

Schenk zu Schweinsberg 1943 – Eberhard Freiherr Schenk zu Schweinsberg, "Kopernikus-Bildnisse", in Fritz Kubach (ed.), *Nikolaus Kopernikus: Bildnis eines großen Deutschen*, Munich – Berlin, 1943, 257–285.

Scheurenbrand/Schwarz/Hahn 2005 – Hans Scheurenbrand, Michael Schwarz, Hermann-Michael Hahn, *Astrolabium, Kalenderuhr und Glockenspiel. Festo Harmonices Mundi. Konstruktion, Funktion, Präzision*, Stuttgart, 2005.

Schieck 1846 – Heinrich August Schieck, "Über die Himmelsgloben des Anaximander und Archimedes: Ein Beitrag zur Aufhellung des Alterthums. Zweite Abtheilung. Die Sphäre des Archimedes", *Jahresbericht über das Kurfürstliche Gymnasium zu Hanau*, 1846, 1–32.

Schiller 1971 – Gertrud Schiller, *Ikonographie der christlichen Kunst*, Gütersloh, 1971.

Schiller 1997 – Peter Schiller, "Die Horoskope von Nikolaus Prugner für Bonifazius und Bruno Amerbach", *Annuaire historique de Mulhouse* 8, 1997, 81–85.

Schilter 1698 – Johann Schilter (ed.). *Die Alteste Teutsche so wol Allgemeine Als insonderheit Elsassische und Straßburgische Chronicke Von Jacob von Koenigshoven Priestern in Straßburg Von Anfang der Welt biß ins Jahr nach Christi Geburth MCCCLXXXVI.*

beschrieben. Anjetzo zum ersten mal heraus und mit Historischen Anmerckungen in Truck gegeben von D. Johann Schiltern, Strasbourg, 1698.

Schindling 1977 – Anton Schindling, *Humanistische Hochschule und freie Reichsstadt: Gymnasium und Akademie in Straßburg 1538–1621*, "Veröffentlichungen des Instituts für Europäische Geschichte Mainz 77", Wiesbaden, 1977.

Schlachter 1927 – Alois Schlachter, *Der Globus: Seine Entstehung und Verwendung in der Antike nach den literarischen Quellen und den Darstellungen in der Kunst*, "Stoicheia: Studien zur Geschichte des antiken Weltbildes und der griechischen Wissenschaft 8", Leipzig – Berlin, 1927.

Schlee 1965 – Ernst Schlee (Ed.), *Gottorfer Kultur im Jahrhundert der Universitätsgründung: Kulturgeschichtliche Denkmäler und Zeugnisse des 17. Jahrhunderts aus der Sphäre der Herzöge von Schleswig-Holstein-Gottorf* (exhibition catalog for 300th anniversary of the Christian-Albrechts-Universität Kiel, Neues Schloß Kiel, 31 May–31 July 1965), Flensburg, 1965.

Schmauch 1940/43 – Hans Schmauch, "Neues über die ärztliche Tätigkeit des Astronomen Kopernikus", *Proteus: Verhandlungsberichte und Abhandlungen der Rheinischen Gesellschaft für Geschichte der Naturwissenschaft, Medizin und Technik* 3 (4), 1940–1943, 115–119.

Schmidt 1842 – Charles Schmidt, *Notice historique sur l'horloge astronomique de la cathédrale de Strasbourg*, Strasbourg 1842; (German tr.: *Bericht über das astronomische Uhrwerk des Straßburger Münsters*, Strasbourg, 1842).

Schmidt 1856 – Charles Schmidt, *Der Antheil der Straßburger an der Reformation von Churpfalz*, Strasbourg, 1856.

Schmidt 1898 – Wilhelm Schmidt, "Heron von Alexandria, Konrad Dasypodius und die Straßburger Astronomische Münsteruhr", *Abhandlungen zur Geschichte der Mathematik* 8-XIII (Suppl. of *Zeitschrift für Mathematik und Physik* 42), 1898, 177–194.

Schmidt 1912 – Max Carl Paul Schmidt, *Die Entstehung der antiken Wasseruhr*, "Kulturhistorische Beiträge zur Kenntnis des griechischen und römischen Altertums 2", Leipzig, 1912.

Schmidt 1927 – Arno Schmidt, "Die astronomische Uhr in der Marienkirche zu Danzig", *Ostdeutsche Monatshefte* 8, 1927, 386–390.

Schmitt 1930 – Otto Schmitt, "Ein unvollendetes 'Straßburger Münsterbüchlein' aus dem 17. Jahrhundert", *Elsaß-Lothringisches Jahrbuch* 9, 1930, 228–253.

Schmitz-Moormann 1989 – Karl Schmitz-Moormann, "Alchemie, Astronomie und Theologie: Astronomische Uhren an Gotteshäusern und die Kenntnis der Naturgesetze des Kosmos in der Theologie des Mittelalters", in Änne Bäumer, Martin Büttner (eds.), *Science and Religion / Wissenschaft und Religion: Proceedings of the Symposium of the XVIII International Congress of History of*

Science at Hamburg-Munich, 1–9 August 1989, "Abhandlungen zur Geschichte der Geowissenschaften und Religion/Umwelt-Forschung 3", Bochum, 1989, 94–101.

Schneegans 1852 – Ludwig Schneegans, *Das Pfingstfest und der Roraffe im Münster zu Straßburg, ein mittelalterliches Sittengemälde und Volksbild*, Mühlhausen, 1851 [1852] (Offprint of a paper in *Alsatia*, 1852, 191–245).

Schoener 1996 – Johannes Schoener, *Opusculum Astrologicum* (ed. and tr. by Robert Hand), "Project Hindsight: Latin Track, 4", Berkeley Springs, 1996^2.

Scholfield 1958 – Peter Hugh Scholfield, *The Theory of Proportions in Architecture*, Cambridge, 1958.

Schöller 1989 – Wolfgang Schöller, *Die rechtliche Organisation des Kirchenbaues im Mittelalter, vornehmlich des Kathedralbaues: Baulast—Bauherrenschaft—Baufinanzierung*, Cologne – Vienna, 1989 (PhD dissertation, University of Marburg, 1985).

Schöner 1964 – Erich Schöner, *Das Viererschema in der antiken Humoralpathologie*, "Sudhoffs Archiv für Geschichte der Medizin und der Naturwissenschaften, Suppl. 4", Wiesbaden, 1964.

Schrade 1984 – Richard Erich Schrade, "Kunst, Literatur und die Straßburger Münsteruhr", in *Spätrenaissance am Oberrhein: Tobias Stimmer 1539–1584* (exhibition catalogue), Basel, 1984, 112–117.

Schramm 1991 – Matthias Schramm, "Symmetrie—zur Geschichte von Wort und Sache", in Inga Hosp, Valentin Braitenberg, Peter Mulsner (eds.), *Bozener Treffen 1990: Symmetrien, Dynamik, Strukturen. Gespräche unter Naturwissenschaftlern*, Bozen, 1991, 17–36.

Schrauf 1894/1904 – Karl Schrauf (ed.), *Acta facultatis medicae universitatis Vindobonensis*, Vienna, 1894–1904.

Schreiber 1868 – Heinrich Schreiber, *Geschichte der Albert-Ludwigs-Universität zu Freiburg im Breisgau*, Freiburg im Breisgau, 1868.

Schulte 1882 – Aloys Schulte, "Zur Geschichte der Straßburger Münsterbaumeister", *Repertorium für Kunstwissenschaft* 5, 1882, 21–32, 270–279.

Schulte 1927 – Aloys Schulte, "Aus dem Leben des Straßburger Domkapitels 1150–1332", *Elsaß-Lothringisches Jahrbuch* 6, 1927, 1–46.

Schulte/Wolfram 1888 – Aloys Schulte, Georg Wolfram (eds.), *Urkundenbuch der Stadt Straßburg*, vl. IV.2, Strasbourg, 1888.

Schulze-Manitius 1950 – Hans Schulze-Manitius, "Die Straßburger Münsteruhren", *Neue Uhrmacherzeitung* 4 (10, 12, 13), 1950, 293–297, 372, 409–411.

Schürmann 1990 – Astrid Schürmann, "Antike Wasseruhren", *Deutsches Museum: Wissenschaftliches Jahrbuch, Abhandlungen und Berichte, new ser.* 7, 1990, 82–96.

Schürmann 1991 – Astrid Schürmann, *Griechische Mechanik und antike Gesellschaft: Studien zur Förderung einer technischen Wissenschaft*, "Boethius: Texte und

Abhandlungen zur Geschichte der Mathematik und der Naturwissenschaften XXVII", Stuttgart, 1991.

Schwarz 1943a – Friedrich Schwarz, "Kopernikus-Bildnisse", in Johannes Papritz, Hans Schumauch (eds.), *Kopernikus-Forschungen*, "Deutschland und der Osten: Quellen und Forschungen zur Geschichte ihrer Beziehungen 22", Leipzig, 1943, 143–171.

Schwarz 1943b – Friedrich Schwarz, "Wie sah Kopernikus aus?", in August Georgens (ed.), *Nikolaus Kopernikus: Persönlichkeit und Werk*, "Kulturpolitische Schriftenreihe für den Reichsgau Danzig-Westpreußen 4", Danzig, 1943, 110–132.

Schweighäuser 1876 – Theodor Schweighäuser, *Historische Notizen über die Astronomische Münsteruhr zu Strassburg nebst einer ausführlichen Beschreibung des mechanischen Werkes*, Strasbourg, 1876; French ed.: *Notices historiques sur l'horloge astronomique de la cathédrale de Strasbourg*, Strasbourg, 1876.

Schwilgué 1827 – Jean-Baptiste Schwilgué, "Projet pour la réparation, le perfectionnement ou la construction à neuf de l'horloge astronomique de la cathédrale de Strasbourg", *Journal de la Société des Sciences, Agriculture et Arts du Département du Bas-Rhin* 4, 1827, 200–206.

Schwilgué 1857 – Charles Schwilgué, *Notice sur la vie, les travaux et les ouvrages de mon père J.B. Schwilgué*, Strasbourg, 1857.

Schwilgué 1861 – Charles Schwilgué, "A Brief Description of the Astronomical Clock of the Cathedral of Strasbourg", *Horological Journal* 3, 1861, 111–113, 124–127, 137–139, 144–148.

Sebitz 1641 – Melchior Sebitz, *Straßburgischen Gymnasii Christliches Jubelfest; Gott dem Allerhöchsten zu schuldigen Ehren, beedes in Kirch und Schulen mit unterschiedenen Predigten und Orationibus, auch Gebet und Gesang danckbarlich und andächtig im Jahr nach Christi Geburt 1638 celebrirt und begangen*, Strasbourg, 1641.

Seel 1966 – Otto Seel (ed.), *Physiologus*, Zurich – Munich, 1960.

Segonds 1981 – Alain P. Segonds (ed.), *Jean Philopon: Traité de l'Astrolabe*, "Astrolabica 2", Paris, 1981.

Sextus Empiricus 1987 – Sextus Empiricus, *Against the Physicists* (tr. by Robert Gregg Bury), "Loeb Classical Library 311", Cambridge (MA), 1987.

Shamos 2013 – Geoffrey Shamos, "Astrology as a Social Framework: The 'Children of Planets', 1400–1600", *Journal for the Study of Religion, Nature and Culture* 7, 2013, 434–460.

Sig 1906 – Ludwig Sig, *Das geistliche Schauspiel im Elsaß: Ein Beitrag zur Geschichte des mittelalterlichen Theaters*, Strasbourg, 1906.

Signori 2006 – Gabriela Signori, "Sakral oder profan? Der Kommunikationsraum Kirche", in Marjan De Smet, Paul Trio (eds.), *The Use and Abuse of Sacred Places in Late Medieval Towns* (Colloquium at the Kortrijk Campus of The Katholieke

University of Leuven, 4–5 December 2003), "Mediaevalia Lovaniensia, Ser. 1, Studia 38", Leuven, 2006, 117–134.

Siliprandi 1915 – Otello Siliprandi, *Notizie su Gli Orologi Pubblici di Reggio* (*Sec. XIV–XIX*), Reggio Emilia, 1915.

Sitzmann 1909/10 – Edouard Sitzmann, *Dictionnaire de Biographie des hommes célèbres de l'Alsace depuis les temps plus reculès jusqu'a nos jours*, Rixheim, 1909–1910.

Smital 1920 – Ottokar Smital, "Die Hofbibliothek", in O. Smital, Heinrich Zimmermann, Anton Handlirsch (eds.), *Die beiden Hofmuseen und die Hofbibliothek: Der Werdegang der Sammlungen, ihre Eigenart und Bedeutung*, Vienna – Leipzig, 1920, 49–110.

Smital 1926 – Ottokar Smital, "Miszellen zur Geschichte der Wiener Palatina", in *Festschrift der Nationalbibliothek in Wien*, Vienna, 1926, 771–794.

Sommerfeld 1981 – Erich Sommerfeld, "Die Selbstbildnisse des Nicolaus Copernicus", *Mitteilungen der Archenhold-Sternwarte Berlin-Treptow* 6 (134), 1981, reprint.

Späth 1909 – C.J. Späth, "Der mechanisch bewegte Hahn der alten Straßburger Münsteruhr", *Deutsche Uhrmacherzeitung* 33, 1909, 349–351.

Spleiss 1702 – Johann Jacob Spleiss, *Kurze Beschreibung deß astronomischen Uhrwerks auf dem Fronwaag-Thurn zu Schaffhausen*, [Schaffhausen], 1702.

Staub 1992 – Hervé Staub, "Les installations gnomoniques de la cathédrale de Strasbourg", *Cahiers alsaciens d'archéologie, d'àrt et d'histoire* 35, 1992, 99–112.

Stechow 1938 – Wolfgang Stechow, "Homo Bulla", *The Art Bulletin* 20, 1938, 227–228.

Steinhauser 1939 – August Steinhauser, *Rottweiler Künstler und Kunstwerke des 15. und 16. Jahrhnderts*, Rottweil, 1939.

Steinschneider 1865 – Moritz Steinschneider, "Die 'mittleren' Bücher der Araber und ihre Bearbeiter", *Zeitschrift für Mathematik und Physik* 10, 1865, 456–498.

Stimmer 1576 – Tobias Stimmer, *Neue Künstliche Figuren Biblischer Historien grüntlich von Tobia Stimmer gerissen: Und zu Gotsfoerchtiger ergetzung andaechtiger hertzen mit artigen Reimen begriffen durch J.F.G.M.* [Johann Fischart], Basel, 1576; reprint Munich, 1923.

Stöcklein 1969 – Ansgar Stöcklein, *Leitbilder der Technik: Biblische Tradition und technischer Fortschritt*, Munich, 1969.

Stolberg 1898 – August Stolberg, *Tobias Stimmers Malereien an der astronomischen Münsteruhr zu Straßburg*, "Studien zur deutschen Kunstgeschichte 13", Strasbourg, 1898.

Stolberg 1901 – August Stolberg, *Tobias Stimmer: Sein Leben und seine Werke. Mit Beiträgen zur Geschichte der deutschen Glasmalerei im sechzehnten Jahrhundert*, "Studien zur deutschen Kunstgeschichte 31", Strasbourg, 1901.

Strabo 1989 – Strabo, *Geography Books 1–2* (tr. by Horace Leonard Jones), "Loeb Classical Library, 49", Cambridge (MA) – London, 1989[6].

Strauss 1856 – David Friedrich Strauss, *Leben und Schriften des Dichters und Philologen Nicodemus Frischlin: Ein Beitrag zur deutschen Culturgeschichte in der zweiten Hälfte des sechszehnten Jahrhunderts*, Frankfurt am Main, 1856.

Stückelberger 1998 – Alfred Stückelberger, "Der Astrolab des Ptolemaios: Ein antikes astronomisches Meßgerät", *Antike Welt: Zeitschrift für Archäologie und Kulturgeschichte* 29, 1998, 377–383.

Sudhoff 1902 – Karl Sudhoff, *Iatromathematiker vornehmlich im 15. und 16. Jahrhundert*, "Abhandlungen zur Geschichte der Medizin 2", Breslau, 1902.

Sumner 1989 – Cheryl Lynn Sumner, *The Artistic Observation of the Copernican Universe 1543–1750*, PhD dissertation, Florida State University, 1989.

Tabulae codicum Vindobonensi 1864/68 – *Tabulae codicum manu scriptorum praeter graecos et orientales in Bibliotheca Palatina Vindobonensi asservatorum*, Vienna, 1864–1868; reprint Graz, 1965.

Tardy 1980 – [Henri-Gustave Lengellé] Tardy, *Bibliographie générale de la mesure du temps suivi d'un essai de classification technique et géographique*, Paris, 1980².

Teissier 1715 – Antoine Teissier, *Les Eloges des Hommes Savans, Tirez de l'Histoire de M. de Thou, avec des Additions contenant l'Abbrégé de leur Vie, le Jugement et le Catalogue de leurs Ouvrages, par Antoine Teissier*. Leiden, 1715.

Thomann 1996 – Marcel Thomann, "Notice biographique de François Joseph Fuchs", *Bulletin de la Cathédrale de Strasbourg* 22, 1996, 11–18.

Thomann 2013 – Marcel Thomann, "François Joseph Fuchs (1921–2012)", *Revue d'Alsace* 139, 2013, 529–534.

Thomson 1978 – Ron B. Thomson, *Jordanus de Nemore and the Mathematics of Astrolabes: De plana spera*, "Pontifical Institute of Mediaeval Studies: Studies and Texts 39", Toronto, 1978.

Thöne 1935 – Friedrich Thöne, "Christoph Murers Holzschnitte", *Kunst- und Antiquitäten-Rundschau* 43, 1935, 25–31.

Thöne 1936 – Friedrich Thöne, *Tobias Stimmer: Handzeichnungen. Mit einem Überblick über sein Leben und sein gesamtes Schaffen*, Freiburg im Breisgau, 1936.

Thorndike 1923/41 – Lynn Thorndike, *A History of Magic and Experimental Science*, New York, 1923–1941.

Treweek 1957 – Athanasius Pryor Treweek, "Pappus of Alexandria: The Manuscript Tradition of the Collectio Mathematica", *Scriptorium: Revue Internationale des Etudes relatives aux Manuscrits* 11, 1957, 195–233.

Trinkert 2014 – Julia Trinkert, "Weltweise an der Peripherie? Darstellungstraditionen von Gelehrten in Zwickeln als kompositorische Bildlösungen", in Manfred Schukowski, Uta Jahnke, Wolfgang Fehlberg (eds.), *Mittelalterliche astronomische Großuhren: Internationales Symposium in Rostock 25. bis 28. Oktober 2012*, "Acta Historica Astronomiae 49", Leipzig, 2014, 201–233.

Truitt 2013 – Elly Rachel Truitt, "The Incarnation of Time", in Aurélia Gaillard, Jean-Yves Goffi, Bernhard Roukhomovsky, Sophie Roux (eds.), *L'Automate: Modèle, métaphore, machine, merveille. Actes du colloque internationale de Grenoble* (*19–21 mars 2009*), Bordeaux, 2013, 365–378.

Truitt 2015 – Elly Rachel Truitt, *Medieval Robots: Mechanism, Magic, Nature, and Art*, Philadelphia, 2015.

Turner 1989 – Anthony John Turner, "Sun-Dials: History and Classification", *History of Science* 27, 1989, 303–318.

Uhlemann 1857 – Max Uhlemann, *Grundzüge der Astronomie und Astrologie der Alten besonders der Aegypter*, Leipzig, 1857.

Ungerer 1922 – Alfred Ungerer, Théodore Ungerer, *L'horloge astronomique de la Cathédrale de Strasbourg*, Strasbourg, 1922.

Ungerer 1924 – Théodore Ungerer, "Le coq de Strasbourg", *La vie en Alsace* 2, 1924, 127–128.

Ungerer 1925 – Théodore Ungerer, "Les Habrecht: Une dynastie d'horlogers strasbourgeois au XVIe et au XVIIe siècle", *Archives alsaciennes d'histoire de l'art* 4, 1925, 95–146.

Ungerer 1927a – Théodore Ungerer, "L'œuvre de maitrise d'Isaac Habrecht (1578)", *Archives alsaciennes d'histoire de l'art* 6, 1927, 43–50.

Ungerer 1927b – Théodore Ungerer, *Die Astronomische Uhr im Straßburger Münster*, Strasbourg, 1927.

Ungerer 1928 – Théodore Ungerer, "Un dessin original d'Isaac Habrecht Horloger à Strasbourg (1580)", *Archives alsaciennes d'histoire de l'art* 7, 1928, 79–82.

Ungerer 1930 – Théodore Ungerer, "Une Horloge Astronomique disparue d'Isaac Habrecht (1583)", *Archives alsaciennes d'histoire de l'art* 9, 1930, 101–111.

Ungerer 1931 – Alfred Ungerer, *Les horloges astronomiques et monumentales les plus remarquables de l'Antiquité jusqu'à nos jours*, Strasbourg, 1931.

Ungerer/Glory 1933 – Théodore Ungerer, André Glory, "L'astrologue au cadran solaire de la Cathédrale de Strasbourg (1493)", *Archives alsaciennes d'histoire de l'art* 12, 1933, 73–108.

Unterkircher 1968 – Franz Unterkircher, "Hugo Blotius und seine ersten Nachfolger (1575–1663)", in Josef Stummvoll (ed.), *Geschichte der Österreichischen Nationalbibliothek*, "Museion, Veröffentlichungen der Österreichischen Nationalbibliothek, new ser. 2, 3 (1)", Vienna, 1968, 79–127.

Veltman 1890 – Hermann Veltman, "Handschriftliche Aufzeichnungen über einige alte, jetzt verschwundene Uhrwerke der Stadt Osnabrück, insbesondere über die vormalige astronomische Uhr im Dome daselbst", *Mittheilungen des Vereins für Geschichte und Landeskunde von Osnabrück* (*Historischer Verein*) 15, 1890, 232–302.

Vincent 1858 – Alexandre Joseph Hidulphe Vincent, "Extraits des manuscrits relatifs à la géometrie pratique des Grecs", *Académie des Inscriptions et Belles-Lettres: Notices et extraits des manuscrits de la Bibliothèque Impériale et autres bibliothèques* 19 (2), 1858, 157–431.

Vitruvius 1969 – Vitruvius, *De l'Architecture, Livre IX* (ed. and tr. by Jean Soubiran), Paris, 1969.

Vitruvius 1983 – Vitruvius, *De Architectura* (ed. and tr. by Frank Granger), "*Loeb Classical Library, 251, 280*", Cambridge (MA) – London, 1983^{6}.

Vogeleis 1911 – Martin Vogeleis, *Quellen und Bausteine zu einer Geschichte der Musik und des Theaters im Elsaß 500–1800*, Strasbourg, 1911.

Vossius 1650 – Gerardus Joannes Vossius, *De universae mathesios natura et constitutione liber: Cui subjungitur chronologia mathematicorum*, Amsterdam, 1650.

Wåhlin 1923 – Theodor Wåhlin, *Horologium mirabile Lundense: Det astronomiska uret i Lunds domkyrka*, Lund, 1923.

Walter 1929 – Joseph Walter, "Le Mystère 'Stella' des trois Mages joué à la Cathédrale de Strasbourg au XIIe siècle", *Archives alsaciennes d'histoire de l'art* 8, 1929, 39–50.

Walther 1877 – Philipp Alexander Ferdinand Walther (ed.), *Hans Georg Ernstingers Raisbuch*, "Bibliothek des Litterarischen Vereins in Stuttgart 135", Tübingen, 1877.

Wardeska 1973 – Zofia Wardeska, "Copernicus und die deutschen Theologen des 16. Jahrhunderts", in Friedrich Kaulabach, Udo Wilhelm Bsrgenda, Jürgen Blühdorn (eds.), *Nicolaus Copernicus zum 500. Geburtstag*, Cologne – Vienna, 1973, 155–184.

Webster 1938 – James Carson Webster, *The Labors of the Months in Antique and Medieval Art to the End of the 12th Century*, Princeton, 1938.

Wehrli 2001 – Ruedi Wehrli, "Jost Bürgi (1552–1632): Ein biografischer Versuch", *Chronométrophilia* 51, 2001, 42–72.

Weichenhan 2005 – Michael Weichenhan, "Caspar Peucers Astronomie zwischen christlichem Humanismus und Nicolaus Copernicus", in Hans-Peter Hasse, Günther Wartenberg (eds.), *Caspar Peucer (1525–1602): Wissenschaft, Glaube und Politik im konfessionellen Zeitalter*, Leipzig, 2005, 91–110.

Weidler 1741 – Johann Friedrich Weidler, *Historia astronomiae, sive De ortu et progressu astronomiae liber singularis*, Wittenberg, 1741.

Weitenmühlen 1950 – Hans Weitenmühlen [Charles Wittmer], "Zur elsass-schweizerischen Genealogie: Die Uhrmacherfamilie Habrecht", *Honneur et Patrie* 23 November, 30 November and 7 December 1950.

Weitmann 2011 – Pascal Weitmann, *Technik als Kunst: Automaten in der griechisch-römischen Antike und deren Rezeption in der frühen Neuzeit als Ideal der Kunst oder Modell für Philosophie und Wirtschaft*, Tübingen – Berlin, 2011 (Postdoctoral thesis, University of Kiel).

Werkmeister 1912 – P. Werkmeister, "Über die Zeitmesser des Strassburger Münsters insbesondere die Sonnenuhren am Giebel der Südseite", *Strassburger Münsterblatt: Organ des Straßburger Münsterbauvereins* 6, 1912, 62–74.

Werland 1926 – Peter Werland, "Die Uhr im Dom zu Münster", *Deutsche Uhrmacherzeitung* 50, 1926, 62–65, 98–101.

Wescher 1867 – Carle Wescher, *Poliorcétique des Grecs: traités théoriques, récits historiques; textes restitués d'après les manuscrits de Paris, du Vatican, de Vienne, de Bologne, de Turin, de Naples, d'Oxford, de Leyde, de Munich, de Strasbourg, augmentés de fragments inédits, et accompagnés d'un commentaire paléographique et critique*, Paris, 1867.

Wescher 1871 – Carle Wescher, "Note relative aux Manuscrits grecs de la Bibliothèque de Strasbourg", *Académie des Inscriptions et Belles-Lettres: Comptes rendus, new ser.* 7, 1871, 182–183.

Westermayer/Wagner/Demmler 1912 – Albert Westermayer, Emil Wagner, Theodor Demmler, *Die Grabinschriften der Stiftskirche zu St. Georg in Tübingen*, Tübingen, 1912.

Westfall 1958 – Richard Samuel Westfall, *Science and Religion in Seventeenth-Century England*, "Yale Historical Publications 67", New Haven (CT) – London, 1958; reprint Hamden (CT), 1970.

Westman 1975a – Robert S. Westman, "The Melanchthon Circle, Rheticus, and the Wittenberg Interpretation of the Copernican Theory", *Isis: An International Review Devoted to the History of Science and its Cultural Influences* 66, 1975, 165–193.

Westman 1975b – Robert S. Westman, "Three Responses to the Copernican Theory: Johannes Praetorius, Tycho Brahe, and Michael Maestlin", in R.S. Westman (ed.), *The Copernican Achievement*, "UCLA Center for Medieval and Renaissance Studies, Contribution 7", Berkeley – Los Angeles – London, 1975, 285–345.

Werlman 2012 – Robert S. Westman, *The Copernican Question: Prognostication, Skepticism, and Celestial Order. University of California Press*, Berkeley – Los Angeles – London, 2012.

Wickersheimer 1914 – Ernest Wickersheimer, "Figures médico-astrologiques des IXe, X^{e} et XIe siècles", *Janus: Archives internationales pour l'Histoire de la Médecine et la Géographie Médicale* 19, 1914, 157–177.

Wickersheimer 1923 – Ernest Wickersheimer, *Catalogue Général des Manuscrits des Bibliothèques publiques de France: Strasbourg. Bibliothèque Universitaire et Régionale*, Paris, 1923.

Wiek 1958 – Peter Wiek, *Untersuchungen über die Mitwirkung des Stadtbürgertums am Bau bischöflicher Kathedralkirchen im Spätmittelalter*, PhD dissertation, University of Hamburg, 1958.

Wiek 1959 – Peter Wiek, "Das Straßburger Münster: Untersuchungen über die Mitwirkung des Stadtbürgertums am Bau bischöflicher Kathedralkirchen im Spätmittelalter", *Zeitschrift für die Geschichte des Oberrheins* 107 (new ser. 68), 1959, 40–113.

Wieschebrink 1983 – Theodor Wieschebrink, *Die astronomische Uhr im Dom zu Münster*, Münster, 1983[2].

Wilmart 1928 – André Wilmart (ed.), *L'ancien cantatorium de l'Eglise de Strasbourg: Manuscrit additionnel 23.922 du Musée Britannique*, Colmar, 1928.

Wimpfeling 1505 – Jacob Wimpfeling, *Hic subnotata continentur* [...] *Epithoma rerum Germanicarum vsque ad nostra tempora*, Strasbourg, 1505.

Winckelmann 1907 – Otto Winckelmann, "Zur Kulturgeschichte des Strassburger Münsters", *Zeitschrift für die Geschichte des Oberrheins* 61 (new ser. 22), 1907, 247–290.

Winckelmann 1909 – Otto Winckelmann, "Nachträge zur Kulturgeschichte des Straßburger Münsters im 15. Jahrhundert", *Zeitschrift für die Geschichte des Oberrheins* 63 (new ser 24), 1909, 302–323.

Winckelmann/Schmitz 1894 – Otto Winckelmann, Th. Schmitz, "Die Profanbauten des Mittelalters und der Renaissance", in *Strassburg und seine Bauten*, Strasbourg, 1894, 255–314.

Winter 1980 – Robert Winter, *Paulus Fabricius: Ein Wiener Universitätsprofessor des 16. Jahrhunderts*, PhD dissertation, University of Vienna, 1980.

Witte 1677 – Henning Witte, *Memoriae philosophorum, oratorum, poetarum, historicorum, et philologorum: Nostri seculi clarissimorum renovatae Decas prima*, Frankfurt am Main, 1677.

Witte 1900 – Hans Witte (ed.), *Urkundenbuch der Stadt Straßburg, Bd. VII: Privatrechtliche Urkunden und Rathslisten von 1332–1400*, Strasbourg, 1900.

Wittmer/Meyer 1948/61 – Charles Wittmer, Charles Meyer (eds.), *Le Livre de Bourgeoisie de la ville de Strasbourg 1440–1530*, Strasbourg, 1948–1961.

Wohlwill 1904 – Emil Wohlwill, "Melanchthon und Copernicus", *Mitteilungen zur Geschichte der Medizin und der Naturwissenschaften* 3, 1904, 260–267.

Wolf 1845/54 – Rudolf Wolf, "Notizen zur Geschichte der Mathematik in der Schweiz III: Conrad Dasypodius", *Mittheilungen der Naturforschenden Gesellschaft in Bern*, 1845, 137–138; 1854, 69.

Wolf 1860 – Rudolf Wolf, "Konrad Dasypodius von Frauenfeld 1531–1600", in R. Wolf, *Biographien zur Kulturgeschichte der Schweiz*, Zürich, 1860, v. III, 51–62.

Wolf 1877 – Rudolf Wolf, *Geschichte der Astronomie*, Munich, 1877.

Zanetti 2014 – Cristiano Zanetti, "The Microcosm: Innovation and Technological Transfer in the Habsburg Empire of the Sixteenth Century", *History of Technology* 32, 2014, 35–65.

Zanetti 2016 – Cristiano Zanetti (ed.), *Janello Torriani: A Renaissance Genius* (catalog of the exhibition at the Museo del Violino, Cremona, 10 September 2016–29 January 2017), Cremona, 2016.

Zanetti 2017 – Cristiano Zanetti, *Janello Torriani and the Spanish Empire: A Vitruvian Artisan at the Dawn of the Scientific Revolution*, "Nuncius Series: Studies and Sources in the Material and Visual History of Science 2", Leiden – Boston, 2017 (revised version of the PhD dissertation, *Janello Torriani (Cremona c. 1500–1585 Toledo): A Social History of Invention between Renaissance and Scientific Revolution*, European University Institute, Florence, 2012).

Zerubavel 1985 – Eviatar Zerubavel, *The Seven Day Circle: The History and Meaning of the Week*, New York, 1985.

Ziegler 1536 – Jacob Ziegler, *Sphaerae atque astrorum coelestium ratio, natura, et motus: ad totius mundi fabricationis cognitionem fundamenta*, s.l. [Basel], 1536.

Zinner 1925 – Ernst Zinner, *Verzeichnis der astronomischen Handschriften des deutschen Kulturgebietes*, München, 1925.

Zinner 1939 – Ernst Zinner, "Die ältesten Räderuhren und modernen Sonnenuhren: Forschungen über den Ursprung der modernen Wissenschaft", *Bericht der Naturforschenden Gesellschaft Bamberg* 28, 1939, 1–148.

Zinner 1941 – Ernst Zinner, *Geschichte und Bibliographie der astronomischen Literatur in Deutschland zur Zeit der Renaissance*, Leipzig, 1941.

Zinner 1953 – Ernst Zinner, "Die astronomische Bestimmung der Empfängniszeit und der Schwangerschaftsdauer", *Sudhoffs Archiv für Geschichte der Medizin und der Naturwissenschaften* 37, 1953, 432–437.

Zinner 1964 – Ernst Zinner, *Alte Sonnenuhren an europäischen Gebäuden*, "Boethius: Texte und Abhandlungen zur Geschichte der Mathematik und der Naturwissenschaften, 3", Wiesbaden, 1964.

Zinner 1968 – Ernst Zinner, *Leben und Wirken des Johannes Müller von Königsberg genannt Regiomontanus*, "Milliaria: Faksimiledrucke zur Dokumentation der Geistesentwicklung, x.1", Osnabrück, 1968.

Zinner 1979 – Ernst Zinner, *Deutsche und niederländische astronomische Instrumente des 11.–18. Jahrhunderts*, München, 1979[2].

Zinner 1988 – Ernst Zinner, *Entstehung und Ausbreitung der copernicanischen Lehre*, Munich, 1988[2].

Zülch 1935 – Walther Karl Zülch, *Frankfurter Künstler 1223–1700*, Frankfurt am Main, 1935.

Index

www.ingramcontent.com/pod-product-compliance
Lightning Source LLC
LaVergne TN
LVHW010640300926
849161LV00005B/995
9789004423466